Plant Gene Research

Basic Knowledge and Application

Edited by

E. S. Dennis, Canberra

B. Hohn, Basel

Th. Hohn, Basel (Managing Editor)

P. J. King, Basel

J. Schell, Köln

D. P. S. Verma, Montreal

Springer-Verlag Wien New York

A Genetic Approach to Plant Biochemistry

Edited by A. D. Blonstein and P. J. King

Springer-Verlag Wien New York

Dr. Anne D. Blonstein
Friedrich Miescher-Institut, Basel

Dr. Patrick J. King
Friedrich Miescher-Institut, Basel

With 30 Figures

Softcover reprint of the hardcover 1st edition 1986

Library of Congress Cataloging in Publication Data

A Genetic approach to plant biochemistry.

(Plant gene research)
Includes bibliographies and index.
1. Plant biochemical genetics. I. Blonstein, A. D. (Anne D.), 1958– . II. King, P. J. (Patrick J.), 1941– III. Series.
QK981.3.G46 1986 581.19′2 86-24865

ISSN 0175-2073

ISBN-13: 978-3-7091-7463-0 e-ISBN-13: 978-3-7091-6989-6
DOI: 10.1007/978-3-7091-6989-6

Preface

Biologists ask how the growth, development and behaviour of organisms happen, how these processes are co-ordinated and how they are regulated by the environment. Today the questions are phrased in terms of the genes involved, their structure and the control of their expression. Mutations (recognised by a change in phenotype) label genes and can be used to study gene structure, gene function and the organisation of the genome. This is "Genetics". Study of phenotypes down to the level of the enzymes and structural proteins coded for by genes is "Biochemistry". It is self evident that only by studying phenotype ("Biochemistry") can we do "Genetics" and that "Genetics" (perturbation of the phenotype) is the key to understanding the "Biochemistry". There can surely be no better arguments for a more holistic approach to biology than the massive output of knowledge from microbial "Biochemical Genetics" and the more recent revelations from "Molecular Genetic" studies of development in *Drosophila.*

When one remembers that most of the important conceptual developments in genetics (the discovery of the nucleus, the hereditary mechanism, cytoplasmic inheritance, mutation, the one gene-one enzyme hypothesis, semiconservative replication of DNA, heterochromatin, transposable elements) were discovered by biologists working with plants, it is surprising that the genetic approach to the study of basic plant properties is so underdeveloped when compared to all other biological systems. Because the advantages, the directness and reduced ambiguity, of studying normal functions by perturbing or eliminating single genes are so obvious when compared to the circumstantial approach of pure physiology and biochemistry, there must exist strong reasons why the biochemical genetics of plants is still an emerging field. Undoubtedly there are aspects of biochemistry or molecular biology common to several organisms that one would not choose to study with a complex higher plant. There are not that many efficient cytogenetic systems in plants, like those of maize, *Arabidopsis* or tomato, which are at the same time suited for particular biochemistry. There are also various physical criteria for the isolation of mutants that must be met: Where traits must be screened for by examination of individual plants, the investment of the necessary time and space may be inhibiting. The selection techniques applicable to whole plants required for the rescue of specific mutants may not be available. Lethal nutritional mutants isolated with ease from *Neurospora* may be impossible to find given the complex life cycle of the higher plant.

The "Princes of Serendip" were probably not very far away when studies of genetic aspects of gibberellins (Chapter 1), abscisic acid (Chapter 2), photosynthesis (Chapter 3) and endosperm proteins (Chapter 7) got under way. Some areas of investigation were set off initially by spontaneous variation in a field and a passing, curious biologist. Others started with the analysis of randomly generated variation following mutagenesis or a dip into a local seed collection to look for a useful variant. The rapid success of the alcohol dehydrogenase systems (Chapter 4) and the growth of nitrate reductase studies (Chapter 5) are clearly due to powerful and simple chemical selection systems being available for mutants deficient in the enzymes, and to the conditional lethality of the traits. There are, unfortunately, very few positive selection systems for deficient mutants. Cell and protoplast culture offer both a method for conveniently selecting or screening for mutants amongst very large cell populations and a way of rescuing lethal nutritional mutants. Isolation of variation is carried out at the cell level and thus many mutants may be found that would not otherwise be recognised by plant selection. After selection, plants may be regenerated and normal genetics and biochemistry undertaken. Cell culture was central to the isolation of the majority of nitrate reductase deficient mutants (Chapter 5). An account of the recent successes in the isolation of plant auxotrophs via cell and protoplast culture is given in Chapter 10.

The life cycle of the majority of higher plants includes the liberation of enormous numbers of semi-autonomous haploid gametophytes (pollen grains). The realisation that many important sporophytic genes are expressed in pollen grains and that the selection for new alleles may occur in natural populations at this level has stimulated interest in gametophytic gene expression (Chapter 9) and the possibility that mutant selection using pollen could increase the variation available to the breeder and the biochemist. Plant breeding rather than biochemistry is probably the greater stimulus to research in nodulation and nitrogen fixation in legumes (Chapter 6) and in plant/pathogen genes (Chapter 8) but, as the authors point out, in these and so many other areas of plant breeding there is an increasing need to understand the mechanisms and the molecular events involved. The protection and stabilisation of our plant resources is a worthy target for plant biochemical genetics.

Basel, August 1986 A. D. Blonstein and P. J. King

Contents

Chapter 1

Gibberellin Mutants

James B. Reid

Department of Botany, University of Tasmania, Hobart, 7001, Australia

With 3 Figures

Contents

I. Introduction

An enormous amount has been written on the gibberellins (GAs) during the last decade (for reviews see Graebe and Ropers, 1978; Hedden *et al.*, 1978; MacMillan, 1980) culminating in the extensive work by Crozier (1983). During this time considerable advances have been made regarding their biosynthesis and metabolism due to the use of GC-MS techniques and the production of both radio- and stable-isotope labelled compounds (see Hedden, 1983). However, information is still lacking on the developmental processes controlled by endogenous GAs, the site of action of the GAs and the mechanism(s) by which GAs elicit a physiological response. Indeed it is still debated if endogenous GA levels are responsible for controlling developmental processes (Trewavas, 1981; Ingram *et al.*, 1983; Phinney, 1984) and which of the over sixty identified GAs are biologically active (e. g. Hoad, 1983; Phinney, 1984).

A range of mutants in the fungus *Gibberella fujikuroi* has been used in biosynthetic studies of the GAs. For example, mutant Bl-41 a, which blocks GA biosynthesis prior to *ent*-kaurenoic acid, has been used extensively since the virtual absence of endogenous GAs allows the use of unlabelled compounds in metabolism studies (see Bearder, 1983). Unfortunately, similar use has not been made of GA synthesis mutants to help elucidate the GA biosynthetic pathways in higher plants. However, GA synthesis mutants in higher plants do provide some of the strongest evidence available to show which developmental processes are under the control of endogenous GAs and which of the multitude of GAs are biologically active. They are also the basis for many of the most widely used bioassays for GA-like substances. For example, the dwarf mutants d_5 in maize, *dx* in rice and *le* in peas have provided useful information on endogenous levels of biologically active GAs in many physiological studies. Mutants may also be useful as probes for examining the site and mode of action of the GAs, especially as this has proven difficult using other techniques (Stoddart, 1983).

The present article concentrates on the mutants which are available in higher plants, how they have been used to show which developmental responses are controlled by endogenous GAs and how mutants may be used in the future to determine the mode and site of action of the GAs. It has not been possible to cover all areas in which mutants have been used in the study of GAs in higher plants. For example, the work using GA synthesis mutants as bioassays, and the extensive metabolic work in which often non-native radio-labelled compounds have been fed to mutants such as d_5 maize or dwarf *(le)* peas have not been examined in detail. Detailed reviews of this work can be found elsewhere (e. g. Graebe and Ropers, 1978; Sponsel, 1983 a).

II. Selection and Identification of Mutants

The selection of mutants influencing GA metabolism or sensitivity has usually been based upon phenotypic changes in characteristics associated with the GAs in application experiments. These characters include internode length (e. g. Brian and Hemming, 1955; Phinney, 1956), apical senescence (Davies *et al.*, 1977; Proebsting *et al.*, 1978), sex expression (Shifriss, 1973) and seed dormancy (Koornneef and van der Veen, 1980). A list of proposed GA mutants is included in Table 1. Many of these mutants arose spontaneously and have subsequently been selected for research or agricultural purposes. For example the *le* gene in peas confers the dwarf habit and this characteristic was mentioned at least as early as 1597 by Gerarde (see Blixt, 1972). It was presumably selected because it reduced lodging of the crop and is today incorporated in many green pea varieties. Other mutants have been produced by either radiation or chemical mutagens (e. g. Sidorova, 1970; Wellensiek, 1969; Koornneef and van der Veen, 1980). Unfortunately the selection of mutants involves the growth of large numbers of plants and no efficient self-selection techniques are available. However, selection based upon characters evident early in the life-cycle may be beneficial. For example, Koornneef and van der Veen (1980) have shown that the more extreme GA synthesis mutants of *Arabidopsis thaliana* L. have increased dormancy in addition to a dwarf habit.

Table 1. The phenotypic effect and proposed mode of action of certain genes suggested to influence GA-synthesis or sensitivity

Species	Gene	Phenotype	Proposed Gene Action	Key Reference(s)
Zea mays	d_1	dwarf	blocks GA_{20} to GA_1	Spray *et al.*, 1984
	d_2	dwarf	blocks C-7-oxidation	Phinney and Spray, 1982
	d_3	dwarf	blocks C-13-hydroxylation	Phinney and Spray, 1982
	d_5	dwarf	blocks B activity of *ent*-kaurene synthetase	Hedden and Phinney, 1979
	an_1	strongly male, dwarf	blocks GA synthesis	Phinney, 1961
	na_1	dwarf	GA-insensitive	Phinney, 1961
	na_2	dwarf	GA-insensitive	Phinney, 1961
	D_8	dwarf	GA-insensitive	Phinney, 1961
	pe_1	dwarf	GA-insensitive	Phinney, 1961
	mi_2	dwarf	GA-insensitive	Phinney, 1961
Pisum sativum	*le*	dwarf	blocks GA_{20} to GA_1	Ingram *et al.*, 1984
	$la\ cry^s$	slender	saturates GA receptors	Potts *et al.*, 1985
	$la\ cry^c$	crypto	partially saturates GA receptors	Potts *et al.*, 1985
	lk	erectoides	GA insensitive	Reid and Potts, 1986
	lh	dwarf	blocks GA synthesis	Reid and Potts, 1986
	ls	nana	blocks GA synthesis	Reid and Potts, 1986
	na	nana	blocks *ent*-7α-hydroxykaurenoic acid to GA_{12}-aldehyde	Potts and Reid, 1983; Ingram and Reid (unpub.)
	Sn Hr	large response to photoperiod	produces polar GA	Proebsting *et al.*, 1978
Arabidopsis thaliana	*ga-1*	non-germinating dwarf	blocks GA synthesis	Koornneef and van der Veen, 1980

Species	Gene	Phenotype	Proposed Gene Action	Key Reference(s)
	ga-2	non-germinating dwarf	blocks GA synthesis	Koornneef and van der Veen, 1980
	ga-3	non-germinating dwarf	blocks GA synthesis	Koornneef and van der Veen, 1980
	ga-4	dwarf	blocks GA interconversions	Koornneef and van der Veen, 1980
	ga-5	dwarf	blocks GA synthesis	Koornneef and van der Veen, 1980
Hordeum vulgare	*sln*	slender	GA synthesis or utilisation	Foster, 1977
	br_2	semi-dwarf	blocks GA interconversions	Suge, 1983
	uz	semi-dwarf	blocks GA synthesis	Suge, 1972
	GA-*less*	dwarf lethal	blocks GA synthesis	Favret *et al.*, 1975
	GA-*ins*	semi-dwarf	alters GA receptor	Favret *et al.*, 1975
	gigas	slender	GA overproducer	Favret *et al.*, 1975
	–	aleurone mutants	GA insensitive and supersensitive	Ho *et al.*, 1980
Lathyrus odoratus	e_l	dwarf	blocks GA synthesis	Magara, 1963
Oryza sativa	*dx*	dwarf	blocks GA synthesis prior to *ent*-kaurene	Murakami, 1972
	dy	dwarf	blocks GA_{20} to GA_1	Murakami, 1972
	–	dwarfs	GA-insensitive	Harada and Vergarra, 1971
Triticum aestivum	*Rht 1*	semi-dwarf	GA insensitive for stem elongation	Radley, 1970; Gale and Marshall, 1973
	Rht 2	semi-dwarf	GA insensitive for stem elongation	Radley, 1970; Gale and Marshall, 1973
	Rht 3	dwarf	reduces GA receptor concentration or affinity in both stem and aleurone	Ho *et al*., 1981; Stoddart, 1984
Pharbitis nil	–	dwarf	blocks GA synthesis	Ogawa, 1962; Barendse and Lang, 1972
Citrullus lanatus	*dw-2*	dwarf	blocks GA synthesis	Loy and Liu, 1974
Lycopersicon esculentum	*ga* (Ve-335)	non-germinating dwarf	blocks synthesis prior to kaurene	Koornneef *et al*., 1981; Zeevaart, 1984

Species	Gene	Phenotype	Proposed Gene Action	Key Reference(s)
	Ve-182	semi-dwarf	partially blocks GA synthesis	Zeevaart, 1984
	Ve-270	dwarf	blocks GA-synthesis between kaurenoic acid and GA_{12}	Zeevaart, 1984
	sl_2	stamenless	blocks GA synthesis	Sawhney, 1974
	yg_6	long-stem, reduced chlorophyll	GA overproducer	Perez *et al.*, 1974
Phaseolus vulgaris	dwarf-1	dwarf	blocks GA synthesis	Moh and Alan, 1967; Proano and Greene, 1968
	–	dwarf	blocks GA synthesis	Goto and Esashi, 1973
Prunus persica	–	dwarf	GA insensitive	Wylie and Ryugo, 1971
Silene armeria	*s*	dwarf	GA insensitive	Wellensiek, 1976
	ef	late flowering	GA insensitive	Wellensiek, 1976
Lolium perenne	*d*	dwarf	blocks GA synthesis	Cooper, 1958
Ipomoea batatas	–	dwarf	GA insensitive	Suge, 1979
Trifolium pratense	dw_1	dwarf	GA sensitive	Smith, 1974
Fragaria vesca	*arb*	long-stemmed	removes block to GA synthesis	Guttridge, 1973
Ricinus communis	*dr*	slow germinating, dwarf, enhanced maleness	GA sensitive	Shifriss, 1973

Selection based upon changes in morphological characters can provide only circumstantial evidence that a mutation is involved directly with GA metabolism or sensitivity. For example, a GA metabolism mutant may be indicated if an accurate phenocopy of the wild type is produced after treatment of the mutant type with an active GA such as GA_3 or GA_1. The hypothesis is further supported if treatment of the wild type with a GA synthesis inhibitor such as AMO 1618 or PP 333 results in a phenocopy of the mutant type. However, even where good application data are available, direct proof that a GA metabolism mutant exists is difficult to obtain because of the extremely low levels of GAs in vegetative tissue. Early workers were confined to bioassays to quantify GA levels. While bioassays have provided some of the classic results regarding the role of GAs (e. g. Phinney, 1961) and are still the most appropriate starting point, they have at times been misleading and their use has been generally criticised (Graebe and Ropers, 1978). This is partly due to their lack of specificity and repeatability, although equally at fault has been the poor separation techniques employed prior to the bioassay (Crozier and Durley, 1983) and the poor choice of tissue extracted (i. e. the tissues extracted had not previously been shown to be a site in which the mutation was expressed). Similar problems in comparing endogenous GA levels can occur with the more definitive techniques such as GC-MS unless losses during purification are accounted for and the correct tissue is used.

Proof that a GA metabolism mutant exists requires not only the identification and quantification of the endogenous GAs but a demonstration of differences in metabolism between the mutant and wild-type. For this purpose GC-MS techniques and a supply of suitably labelled compounds for use in metabolism studies are required. Even these techniques are extended to their limits by the low levels of endogenous GAs in vegetative tissues of wild types and are not sufficiently sensitive to examine the "leakiness" of certain mutant types (e. g. Ingram *et al.*, 1984). With these types it may be necessary to turn to the more sensitive techniques of radio- or enzyme-immunoassays. As with bioassays, problems may arise when mass spectrometry (MS) or immunoassay techniques are employed. For example, a comparison of tall *(Le)* and dwarf *(le)* peas using MS was reported to show substantially more GA_1 in dwarf plants than in tall plants (Keller and Coulter, 1982). The levels described were also far higher than other reports (e. g. Ingram *et al.*, 1984). The cause of these misleading results is not completely clear but poor separation and purification were partly at fault (see Gaskin *et al.*, 1985). Even results from immunoassay techniques may be at variance with results obtained from GC-MS techniques (Atzorn and Weiler, 1983; Gilmour and MacMillan, 1984) suggesting that further refinement of the specificity is required.

GA metabolism mutants may be subdivided into GA synthesis mutants and GA breakdown or utilization mutants. Altered GA levels are indicative of both groups. However, to date no mutants falling into the breakdown or utilization categories have been proven but some of the proposed GA synthesis mutants listed in Table 1 may well fall into these categories when further work is done.

Designation as a GA insensitive mutant has up to now rested upon an altered response to applied GA. They are a large and heterogeneous group of mutants since they may influence any of the steps between reception of the GA signal and the manifestation of the GA response. This group may therefore include mutants which are not directly involved with the GAs. For example, they include any mutant which makes the GAs non-limiting for the response(s) being examined. These, at least potentially, include factors such as insufficient substrates, limiting levels of other plant hormones and limited uptake or transport systems within the plant. Many of the GA insensitive mutants listed in Table 1 have not been examined in sufficient detail to determine whether the GA insensitivity is of a direct, primary nature or is an indirect effect via one of these mechanisms.

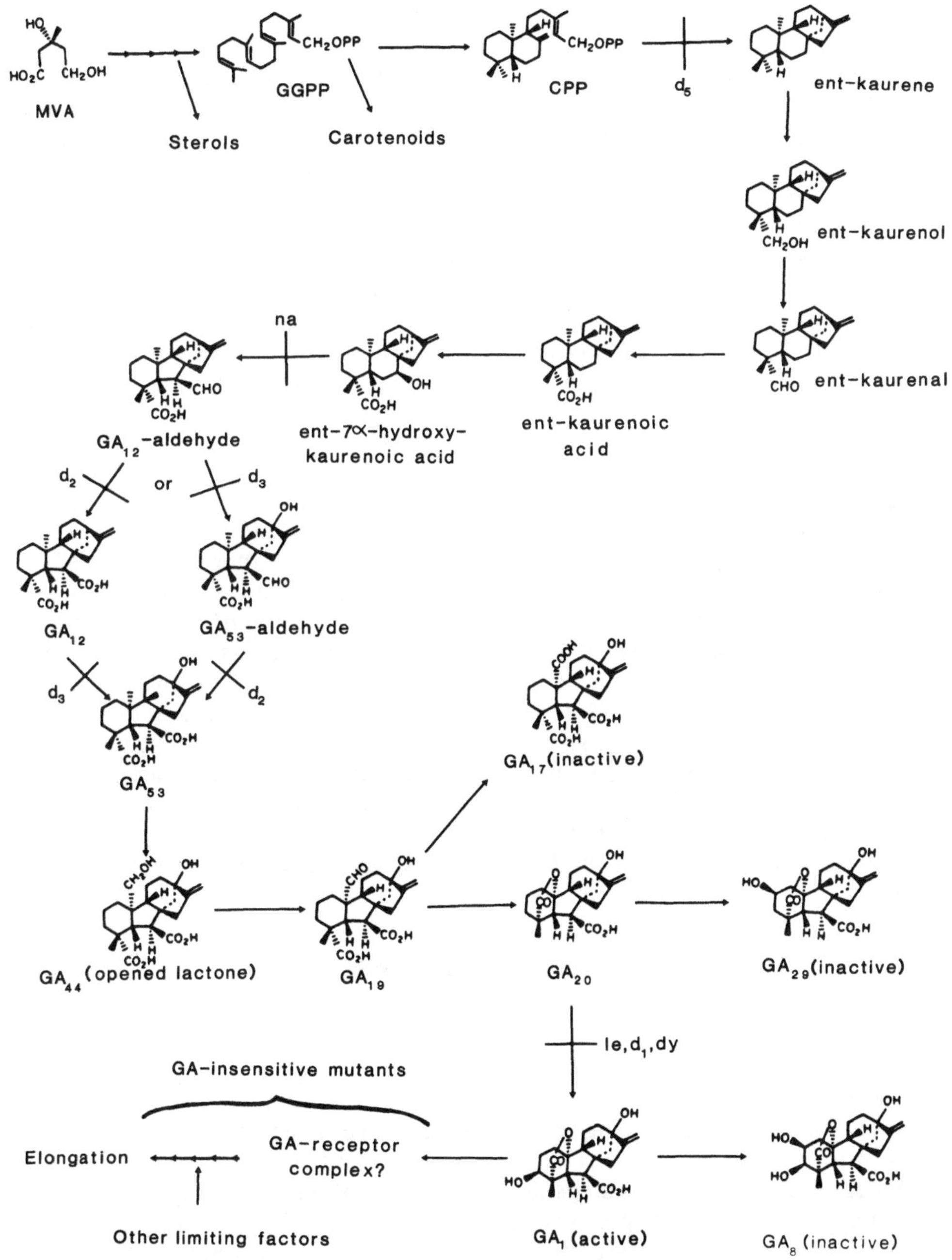

MVA → GGPP → CPP → (d_5) ent-kaurene → ent-kaurenol → ent-kaurenal → ent-kaurenoic acid → ent-7α-hydroxy-kaurenoic acid → (na) GA_{12}-aldehyde

Sterols

Carotenoids

GA_{12}-aldehyde → (d_2) GA_{12} or (d_3) GA_{53}-aldehyde → (d_3 / d_2) GA_{53} → GA_{44} (opened lactone) → GA_{19} → GA_{20} → (le, d_1, dy) GA_1 (active)

GA_{19} → GA_{17} (inactive)

GA_{20} → GA_{29} (inactive)

GA_1 → GA_8 (inactive)

GA_1 → GA-receptor complex? → Elongation

GA-insensitive mutants

Other limiting factors

Fig. 1. The proposed sites of action of GA mutants in maize *(d_1, d_2, d_3 and d_5)*, peas *(le, na)* and rice *(dy)* in the early 13-hydroxylation gibberellin biosynthetic pathway from mevalonic acid (MVA) to the biologically active product GA_1, and the possible site(s) of action of GA insensitive mutants.

III. GA Synthesis

An important impetus in determining the action of GA synthesis mutants has been the determination of GA biosynthetic pathways in certain higher plants. These have been reviewed recently by Sponsel (1983a) and MacMillan (1984). The pathway from mevalonic acid (MVA) to GA_{12}-aldehyde appears common to all the plants examined (see Hedden, 1983) (Fig. 1) while after GA_{12}-aldehyde several pathways differing largely in the pattern and timing of hydroxylation occur (MacMillan, 1984). The one of most significance to the discussion of GA mutants is the early 13-hydroxylation pathway which has been demonstrated most extensively by Kamiya and Graebe (1983) in cell free systems from pea seeds (Fig. 1). The point of 13-hydroxylation is not resolved but is likely to be between GA_{12} and GA_{53} or GA_{12}-aldehyde and GA_{53}-aldehyde (shown as alternatives in Fig. 1). This pathway probably occurs in the seeds of many higher plants since 13-hydroxylated GAs are native to a number of species (see Bearder, 1980). It is also the probable pathway in other tissues including the shoots of peas and maize (Davies *et al.*, 1982; Phinney and Spray, 1982; Gaskin *et al.*, 1985) where the pathway after GA_{20} splits to produce either the inactive 2β-hydroxylated GA, GA_{29}, or the highly active 3β-hydroxylated GA, GA_1 (Phinney and Spray, 1982; Ingram *et al.*, 1984; Spray *et al.*, 1984). Further metabolism of GA_1 to GA_8 has been demonstrated (Davies and Rappaport, 1975). GA_{17} appears to be a side branch from GA_{19} in this pathway (Davies *et al.*, 1982; Kamiya and Graebe, 1983). A knowledge of the endogenous GAs and the probable biosynthetic pathway is essential if application studies are to provide evidence regarding the steps controlled by GA mutants and the order in which they act. Early work in which the metabolism of, or response to, non-native GAs was examined did not provide clear results (e. g. Crozier *et al.*, 1970; Musgrave and Kende, 1970).

IV. Internode Length

A. Synthesis Mutants

i) Maize

Thirty mutants influencing plant height have been reported in maize (Phinney and Spray, 1982). The five dwarf mutants d_1, d_2, d_3, d_5 and an_1 are reported to block various steps in the GA biosynthetic pathway prior to the formation of the active product, GA_1, since the application of appropriate quantities of GA_3 to any of these mutants results in a phenocopy of the wild-type plant (Phinney, 1956) and all have reduced levels of GA-like substances when extracts from the shoots are bioassayed (Phinney, 1961).

Gene d_5 reduces the production of *ent*-kaur-16-ene (*ent*-kaurene) from MVA to less than one quarter of that found in normals by reducing the B activity of *ent*-kaurene synthetase (Fig. 1). This was demonstrated using a cell-free system developed from young etiolated shoots of d_5 and normal maize seedlings (Hedden and Phinney, 1979). The major diterpene hydrocarbon synthesised by the d_5 system was *ent*-kaur-15-ene (*ent*-isokaurene) regardless of whether [C^{14}]-MVA, [C^{14}]-geranylgeranylpyrophosphate (GGPP) or [^3H]-copalylpyrophosphate (CPP) was used as substrate. In comparison, cell-free extracts from normal seedlings predominantly produced *ent*-kaurene (Hedden and Phinney, 1979). *Ent*-kaurene is a known precursor to the GAs (Cross *et al.*, 1964) but the isomer *ent*-isokaurene is biologically inactive (Phinney and Spray, 1982) and not a precursor to the GAs (Hedden *et al.*, 1977). Both systems produce the other isomer as a minor product. The reduction in *ent*-kaurene synthesis is almost certainly the cause of the dwarf habit in d_5 plants. The relatively early block in the GA biosynthetic pathway by d_5 is supported by the fact that the GA-precursors, *ent*-kaurene, *ent*-kaurenol, *ent*-kaurenoic acid as well as GA_{12}-aldehyde, GA_{53} and GA_{20} elicit a GA-like growth response when applied to d_5 seedlings (Katsumi *et al.*, 1964; Phinney and Spray, 1982).

Gene d_1 blocks the 3β-hydroxylation of GA_{20} to GA_1 (Fig. 1) (Spray *et al.*, 1984). Application of GA_{20} to d_1 plants results in a weak promotion of elongation compared to other mutants (e. g. d_2 , d_3 or d_5) and GA_{20} is only one per cent as active as GA_1 (Phinney and Spray, 1982). When d_1 seedlings were treated with [^{13}C, ^3H]GA_{20} and the metabolites in the shoot identified by GC-MS, only [^{13}C, ^3H]GA_{29} and [^{13}C,^3H]GA_{29}-catabolite were found (Spray *et al.*, 1984). No evidence for the 3β-hydroxylated metabolites [^{13}C,^3H]GA_1 or [^{13}C,^3H]GA_8 was found. In comparable feeds to normal (tall) seedlings [^{13}C,^3H]GA_{29}-catabolite and [^{13}C,^3H]GA_1 were identified but no [^{13}C,^3H]GA_8 or [^{13}C,^3H]GA_{29}. Dilution of the [^{13}C] label by endogenous [^{12}C]GAs was found in metabolites from both the normal and d_1 seedlings. The cause of the absence of [^{13}C,^3H]GA_{29} from the tall seedlings is unclear but it was identified in similar feeds to d_5 seedlings. No dilution of [^{13}C] metabolites occurred in d_5 seedlings confirming that the gene d_5 blocks a step prior to the formation of C_{19}-GAs. The late nature of the d_1 block is supported by the fact that extracts from d_1 plants show several zones of GA-like activity when bioassayed on d_5 plants (Phinney and Spray, 1982). Furthermore, in $d_5 — d_1$ grafts the d_5 shoots show a small but significant increase in elongation compared with shoots in $d_5 — d_5$ control grafts (Katsumi *et al.*, 1983).

The steps in the GA-biosynthetic pathway blocked by genes d_2 , d_3 and an_1 have not been determined with the same degree of certainty. Gene d_2 appears to block the carbon-7-oxidation step from GA_{53}-aldehyde to GA_{53} and/or GA_{12}-aldehyde to GA_{12} (Fig. 1) since *ent*-kaurene, GA_{12}-aldehyde and GA_{53}-aldehyde show no activity when assayed on d_2 plants while GA_{53} and GA_{20} result in substantial elongation (Phinney and Spray, 1982; Phinney, 1984). Similarly, gene d_3 appears to block 13-hydroxylation from GA_{12}-aldehyde to GA_{53}-aldehyde and/or GA_{12} to GA_{53} since d_3 plants show no response to kaurene or GA_{12}-aldehyde but respond well to GA_{53}-aldehyde, GA_{53} and GA_{20} (Phinney and Spray, 1982; Phinney, 1984). The site of action of an_1 has not been reported but it is possibly before kaurenol since Katsumi *et al.,* (1964) reported that kaurenol could stimulate leaf sheath elongation in an_1 plants.

ii) Rice

Of the wide range of GA sensitive mutants reported in rice, two recessive dwarfs, *dx* and *dy*, have been examined in detail (Suge and Murakami, 1968; Murakami, 1972; Suge, 1978). Plants possessing gene *dx* (e. g. cv. Tan-ginbozu) possess no GA-like substances in the shoots and show elongation when treated with kaurene, kaurenol, kaurenoic acid or a wide range of GAs (Murakami, 1972; Murofushi, 1983). This suggests that gene *dx* blocks an early step (prior to kaurene) in the biosynthetic pathway leading to the GAs (Fig. 1) (Suge, 1978). Extracts from plants possessing gene *dy* possess GA-like substances. The pattern found showed little difference from extracts from normal *(Dy)* plants (Suge and Murakami, 1968; Murakami, 1972). Unlike *dx* plants, *dy* plants did not respond to kaurene, kaurenol or kaurenoic acid and showed little response to non-3β-hydroxylated GAs such as GA_{19} and GA_{20} (Murakami, 1972). Both *dx* and *dy* plants respond well to 3β-hydroxylated GAs such as GA_1 and GA_3. The biosynthetic pathway for GAs has not been determined in rice although an early 13-hydroxylation pathway (Fig. 1) seems likely since GA_{19} has been identified by GC-MS as the major GA in the shoots, roots and ears of rice plants and GA_1 has been identified by GC-SICM (Kurogochi *et al.,* 1979; Suzuki *et al.,* 1981; Murofushi, 1983). If this is the case, it seems likely from application data that gene *dy* inhibits the 3β-hydroxylation of GA_{20} to GA_1 (Fig. 1) (Murakami, 1972; Kurogochi *et al.,* 1979). The genes *dx* and *dy* are therefore closely comparable to the d_5 and d_1 genes in maize, respectively, although detailed metabolic studies are required to confirm these suggestions. Like the d_5 and d_1 maize mutants the *dx* and *dy* dwarf rice mutants have been used extensively as bioassays and the joint use of both mutants has proved most useful in suggesting if GA-like fractions are 3β-hydroxylated (Murakami, 1972; Potts *et al.,* 1982).

iii) Barley

Many dwarf mutants occur in barley *(Hordeum vulgare)* although they often differ in their cause (e. g. cell size, cell number, stem architecture, etc., Blonstein, 1981). Suge (1983) discusses three non-allelic dwarf or semi-dwarf mutants, *uz, br* and br_2. Gene br_2 has been suggested to be equivalent to the d_1 mutant of maize and the *dy* mutant in rice. However, the small differences in the bioassay results and the crude separation techniques used mean that firm evidence is still required. Gene *uz* has been reported to reduce the level of GA-like substances in the shoot (Suge, 1972) but reports by Kuraishi (1974) and Inouke *et al.,* (1982) suggest that this gene possibly has its primary action by inhibiting IAA synthesis, at least in dark-grown coleoptiles. The levels of GA-like substances in *br* plants were similar to those of normal varieties when bioassayed using the dwarf rice cv. Tanginbozu. The results for these three mutants are at best inconclusive and further detailed study is required.

Favret *et al.*, (1975) and Hopp *et al.*, (1981) have examined two further synthesis mutants. Their dwarf "GA-*less*" mutant becomes a phenocopy of the normal type when treated with GA_3 and contains no detectable GA-like activity. The metabolism of GA_1 is not altered and it therefore appears to be a synthesis mutant but its relationship to the mutants *uz* and br_2 is not clear. Of more interest is the *gigas* mutant (Favret *et al.*, 1975). This giant mutant is phenotypically similar to a normal plant treated with a large quantity of an active GA. It appears to be a synthesis mutant since the GA synthesis inhibitor CCC can reduce its growth and the "GA-less" mutant is epistatic to the *gigas* mutant. While many GA-deficient mutants have been described, few GA-overproducing mutants have been proposed and further examination is warranted as it may allow the mechanisms controlling GA synthesis to be determined.

Recent work with developing seeds of barley has shown they contain a wide array of gibberellins (Gaskin *et al.*, 1984) and a tentative scheme for the metabolism of GAs in this species has been proposed (Gilmour *et al.*, 1984). This offers the hope that further progress on the action of the mutants may be forthcoming.

iv) Peas

In the garden pea (*Pisum sativum* L.) over 40 mutants influencing plant height have been reported (e. g. de Haan, 1927; Wellensiek, 1969; Sidorova, 1970; Blixt, 1972; Weber and Gottschalk, 1973). Eight non-allelic mutants which have a marked influence on internode length have been examined in some detail and four of these, *le, na, lh* and *ls,* have been suggested to influence gibberellin synthesis (Potts *et al.*, 1982; Potts and Reid, 1983; Ingram *et al.*, 1983, 1984; Reid and Potts, 1986).

Gene *le* has been subjected to many investigations (e. g. Brian and Hemming, 1955; Lockhart, 1956; Jones and Lang, 1968; Köhler, 1970) and results in a 70 per cent reduction in internode length compared to comparable *Le* (tall) plants (Blixt, 1972; Reid *et al.*, 1983). Extracts from shoots of tall *(Le)* plants contain the highly active GA_1, a GA not found in extracts from the shoots of dwarf *(le)* peas (Potts *et al.*, 1982; Davies *et al.*, 1982; Potts and Reid, 1983; Ingram *et al.*, 1983). Application of GA_1 can mask the difference between tall *(Le)* and dwarf *(le)* pea plants while only plants possessing *Le* respond as well to GA_{20} as to GA_1 (Ingram *et al.*, 1983). This suggests that the gene *le* is equivalent to the gene d_1 in maize and *dy* in rice and probably acts by controlling the production of a 3β-hydroxylase which allows the conversion of GA_{20} to GA_1 (Fig. 1). This was confirmed by Ingram *et al.*, (1984) who showed that $[^{13}C,^{3}H]GA_{20}$ is metabolised to $[^{13}C,^{3}H]GA_1$, $[^{13}C,^{3}H]GA_8$ and $[^{13}C,^{3}H]GA_{29}$ in the immature, shoot tissue of *Le* plants. In contrast, $[^{13}C,^{3}H]GA_{29}$ and $[^{13}C,^{3}H]GA_{29}$-catabolite were the only GAs produced in measurable quantities by plants homozygous for *le*. However, *le* is probably a leaky mutant, as *le* plants show moderate elongation if GA_{20} is applied (Ingram *et al.*, 1983) and a trace of GA_8 was detected in one dwarf *(le)* line by Ingram *et al.* (1984). A comparison of the molecular ion currents of methyl TMS GA_8 and methyl TMS GA_{29} in the

Fig. 2. The response of the GA sensitive nana phenotype (Weibullsholm line 1766, type line for gene *na*) and the GA-insensitive erectoides phenotype (John Innes line 1420, type line for gene *lk*) of peas to 10 μg of GA_1 applied to the third leaf after 11 days growth. Line 1766 was produced by Professor S. J. Wellensiek and line 1420 was selected by Dr. P. Matthews.

dwarf gave a ratio of approximately 0.02 compared with a ratio of approximately 1.6 in the two tall *(Le)* lines examined (Ingram *et al.*, 1984).

Gene *na* in peas has a more extreme effect on internode length than *le* and results in the extremely short nana phenotype (Fig. 2) (Wellensiek, 1969; Reid *et al.*, 1983). The young shoots from plants homozygous for *na* do not possess detectable quantities of C_{19}-GAs since the [^{13}C]-metabolites from [^{13}C,^{3}H]GA_{20} feeds to *na* plants show no dilution with endogenous [^{12}C]GAs. In contrast, significant dilution was found in *Na* plants (Ingram *et al.*, 1984). The block caused by gene *na* is possibly prior to GA_{19} and GA_{44} since in the rice seedling bioassay (cv. Tan-ginbozu) no GA-like substances were detected in extracts from the shoots of *na* plants (Potts and Reid, 1983). GA_{44} and GA_{19} have been indicated by GC-MS from peaks in the rice seedling bioassay from *Na* plants (Davies *et al.*, 1982; Potts, unpub. in Potts *et al.*, 1985). Application data suggest that the block may be between *ent*-7α-hydroxykaurenoic acid and GA_{12}-aldehyde since *na* plants showed no response to 1, 10 or 100 μg of *ent*-kaurene, *ent*-kaurenol, *ent*-kaurenal, *ent*-kaurenoic acid or *ent*-7α-hydroxykaurenoic acid but showed a 90, 280 and 380 per cent increase in internode length when treated with 1, 10 or 100 μg of GA_{12}-aldehyde, respectively (Ingram and Reid, unpub.). However, further direct evidence is required as there is no evidence that *ent*-kaurene, etc. were reaching a site within the plant where they could

Fig. 3. The phenotype of two short-internode GA synthesis mutants of pea, *lh* (in its type line K511) and *ls* (in its type line M26), compared with the parental cultivars Torsdag *(Lh)* and Dippes gelbe Viktoria *(Ls)*, respectively. K511 and Torsdag were provided by Dr. K. K. Sidorova and M26 and Dippes gelbe Viktoria by Professor W. Gottschalk.

elicit a biological response. In addition, GA_{12}-aldehyde was not as effective as GA_1 or GA_{20} in inducing elongation in *na Le* plants, the latter two compounds inducing true phenocopies of isogenic *Na* plants.

Two further recessive dwarfing mutations in peas, *lh* and *ls* (Fig. 3) (Reid, 1986), appear to be GA synthesis mutants since phenocopies of normal tall plants can be produced by appropriate treatment with the native GA, GA_1 (Reid and Potts, 1986). Both *lh* and *ls* plants contain little or no GA-like substances when extracts from the shoots of 21 to 23 day old plants were examined using the rice seedling bioassay (cv. Tan-ginbozu) suggesting a block prior to GA_{44} (Reid and Potts, 1986). The precise steps in the GA biosynthetic pathway being affected will need to await the availability of suitably labelled intermediates.

Double recessive types, *na ls, na lh* and *ls lh,* possess shorter internodes than the single recessive types (Reid, 1986). This suggests that the mutant alleles are "leaky". However, it has not been possible to show the presence of C_{19}-GAs in the shoots of *na* plants by bioassays (Potts and Reid, 1983) or by dilution of $[^{13}C,^{3}H]GA_{20}$ metabolites by endogenous $[^{12}C]$GAs using GC-MS techniques (Ingram *et al.,* 1984). Consequently, the genetic results suggest that even though the levels of biologically active GAs are below the levels of detection by these techniques they are still of biological significance.

v) Other Species

Recent work by Zeevaart (1984) with three tomato mutants, Ve-182 (semi-dwarf) and Ve-270 and Ve-335 (very short internodes) has produced similar results to those found in maize, rice and peas. The semi-dwarf mutant Ve-182 has qualitatively similar zones of GA-like activity to the parental cv. "Money Maker" but the levels are much lower when determined by the d_5 maize bioassay. Mutant Ve-182 is therefore a leaky mutant. The mutants Ve-270 and Ve-335 possess no detectable GA-like activity. All mutants become phenocopies of the wild-type if treated with biologically active GAs. Feeds with GA precursors suggest that mutant Ve-335 may block GA synthesis prior to kaurene while Ve-270 may block GA biosynthesis between kaurenoic acid and GA_{12}. Perez *et al.*, (1974) have suggested that the long-stemmed tomato mutant yg_6 is a GA overproducer and it may be similar to the *gigas* mutant in barley.

In many other species dwarf mutants containing reduced levels of GAs have been found (e. g. *Lathyrus odoratus, Pharbitis nil, Phaseolus vulgaris, Arabidopsis thaliana,* see Table 1). Such mutants usually show a strong growth response when treated with biologically active GAs. These data provide circumstantial evidence that these mutants should be considered GA synthesis mutants analogous to those described in detail for maize, rice, peas and tomatoes. However, many of these results were collected before the GA biosynthetic pathway was fully elucidated in any higher plant and consequently the biochemical steps involved have not been determined. Further work with these mutants is justified since it may help to elucidate the GA biosynthetic pathway in these plants as well as provide information regarding the control of elongation.

B. Insensitive Mutants

GA insensitive internode length mutants are common (see Table 1) and are economically important since they are used extensively in agriculture (e. g. wheat). The majority of insensitive mutants exhibit reduced internode length [e. g. wheat (Allan *et al.*, 1959), maize (Phinney, 1956)], but a few exhibit increased internode length such as the slender phenotypes in peas (de Haan, 1927) and barley (Foster, 1977). The two types will be considered within the one group since the mode of action of both types of mutants appears similar (e. g. Ho *et al.*, 1981; Potts *et al.*, 1985).

i) Wheat

Owing to their economic significance, the dwarfing genes in hexaploid wheat *(Triticum aestivum), Rht 1, Rht 2* and *Rht 3* have received the most attention. *Rht 3* displays either little or partial dominance over the wild-type allele *rht 3* (Morris *et al.*, 1972; Fick and Qualset, 1973) and is more potent than the recessive genes *Rht 1* and *Rht 2* (Fick and Qualset, 1973; McVittie *et al.*, 1978). *Rht 3* and *Rht 1* are thought to be allelic and located on chromosome 4A (Morris *et al.*, 1972; Gale and Marshall, 1976;

Gale and Law, 1977). *Rht 2* is on chromosome 4D (Gale *et al.*, 1975) and is probably homoeologous with the *Rht 3/Rht 1* locus (McVittie *et al.*, 1978). All three dwarfs are insensitive to GA in terms of elongation (Allan *et al.*, 1959; Gale and Marshall, 1973). At first this insensitivity was considered an independent response (Hu and Konzak, 1974) controlled by the three genes *Gai 1, Gai 2* and *Gai 3* closely linked to *Rht 1, Rht 2* and *Rht 3*, respectively (Gale and Law, 1977). However, it now seems likely that these two responses may be pleiotropic effects of the same genes since no recombinants have been found even in large progenies (Gale *et al.*, 1975; Fick and Qualset, 1975; Gale and Law, 1977; McVittie *et al.*, 1978). The GA insensitivity of *Rht 3* plants extends to all GA responses including α-amylase production by the aleurone layer (Gale and Marshall, 1973; Fick and Qualset, 1975; Ho *et al.*, 1981). In contrast *Rht 1* and *Rht 2* plants do not show GA insensitivity for α-amylase production (Radley, 1970; Gale and Marshall, 1973; Fick and Qualset, 1975). The differences for *Rht 1* and *Rht 3* are unexpected since alleles normally show similar sites of action although differences may occur if the threshold for sensitivity differs between tissues. The dominance relationships of *Rht 1, Rht 3* and the wild-type allele are also unexpected if *Rht 1* and *Rht 3* are allelic.

Radley (1970) examined the GA-like activity of Norin 10 type dwarf wheat cultivars (probably possessing genes *Rht 1* and *Rht 2*) using the barley endosperm bioassay and found that the dwarfs contained more GA-like activity than talls regardless of whether germinating grains or green seedlings were examined. Most of the activity was thought to be due to GA_1 and it was suggested that there was a block to the utilisation of GA in the dwarfs which resulted in its accumulation. Stoddart (1984) has shown that the leaves of tall seedlings metabolise $[^3H]GA_1$ rapidly to compounds co-chromatographing on HPLC with $[^3H]GA_8$ and a conjugate of $[^3H]GA_8$. *Rht 3* dwarf segregates from the same population metabolised $[^3H]GA_1$ more slowly although the same products were produced. In both cases 2β-hydroxylation of the active $[^3H]GA_1$ to the inactive $[^3H]GA_8$ was the initial metabolic modification. Radioimmunoassay indicated a 12—15 times higher level of GA in the dwarf leaves, similar to the earlier findings of Radley (1970). When the effects of disparate endogenous GA pool sizes were taken into account the rates of 2β-hydroxylation were essentially comparable in tall and dwarf immature leaf samples, suggesting that insensitivity in *Rht 3* dwarfs is unrelated to enzymatic modification of the active GA_1. Instead, insensitivity probably relates to the ability to interact with GA_1 (Stoddart, 1984). Ho *et al.* (1981) came to an essentially similar conclusion after comparing the physiology of the diverse GA responses in an *Rht 3* dwarf and a normal tall variety. In the dwarf, elongation of the leaves, synthesis and release of hydrolytic enzymes and secretion of phosphate ions and reducing sugars by the aleurone layers were retarded. The dwarf did not show a general slowdown in cellular metabolism, differ in the uptake of GA_3 or possess a different level of endogenous inhibitors from the tall type. Differences between the cultivars in α-amylase production were relatively greater at higher concentrations of GA_3. This was used to suggest that in *Rht 3* plants the level or activity of the GA receptor for all responses was reduced (Ho *et al.*, 1981).

ii) Other Species

The physiology of GA-insensitive dwarfs in maize, barley and peas (Fig. 2) has not been examined in as much detail as in the dwarf wheats. However, in barley and peas the levels of endogenous GA-like substances have been examined and no qualitative changes have been found (Hopp *et al.*, 1981; Reid and Potts, 1986). Small quantitative differences do occur in peas but these may be due to inherent differences in the ratios of leaf to stem tissue extracted (Reid and Potts, 1986). All the available information is consistent with the view that in the GA-insensitive dwarfs some step leading to the GA response(s) at or after the GA-receptor site is limiting.

The GA insensitive slender phenotype in peas possesses long, thin internodes, rapid germination, pale foliage, reduced branching, malformed and abortive flowers, reduced seed set and the production of parthenocarpic pods (de Haan, 1927; Reid *et al.*, 1983). This set of pleiotropic characteristics can be mimicked by the application of non-limiting quantities of GA_3 (Dalton and Murfet, 1975; Potts *et al.*, 1985). The slender phenotype is conferred by the combination of recessive genes *la* and cry^s (see Reid *et al.*, 1983) and is not altered by application of the GA synthesis inhibitors AMO 1618 or PP 333 (McComb and McComb, 1970; Potts *et al.*, 1985). Further, the slender gene combination *la* cry^s is epistatic to the GA synthesis mutants *le* (de Haan, 1927) and *na* (Potts *et al.*, 1985). These results suggest that the slender phenotype is not dependent on the level of endogenous GAs (Brian, 1957; Potts *et al.*, 1985). Direct support for this view was obtained when the endogenous levels of GA-like substances were examined using the rice seedling (cv. Tan-ginbozu) and lettuce hypocotyl bioassays. Slender segregates (genotype *le la* cry^s *Na*) were found to possess qualitatively similar zones of GA-like substances to the dwarf segregates *(le la Cry Na)* although quantitatively the levels were lower in the slender plants. Slender plants possessing the "GA-less" allele *na* possessed no detectable GA-like substances even though they were phenotypically the same as slender plants possessing *Na* (Potts *et al.*, 1985). The slender gene combination *la* cry^s therefore allows the plant to act as if it is fully saturated with GAs for growth regardless of the level of endogenous GAs, perhaps by influencing a normally rate limiting step between the primary site of perception of the GAs and the phenotypic response (Potts *et al.*, 1985). The exact mode of action is unknown but any hypothesis must take into account the genetic evidence that *La* and *Cry* are near duplicate genes and that the GA insensitive dwarfing gene *lk* is essentially epistatic to the gene combination *la* cry^s (Reid, 1986). This may suggest *la* and cry^s are operating prior to the step controlled by *lk* and that *la* and cry^s may act to release some step which is normally repressed.

Very little is known about GA receptor sites or the steps between reception and the production of a phenotypic response (see Stoddart and Venis, 1980; Stoddart, 1983). This has been suggested as one of the major problems facing plant hormone research (Vanderhoef and Kosuge, 1984). The GA insensitive mutants may provide the necessary probes to examine this problem especially where several such mutants occur in the same species (e. g. maize, wheat and peas). Genetic evidence may allow the order in which the mutants operate to be deduced.

V. Seed Dormancy

Recent work by Koornneef and co-workers has led to the isolation of non-germinating dwarf mutants in *Arabidopsis thaliana* and tomato (Koornneef and van der Veen, 1980; Koornneef *et al.*, 1981). In *Arabidopsis* 56 GA sensitive dwarf mutants were isolated and shown to occur at five loci. Thirty seven mutants at three of these loci, *ga-1, ga-2* and *ga-3,* did not germinate unless treated with GA (Koornneef and van der Veen, 1980). All GAs tested, GA_3, GA_{4+7}, GA_7 and GA_9, could induce germination with the GA_{4+7} mixture being the most effective. Without further GA treatment the seedlings developed into dark green dwarfs but with weekly sprays of 10^{-4} M GA_{4+7} phenocopies of the wild-type were produced. Mutants with similar phenotypes, except showing normal germination, also occurred at loci *ga-1, ga-2* and *ga-3.* Dwarf mutants at loci *ga-4* and *ga-5* were generally less extreme than mutants at *ga-1, ga-2* and *ga-3* (Koornneef and van der Veen, 1980). Although the biochemical characterisation of these mutants has not been reported it appears likely, based on the physiological evidence, that they influence steps in GA synthesis. When the various mutants at loci *ga-1, ga-2* and *ga-3* were further investigated a range from an absolute to no GA requirement for germination was found. This suggests that the alleles vary in their degree of "leakiness". The GA requirement for germination is suggested to be much lower than for elongation and flower development since in certain dwarfs germination can be perfectly normal while length growth and flower development are substantially altered (Koornneef and van der Veen, 1980). While this seems likely the effectiveness of the genes in various tissues requires examination since substantial differences between tissues can occur and certain GA-sensitive dwarf mutants have been shown to be organ specific in their action (Potts and Reid, 1983). The response of non-germinating tomato mutants is similar (Koornneef *et al.*, 1981) and indicates the importance of GA for the normal control of germination in these species. Along with ABA (see Chapter 2 in this volume) and phytochrome mutants they should allow the partial processes involved in this complex development process to be examined. Auxotrophic mutants such as the non-germinating mutants are rare in higher plants (Redei, 1975, and Chapter 10 in this volume) and should be of considerable genetic use. Already they have been used to allow the selection of germination revertants which appear to operate via reduced ABA levels (Koornneef *et al.*, 1982, and Chapter 2 in this volume).

VI. Flowering and Senescence

The involvement of GAs in the control of flowering has been the subject of a recent review (Zeevaart, 1983). Reference here will therefore concentrate on the garden pea where the genetic control of flowering and apical senescence (or more correctly, cessation of apical growth) has been suggested to operate via changes in GA metabolism (Proebsting *et al.*, 1978; Proebsting and Heftmann, 1980). The genes at six established loci, *lf, e, sn, dne, hr* and *veg* interact to control the flowering behaviour of peas (see Murfet, 1985). The complementary genes *Sn* and *Dne* confer the photoperiod response (Barber, 1959; Murfet, 1971 a; King and Murfet, 1985). These two genes have a wide range of pleiotropic effects so that the photoperiod response can be recorded by observing any one of a number of characters including the node of first initiated flower, flowering time, apical senescence, number of reproductive nodes, yield, the rate of flower development, vegetative vigour and the production of lateral branches (Marx, 1968; Murfet, 1971 a, 1982; Reid and Murfet, 1984; Murfet and Reid, 1985). The most appropriate variable to use in examining the photoperiod response depends upon the other flowering genes present since in certain combinations some of these pleiotropic effects are masked (Murfet, 1971 a; Reid and Murfet, 1984). Grafting studies suggest that the gene combination *Sn Dne* regulates the production of an inhibitor which delays flower initation (Murfet, 1971 b; Murfet and Reid, 1973; King and Murfet, 1985) and apical senescence (Proebsting *et al.*, 1977). This inhibitor has a direct effect on apical senescence (Reid, 1980) and is produced in both the leaves and cotyledons (Murfet, 1971 a, 1985). The photoperiod response is mediated by phytochrome (Reid and Murfet, 1977; Reid, 1979 a) and probably acts by regulating some step in the *Sn Dne* pathway leading to the production of the inhibitor. The presence of either *sn* and/or *dne* results in essentially day-neutral plants (Murfet, 1971 a; King and Murfet, 1985). Gene *Hr* magnifies the photoperiod responses conferred by *Sn Dne* by prolonging the action of the *Sn Dne* system in the leaves (Murfet, 1973; Reid, 1979 b). Consequently plants possessing the combination *Sn Dne Hr* show pronounced photoperiod responses for at least certain characteristics.

Application studies have shown that GAs can influence the flowering and apical senescence of peas (e. g. Barber *et al.*, 1958; Wellensiek, 1969; Davies *et al.*, 1977). The effects are generally small and do not suggest that the flowering genes are operating directly by altering GA metabolism even though the responses are strongly dependent on the genotype (Barber *et al.*, 1958; Dalton and Murfet, 1975). However, two exceptions have been reported. Davies *et al.*, (1977) reported that both GA_3 and native GA_{20} could substantially delay the onset of apical senescence in the line G2 under long days (LD). Line G2 has genotype *lf E Sn Dne Hr Veg* (Murfet, 1978) and thus possesses a large response to photoperiod for the number of reproductive nodes and time and node of apical senescence (Marx, 1968). Application of GA_3 or GA_{20} under LD therefore has a similar delaying effect to short days (SD) and led to the suggestion that the gene combination *Sn Hr* was responsible for GA production in the leaves in SD which led to the delay in apical senescence (Davies *et al.*, 1977; Proebsting *et al.*, 1978). Secondly, application of GA_3 to line 24 plants (genotype *Lf e Sn Dne hr Veg*) led to a small and possibly indirect delay in the node of first initiated flower under continuous light but to a substantial and direct effect under SD (Reid *et al.*, 1977). To maximise this photoperiod response GA_3 had to be present from an early stage of growth and therefore was operating in an analogous fashion to the gene *Hr* (i. e. it only exerted a large effect under conditions where the *Sn Dne* pathway would be operating). Both pieces of evidence suggest a link between GAs and the genes determining the size of the photoperiod response in peas.

In a more detailed study, Proebsting *et al.*, (1978) reported that the metabolism of $[^3H]GA_9$ by leaves of G2 plants *(Sn Dne Hr)* was rapid in SD and inhibited by LD. In SD a range of GAs more polar than GA_9 was produced. In contrast, cultivars homozygous for *sn* or *hr* (Marx's I types) (Marx, 1968; Murfet, 1978) extensively metabolised $[^3H]GA_9$ in both LD and SD but the metabolites were reported to be different from those found in G2 plants. In particular, one component, designated GA_E, did not occur in I types, being unique to G2 plants. Proebsting and Heftmann (1980) found results similar to those obtained with G2 plants when feeds of $[^3H]GA_9$ were made to line G *(Lf E Sn Dne Hr Veg)* (Murfet, 1978) plants. These plants show a large photoperiod response for the flowering node (Marx, 1968), and can be induced to flower by one LD at the time of treatment (eight weeks). When the levels of GA-like substances were examined by the lettuce hypocotyl bioassay, extracts from the leaves of G2 plants grown under SD contained more GA-like activity in all three zones of activity than similar plants exposed to 12 LD prior to harvest (Proebsting *et al.*, 1978). Transfer back to SD restored the levels of GA-like activity. Extracts from the leaves of line I_2 *(lf E Sn Dne hr Veg)* grown under SD contained a reduced level of GA-like activity compared to G2 plants in the zone which co-chromatographed with GA_E suggesting that the prevention of apical senescence in G2 by SD may occur because the level of GA_E was sufficient to overcome any senescence stimulus associated with fruit development (Proebsting *et al.*, 1978). The level of GA_E was suggested to be controlled by genes *Sn* and *Hr* [*Dne* is present in these lines (King and Murfet, 1985)]. In LD, or plants with *sn* or *hr*, GA_E levels were too low to prevent apical senescence (Proebsting *et al.*, 1978) since the *Sn Dne* pathway would be blocked at some point. GA_E was originally suggested to be GA_1, but subsequent studies have not identified GA_1 from extracts of line G2 or any other light grown dwarf *(le)* line of peas (Davies *et al.*, 1982; Ingram, 1980; Sponsel, 1983a). Identification by GC-MS of the endogenous GAs from G2 plants led Ingram and Browning (1979) and Davies *et al.*, (1982) to suggest that the biological activity of GA_E may be due to GA_{19}. However, GA_{19} is not a likely metabolite of $[^3H]GA_9$. This leaves the nature and particularly the significance of GA_E in some doubt.

Ingram and Browning (1979) have shown that the levels of GA_{19} and GA_{20} are higher in developing seeds in SD than in LD relative to the level of GA_{29}. Ingram (1980) examined the levels of endogenous GAs by both the lettuce hypocotyl bioassay and GCMS-MIM techniques in the developing seeds from nine pure lines with different flowering genotypes. GA_{19} and GA_{44} levels were increased and GA_{17} levels decreased in the two lines carrying genes *Sn* and *Hr* but this response appeared independent of photoperiod. In one line, the gene *Hr* alone appeared able to confer this response. These results therefore fail to show a direct qualitative effect of either *Sn* or *Hr* on GA levels in developing seeds. Further, in seedlings, no effect of gene *Sn* on GA levels could be detected. Unfortunately isogenic lines for genes *Sn/sn* and *Hr/hr* were unavailable during this work.

Potts (1982) and Potts, Reid and Murfet (unpub.) have examined the levels of GA-like substances in approximately 20-day-old seedlings from essentially isogenic lines for *Sn/sn, Dne/dne* and *Hr/hr* using the rice seedling bioassay (cv. Tan-ginbozu). No qualitative differences in the levels of GA-like activity were found and only minor quantitative differences were apparent. In addition, photoperiod had only a minor quantitative effect on the levels of GA-like activity in the *Hr/hr* and *Dne/dne* lines.

The results of Proebsting *et al.* (1978), Ingram (1980) and Potts (1982) on the levels of endogenous GAs or GA-like activity do not provide strong evidence for a direct control over a step in the GA biosynthetic pathway by the genes *Sn, Hr* or *Dne.* The main differences recorded are possibly the consequence of altered growth rates and developmental patterns caused by these genes. The differences recorded in the levels of GA-like activity from shoots possibly reflects the age of the tissue extracted (see Potts *et al.*, 1982) and changes associated with the onset of apical senescence. Potts *et al.*, (unpub.) have shown that the level of GA-like activity drops in the shoot as it approaches senescence.

This view is supported by evidence from the interaction of internode length mutants, which are known to influence GA levels, and the flowering genes. For example, the flowering phenotype of genotype *Lf E Sn Dne Hr Veg* is the same regardless of whether *Na* or the "GA-deficient" mutant *na* is present (Murfet, 1985) and genotype *Lf E Sn Dne hr Veg* shows normal photoperiod responses regardless of whether *Lh* or *lh* and *Ls* or *ls* are present (Reid, unpub.). Further, segregation of the flowering genes *Hr/hr, Sn/sn* and *Dne/dne* are not masked by the presence of the dwarfing gene *le* (Barber, 1959; Murfet, 1971 a, 1973; King and Murfet, 1985). This is not to suggest that the internode length genes do not have a pleiotropic effect on flowering and vice versa (Barber, 1959; Dalton and Murfet, 1975) but rather that even dramatic changes in the endogenous GA status of the plant do not override the genetic control of flowering or the ability to respond to photoperiod. Likewise, the GA synthesis inhibitor AMO 1618 cannot override the effect of the flowering genes *Sn* and *Hr* although small responses have been reported (Reid, 1976). If any of the flowering genes *Hr, Sn* or *Dne* did operate by increasing the level of GAs it might be expected that some increase in seedling internode length would occur. None of these genes leads to significant increases in internode length (Barber, 1959; Potts, 1982; King, unpub.). These genetic and whole plant physiological results all point to the photoperiod genes *Sn* and *Dne* and the modifying gene *Hr* not operating directly via the control of steps in the GA biosynthetic pathway. The biochemical mode of action of the flowering genes in peas therefore remains a mystery. It is unfortunate that for early studies of GA-levels and metabolism in the flowering genotypes isogenic lines were unavailable and that the detection techniques were insufficient to provide firm answers. In such situations less direct techniques such as application studies and genetic recombinants are perhaps still capable of providing the most valid answers.

Photoperiod control over GA metabolism has been firmly established in other species (see Zeevaart, 1983) although mutants controlling these responses have generally not been isolated. For example, spinach plants exposed to LD possessed increased GA-like activity in the d_5 maize bioassay in one chromatographic zone and decreased activity in two other zones compared with plants maintained in SD. The rate of turnover of GAs appears to be increased under these LD conditions (Zeevaart, 1971). Identification of the native GAs by GC-MS suggested that an early 13-hydroxylation biosynthetic pathway occurs in spinach (Metzger and Zeevaart, 1980a) (Fig. 1). Under SD the level of GA_{19} was high and that of GA_{20} was low as measured by GC-SICM. After transfer to LD, the GA_{19} level decreased and those of GA_{20} and GA_{29} increased, suggesting that photoperiod controls the conversion of GA_{19} to GA_{20} (Metzger and Zeevaart, 1980b). The change in GA metabolism occurred prior to the onset of stem growth suggesting that GA_{20} levels may control stem elongation (Metzger and Zeevaart, 1980b). Application of GA_{20} supports this suggestion (Zeevaart, 1983) as do feeds with $[^2H]GA_{53}$ since in SD only $[^2H]GA_{44}$ and $[^2H]GA_{19}$ were identified while after two LD only $[^2H]GA_{20}$ was identified (Gianfagna *et al.,* 1983). However, the role that this step plays in the regulation of flowering in this long-day plant (LDP) is still unclear.

In *Silene armeria,* another LDP where flower promotion is accompanied by stem elongation, mutants have been reported which affect the time of flowering and the flowering response to GA_3 as well as stem elongation (Wellensiek, 1976). Cleland and Zeevaart (1970) found that AMO 1618 inhibited stem elongation but not flowering in this species suggesting that flower formation is not under GA control. Transfer from SD to LD was shown to increase the levels of GA-like substances in the d_5 maize bioassay. The studies of *Silene* mutants by Wellensiek (1976) suggest responses not found using the single variety used by Cleland and Zeevaart (1970). However, Suttle and Zeevaart (1979), working with the (GA insensitive) stem elongation mutant, provide no evidence implicating the GAs with flowering. If mutants influencing GA metabolism could be found in spinach or *Silene armeria* the relationship between the photoperiod control of GA metabolism, stem elongation and flowering might be resolved finally in these LDP.

VII. Site of Action of GA Mutants

The site of action of GA synthesis mutants has only been examined in a preliminary manner. However, these results suggest that mutants may be of considerable use in examining sites of GA production, transport and action. For example, the pre-fruiting shoots of extremely short *na* pea plants do not contain detectable levels of GAs (Potts and Reid, 1983; Ingram *et al.,* 1984) while the developing seeds contain similar levels of GA-like substances to those found in seeds from normal *Na* plants (Potts and Reid, 1983). Extracts from the pods at the time of contact of the developing seed also contain no GA-like activity corresponding to the major zones of activity in the developing seeds (Potts *et al.,* unpub.). These results suggest little export of GA-like substances occurs from developing seeds, a conclusion consistent with the observation that little increase in elongation of the upper internodes occurs in *na* plants during fruiting. The high level of GA-like substances in the developing seeds of *na* plants is also strong evidence for the developing seeds as a site of GA-synthesis. This has been shown previously in several species by detailed metabolic studies (e. g. Kamiya and Graebe, 1983) but genetic evidence may provide a method for analysing other potential sites of synthesis where the evidence from metabolic studies is not as clear-cut.

Ingram *et al.* (1985) have examined the metabolism of $[^3H^{13}C]GA_{20}$ in the roots following feeds to the shoots of *na* plants and found considerable dilution of the $[^3H^{13}C]GA_8$-catabolite and the $[^3H^{13}C]GA_{29}$-catabolite with endogenous $[^{12}C]$ compounds. In addition low levels of dilution were found in $[^3H^{13}C]GA_8$ and $[^3H^{13}C]GA_{29}$ in contrast to extracts from the shoots. This suggests that *na* may be at least partially inoperative in the roots as well as the developing seed. As a consequence no biologically active GAs would be exported at detectable levels from the roots to the shoots contrary to the hypothesis of Crozier and Reid (1971) that the roots convert biologically inactive GAs from the shoots to active forms before reexport to the shoot. Further, it suggests the roots are a site of GA synthesis. Major differences in the range of metabolites present in the roots and shoots of *na* plants were also observed. Perhaps the most significant difference was that the α, β-unsaturated ketone catabolites of GA_8 and GA_{29} were the predominant metabolites in the roots but were relatively minor metabolites in the shoots suggesting that the roots, through catabolism, may be important in the control of the level of biologically-active GAs (Ingram *et al.,* 1985). Sponsel (1983b) has demonstrated a similar compartmentalisation of GA_{29} catabolism in the testa of maturing seeds. These results suggest that gene *na* may not be an amorph. If this were so *na* might be expected to block GA synthesis in all tissues. Many possibilities exist to explain these results including alternative GA biosynthetic pathways, alternative enzymes (or non-specific enzymes) in the tissues producing GAs or that *na* is a regulator gene which is only operative in certain tissues.

Gene *Le,* which allows the 3β-hydroxylation of GA_{20} to GA_1 (Ingram *et al.,* 1984) also appears to be active in only specific sites within the pea plant since GA_1 is found in the highest concentrations in the young expanding apical portions of tall *(Le)* plants. It is found at much lower levels in mature tissue (Potts *et al.*, 1982; Potts and Reid, 1983) and has not been identified in the developing seed (Eeuwens *et al.*, 1973; Gaskin *et al.*, 1985). This may partially account for the fact that the effect of the *Le* gene is not graft transmissible (e. g. McComb and Mc Comb, 1970; Reid *et al.*, 1983). Even when *le* scions are grafted near actively growing *Le* apices no increase in elongation occurs in the *le* scions suggesting that GA_1 may also be immobilised or compartmentalised in some way within the apex (Reid, unpub.). Jones and Lang (1968) have reported a difference in the diffusable and extractable GA-like substances in the shoots of peas. From metabolism studies using $[^3H]GA_{20}$, the reduced level of GA_1 in mature tissues appears to be at least partly due to a reduction in the rate of 3β-hydroxylation of GA_{20} relative to 2β-hydroxylation of GA_{20} to GA_{29} (Ingram *et al.*, 1985). However, mature tissue can synthesise translocatable precursors of GA_1 which may be of importance to shoot elongation in intact plants since *Na* stocks can promote elongation in *na* scions (Reid *et al.*, 1983). This genetic evidence is consistent with evidence from other sources suggesting that the young leaves are the main source of biologically active GAs within the shoot (see Stoddart, 1983). The evidence from both *na* and *Le* plants clearly shows precise and specific ontogenetic control over GA synthesis.

The lack of GA_1 in developing seeds of *Le* peas (Eeuwens *et al.*, 1973; Gaskin *et al.*, 1985) raises the question of the biological significance of the high levels of other GAs found in developing seeds, especially when in regard to stem elongation GA_1 has been suggested to be the only biologically active GA in peas, maize and rice (Phinney and Spray, 1982; Ingram *et al.*, 1984; Phinney, 1984; Murfet and Reid, 1985). In these plants, other native GAs appear to possess activity only because of conversion to GA_1. AMO 1618 has been shown to dramatically reduce the level of GA-like substances in developing seeds while having only a small effect on seed development (Baldev *et al.*, 1965). Neither Sponsel (1982) nor Ingram and Browning (1979) could identify a clear role for the GAs in seed development. Eeuwens and Schwabe (1975) and Sponsel (1982) suggest that seed GAs are active in controlling pod development while Wareing and Seth (1967) and Stoddart (1983) suggest that the high levels of GAs in developing seeds may be important in attracting nutrients. They do not appear to be associated with the early growth of seedlings in peas since studies with GA-synthesis inhibitors (Sponsel, 1983b) and *na* plants (Reid, 1983) show *de novo* synthesis is responsible for seedling elongation, at least from node 1 onwards (counting from the cotyledons as zero). While it is possible that different GAs are biologically active in the shoot and the developing seed and pod this is perhaps unlikely and unnecessary to propose until direct supportive evidence is available. The lack of a proven precise function for the GAs in developing seeds and the absence at the current levels of detection of GA_1, even in *Le* seeds, might suggest that the GAs in developing seeds are merely secondary metabolites without a key regulatory role. Clearly the physiological significance of GAs in developing seeds requires critical examination especially in view of the large number of metabolic studies that have been based on this tissue.

The tissue specific nature of GA-synthesis mutants does not appear to have been examined in other species. However, in maize comparisons can be made between studies on seedlings and on the young tassels of maturing plants. The levels of GA_1 and GA_8 were found to be very low in tassels from normal plants, and conversion of $[^3H,^{13}C]GA_{20}$ to $[^3H,^{13}C]GA_1$ and $[^3H,^{13}C]GA_8$ could not be demonstrated in this tissue. $[^3H,^{13}C]GA_{29}$ was the only metabolite identified from such feeds in tassels from d_1, d_5 and normal plants (Heupel *et al.*, 1985). However, similar feeds to maize seedlings have shown that GA_{20} is metabolised to GA_1 in the shoots of normal and d_5 seedlings but not in d_1 plants (Spray *et al.*, 1984). This evidence suggests that, like peas, the effect of GA synthesis mutants in maize may also be tissue specific and illustrates the potential pitfalls that exist if the biochemical action of a gene is examined in a tissue other than that in which it exerts its primary effect.

The tissue specific nature of GA insensitive mutants is also well established in wheat and barley. The *Rht 3* gene causes insensitivity in all the GA-mediated physiological processes examined (Gale and Marshall, 1973; Ho *et al.*, 1981) while the dwarfing genes from Norin 10 types (*Rht 1* and *Rht 2*) are reported to influence GA sensitivity only in the shoot and not in the aleurone tissue (Gale and Marshall, 1973; Fick and Qualset, 1975). This specificity of action may well prove to be commercially significant (Gale and Hanson, 1982). In barley the reverse has been demonstrated where GA insensitive aleurone mutants have been isolated in which the shoot appears largely unaffected as no change in height was observed (Ho *et al.*, 1980). This suggests that at least certain steps are essential to all GA-mediated responses while others are less basic and depend on the response and tissue involved.

The influence of light and dark on dwarf mutants is one of the few cases where GA mutants have been used to probe the GA responses to environmental factors. In all cases examined [e. g. d_1, d_3, d_5 in maize (Sembdner and Schreiber, 1965), *le, na,* and *la cry*^{s} in peas (Reid, 1983), and *Rht 3* and *Rht 1* in wheat (Gale *et al.*, 1975; Baroncelli *et al.*, 1984)] the effects of the genes are expressed in darkness as well as in light, regardless of whether the mutants are GA synthesis or GA sensitivity types. This suggests that GA levels or sensitivity are limiting for growth in both the light and dark and that in none of the above cases does darkness allow the steps blocked by these mutants to be completely overcome. The term 'physiological dwarf' was coined to describe the situation in which dwarfness occurred in the light but not in the dark (see Jones, 1973) and was based on results with tall *(Le)* and dwarf *(le)* cultivars of peas (Lockhart, 1956). However the genetic evidence outlined above argues that the effects of light and GAs are, at least partly, independent of each other (Reid, 1983; Baroncelli *et al.*, 1984). Gaskin *et al.* (1985) recently reported that the shoots from a dark grown dwarf *(le)* cultivar of peas (cv. Progress No. 9) contain similar levels of GA_1 to those from a tall *(Le)* cultivar (cv. Alaska) and show similar elongation. Unfortunately, these cultivars are unrelated and differ in many other respects, including the levels of other GAs. The ratio of GA_1 to GA_{20} or GA_{29} is greater in the tall cultivar than in the dwarf cultivar, consistent with the known effect of the *Le* gene in light grown plants. These results therefore do not resolve the question of physiological dwarfism in peas but do highlight the need for the use of isogenic lines or segregating progenies if the effect of a particular gene difference is to be examined.

VIII. Conclusions

The biochemical site of action has now been determined for several GA synthesis mutants influencing internode length (Fig. 1). These results provide some of the strongest evidence available regarding the control of a development process by GA levels and point towards GA_1 being the only active GA in controlling elongation in maize, peas and rice (Phinney, 1984). These mutants provide precise tools for blocking GA synthesis at known steps and they may be of importance in determining the sites of GA synthesis, action and transport since preliminary information suggests that they show precise ontogenetic and tissue specificity. GA synthesis mutants influencing seed dormancy, onset of flowering, apical senescence and sex determination have also been proposed (Table 1) but cover only a minority of the developmental processes and environmental responses suggested to be controlled or influenced by the GAs. Further examination of the pleiotropic effects of proven GA synthesis mutants and the isolation and selection of further mutants may provide the best avenue available to define the regulatory role of endogenous GAs. For example, we presently have detailed information regarding the levels and biosynthetic pathways in developing seeds but lack basic information regarding their biological function.

GA insensitive mutants are common, particularly for internode length (Table 1). Evidence points to at least some of these mutants controlling the GA receptor site or limiting steps between this point and the manifestation of a GA response rather than by influencing GA metabolism (Stoddart, 1984; Potts *et al.*, 1985). The mechanism of action of GAs has proven a difficult area of study (Stoddart, 1983) and has recently been highlighted as an area of potential significance (Vanderhoef and Kosuge, 1984). The GA insensitive mutants offer as yet unexploited probes for examining this process especially where mutants are available to influence a sequence of steps in the one species.

Developmental mutants in higher plants, and GA mutants in particular, have not been exploited to the same degree as mutants in bacteria and lower plants possibly owing to the more complex developmental pathways involved and the partitioning of the problems between disciplines. However, their potential as invaluable research tools in plant biology has been demonstrated (e. g. Phinney, 1961) and provided a multidisciplinary approach, involving genetics, anatomy, physiology and biochemistry is adopted, they offer one of the best methods available to gain a complete understanding of facets of plant development.

Acknowledgements

I wish to thank Drs. T. J. Ingram, I. C. Murfet, W. C. Potts and J. J. Ross for helpful discussions during the development of my views on this topic and/or helpful comments on the manuscript, Mrs. T. Grabek for technical support and the Australian Research Grants Scheme for financial assistance.

IX. References

Allan, R. E., Vogel, O. A., Craddock, J. C., 1959: Comparative response to gibberellic acid of dwarf, semi-dwarf and tall winter wheat varieties. Agron. J. **51,** 737—740.

Atzorn, R., Weiler, E. W., 1983: The role of endogenous gibberellins in the formation of α-amylase by aleurone layers of germinating barley. Planta **159,** 289—299.

Baldev, B., Lang, A., Agatep, A. O., 1965: Gibberellin production in pea seeds developing in excised pods: effect of growth retardant AMO 1618. Science **147,** 155—157.

Barber, H. N., 1959: Physiological genetics of *Pisum.* II. The genetics of photoperiodism and vernalisation. Heredity **13,** 33—60.

Barber, H. N., Jackson, W. D., Murfet, I. C., Sprent, J. I., 1958: Gibberellic acid and the physiological genetics of flowering in peas. Nature (London) **182,** 1321.

Barendse, G. W. M., Lang, A., 1972: Comparison of endogenous gibberellins and the fate of applied radioactive gibberellin A_1 in a normal and a dwarf strain of Japanese morning glory. Plant Physiol. **49,** 836—841.

Baroncelli, S., Lercari, B., Cionini, P. G., Cavallini, A., D'Amato, F., 1984: Effect of light and gibberellic acid on coleoptile and first-foliage-leaf growth in durum wheat (*Triticum durum* L. Desf.). Planta **160,** 298—304.

Bearder, J. R., 1980: Plant hormones and other growth substances — their background, structure and occurrence. In: MacMillan, J. (ed.), Hormonal Regulation of Development. I. Molecular Aspects of Plant Hormones (Encyclopaedia of Plant Physiology, New Series, Vol. 9), pp. 9—112. Berlin - Heidelberg - New York: Springer-Verlag.

Bearder, J. R., 1983: *In vivo* diterpenoid biosynthesis in *Gibberella fujikuroi:* the pathway after *ent*-kaurene. In: Crozier, A. (ed.), The Biochemistry and Physiology of Gibberellins, Vol. 1., pp. 251—387. New York: Praeger.

Blixt, S., 1972: Mutation genetics in *Pisum.* Agri Hortique Genetica **30,** 1—293.

Blonstein, A. D., 1981: Developmental differences between dwarfing genes in barley. In: Asher, M. J. C. (ed.), Barley Genetics IV (Proceedings of the Fourth International Barley Genetics Symposium, Edinburgh 1980), pp. 566—570. Edinburgh: Edinburgh Univ. Press.

Brian, P. W., 1957: The effects of some microbial metabolic products on plant growth. Symp. Soc. Exp. Biol. **11,** 166—181.

Brian, P. W., Hemming, H. G., 1955: The effect of gibberellic acid on shoot growth of pea. Physiol. Plant. **8,** 669—681.

Cleland, C. F., Zeevaart, J. A. D., 1970: Gibberellins in relation to flowering and stem elongation in the long-day plant *Silene armeria.* Plant Physiol. **46,** 392—400.

Cooper, J. P., 1958: The effect of gibberellic acid on a genetic dwarf in *Lolium perenne.* New Phytol. **57,** 235—238.

Cross, B. E., Galt, R. H. B., Hanson, J. R., 1964: The biosynthesis of the gibberellins. Part I. (—)-Kaurene as a precursor of gibberellic acid. J. Chem. Soc. 295—300.

Crozier, A., 1983: The Biochemistry and Physiology of Gibberellins, Vols. 1 and 2, p. 568 and p. 452. New York: Praeger.

Crozier, A., Durley, R. C., 1983: Modern methods of analysis of gibberellins. In: Crozier, A. (ed.), The Biochemistry and Physiology of Gibberellins, Vol. 1, pp. 485—560. New York: Praeger.

Crozier, A., Kuo, C. C., Durley, R. C., Pharis, R. P., 1970: The biological activities of 26 gibberellins in nine plant bioassays. Can. J. Bot. **48,** 867—877.

Crozier, A., Reid, D. M., 1971: Do roots synthesize gibberellins? Can. J. Bot. **49,** 967—975.

Dalton, P. J., Murfet, I. C., 1975: The effect of gibberellic acid and genotype *le la cry^c* on flowering in peas. *Pisum* Newsl. **7,** 5—7.

Davies, L. J., Rappaport L., 1975: Metabolism of tritiated gibberellins in *d-5* dwarf maize. II. [^{3}H] gibberellin A_1 and [^{3}H] gibberellin A_3, and related compounds. Plant Physiol. **56,** 60—66.

Davies, P. J., Emshwiller, E., Gianfagna, T. J., Proebsting, W. M., Noma, M., Pharis, R. P., 1982: The endogenous gibberellins of vegetative and reproductive tissue of G2 peas. Planta **154,** 266—272.

Davies, P. J., Proebsting, W. M., Gianfagna, T. J., 1977: Hormonal relationships in whole plant senescence. In: Pilet, P. E. (ed.), Plant Growth Regulation (9th International Conference on Plant Growth Substances, 1976, Lausanne), pp. 273—280. Berlin - Heidelberg - New York: Springer-Verlag.

de Haan, H., 1927: Length factors in *Pisum.* Genetica **9,** 481—497.

Eeuwens, C. J., Gaskin, P., Macmillan, J., 1973: Gibberellin A_{20} in seed of *Pisum sativum* L. cv. Alaska. Planta **115,** 73—76.

Eeuwens, C. J., Schwabe, W. W., 1975: Seed and pod wall development in *Pisum sativum* L. in relation to extracted and applied hormones. J. Exp. Bot. **26,** 1—14.

Favret, E. A., Favret, G. C., Malvarez, E. M., 1975: Genetic regulatory mechanisms for seedling growth in barley. In: Barley Genetics III (Proc. 3rd Int. Barley Genet. Symp.), pp. 37—42. Garching, W. Germany.

Fick, G. N., Qualset, C. O., 1973: Genes for dwarfness in wheat, *Triticum aestivum* L. Genetics **75,** 531—539.

Fick, G. N., Qualset, C. O., 1975: Genetic control of endosperm amylase activity and gibberellic acid responses in standard-height and short-statured wheats. Proc. Nat. Acad. Sci., U.S.A. **72,** 892—895.

Foster, C. A., 1977: Slender: an accelerated extension growth mutant of barley. Barley Genet. Newsl. **7,** 24—27.

Gale, M. D., Hanson, P. R., 1982: The plant breeding potential of genetic variation in cereal phytohormone systems. In: McLaren, J. S. (ed.), Chemical Manipulation of Crop Growth and Development (Proceedings of Easter School of Agricultural Science, Univ. of Nottingham), pp. 425—449. London: Butterworths.

Gale, M. D., Law, C. N., 1977: The identification and exploitation of Norin 10 semi-dwarfing genes. Plant Breeding Inst., Cambridge. Annu. Report. pp. 21—35.

Gale, M. D., Law, C. N., Marshall, G. A., Worland, A. J., 1975: The genetic control of gibberellic acid insensitivity and coleoptile length in a 'dwarf' wheat. Heredity **34,** 393—399.

Gale, M. D., Marshall, G. A., 1973: Insensitivity to gibberellin in dwarf wheats. Ann. Bot. **37,** 729—735.

Gale, M. D., Marshall, G. A., 1976: The chromosomal location of *Gai 1* and *Rht 1,* genes for gibberellin insensitivity and semi-dwarfism, in a derivative of Norin 10 wheat. Heredity **37,** 283—289.

Gaskin, P., Gilmour, S. J., Lenton, J. R., MacMillan, J., Sponsel, V. M., 1984: Endogenous gibberellins and kaurenoids identified from developing and germinating barley grain. J. Plant Growth Regul. **2,** 229—242.

Gaskin, P., Gilmour, S. J., MacMillan, J., Sponsel, V. M., 1985: Gibberellins in immature seeds and dark-grown shoots of *Pisum sativum.* Gibberellins identified in the tall cultivar Alaska in comparison with those in the dwarf Progress No. 9. Planta **163,** 283—289.

Gianfagna, T., Zeevaart, J. A. D., Lusk, W. J., 1983: Effect of photoperiod on the metabolism of deuterium-labelled gibberellin A_{53} in spinach. Plant Physiol. **72,** 86—89.

Gilmour, S. J., Gaskin, P., Sponsel, V. M., MacMillan, J., 1984: Metabolism of gibberellins in immature barley grain. Planta **161,** 186—192.

Gilmour, S. J., MacMillan, J., 1984: Effect of inhibitors of gibberellin biosynthesis on the induction of α-amylase in embryoless caryopses of *Hordeum vulgare* cv. Himalaya. Planta **162,** 89—90.

Goto, N., Esashi, Y., 1973: Diffusible and extractable gibberellins in bean cotyledons in relation to dwarfism. Physiol. Plant. **28,** 480—489.

Graebe, J. E., Ropers, H. J., 1978: Gibberellins. In: Letham, D. S., Goodwin, P. B., Higgins, T. J. V. (eds.), Phytohormones and Related Compounds — A Comprehensive Treatise. Vol. I. The Biochemistry of Phytohormones and Related Compounds, pp. 107—204. Amsterdam: Elsevier/North-Holland.

Guttridge, C. G., 1973: Stem elongation and runnering in the mutant strawberry, *Fragaria vesca* L. *arborea* Staudt. Euphytica **22,** 357—361.

Harada, J., Vergarra, B. S., 1971: Response of different rice varieties to gibberellins. Crop Sci. **11,** 373—374.

Hedden, P., 1983: *In vitro* metabolism of gibberellins. In: Crozier, A. (eds.), The Biochemistry and Physiology of Gibberellins, Vol. 1, pp. 99—149. New York: Praeger.

Hedden, P., MacMillan, J., Phinney, B. O., 1978: The metabolism of the gibberellins. Annu. Rev. Plant Physiol. **29,** 149—192.

Hedden, P., Phinney, B. O., 1979: Comparison of *ent*-kaurene and *ent*-isokaurene synthesis in cell-free systems from etiolated shoots of normal and dwarf-5 maize seedlings. Phytochemistry **18,** 1475—1479.

Hedden, P., Phinney, B. O., MacMillan, J., Sponsel, V. M., 1977: Metabolism of kaurenoids by *Gibberella fujikuroi* in the presence of the plant growth retardant, N, N, N-trimethyl-l-methyl-(2′,6′,6′-trimethylcyclohex-2′-en-l′-yl)prop-2-enylammonium iodide. Phytochemistry **16,** 1913—1917.

Heupel, R. C., Phinney, B. O., Spray, C. R., Gaskin, P., MacMillan, J., Hedden, P., Graebe, J. E., 1985: Native gibberellins and metabolism of [^{14}C] gibberellin A_{53}

and of [17-^{13}C, 17-$^{3}H_2$] gibberellin A_{20} in tassels of *Zea mays*. Phytochemistry **24,** 47—53.

Ho, T. D., Nolan, R. C., Shute, D. E., 1981: Characterisation of a gibberellin-insensitive dwarf wheat, D6899. Plant Physiol. **67,** 1026—1031.

Ho, T. D., Shih, S., Kleinhofs, A., 1980: Screening for barley mutants with altered hormone sensitivity in their aleurone layers. Plant Physiol. **66,** 153—157.

Hoad, G. V., 1983: Gibberellin bioassays and structure-activity relationships. In: Crozier, A. (ed.), The Biochemistry and Physiology of Gibberellins, Vol. 2, pp. 57—94. New York: Praeger.

Hopp, H. E., Favret, G. C., Favret, E. A., 1981: Control of barley development using dwarf mutants. In: Induced Mutants as a Tool for Crop Improvement, p. 243. Vienna: IAEA.

Hu, M. L., Konzak, C. F., 1974: Genetic association of gibberellic acid insensitivity and semi-dwarfing in hexaploid wheat. Annu. Wheat Newsl. **20,** 184—185.

Ingram, T. J., 1980: Gibberellins and reproductive development in peas. Ph. D. Thesis, University of East Anglia, U. K.

Ingram, T. J., Browning, G., 1979: Influence of photoperiod on seed development in the genetic line of peas G2 and its relation to changes in endogenous gibberellins measured by combined gas chromatography — mass spectrometry. Planta **146,** 423—432.

Ingram, T. J., Reid, J. B., MacMillan, J., 1985: Internode length in *Pisum sativum.* L. The kinetics of growth and [^{3}H] gibberellin A_{20} metabolism in genotype *na Le.* Planta **164,** 429—438.

Ingram, T. J., Reid, J. B., Murfet, I. C., Gaskin, P., Willis, C. L., MacMillan, J., 1984: Internode length in *Pisum.* The *Le* gene controls the 3β-hydroxylation of gibberellin A_{20} to gibberellin A_1. Planta **160,** 455—463.

Ingram, T. J., Reid, J. B., Potts, W. C., Murfet, I. C., 1983: Internode length in *Pisum.* IV. The effect of the *Le* gene on gibberellin metabolism. Physiol. Plant. **59,** 607—616.

Inouke, M., Sakurai, N., Kuraishi, S., 1982: Growth regulation of dark-grown dwarf barley coleoptile by the endogenous IAA content. Plant Cell Physiol. **23,** 689—698.

Jones, R. L., 1973: Gibberellins: their physiological role. Annu. Rev. Plant Physiol. **24,** 571—598.

Jones, R. L., Lang, A., 1968: Extractable and diffusible gibberellins from light- and dark-grown pea seedlings. Plant Physiol. **43,** 629—634.

Kamiya, Y., Graebe, J. E., 1983: The biosynthesis of all major pea gibberellins in a cell-free system from *Pisum sativum.* Phytochemistry **22,** 682—689.

Katsumi, M., Foard, D. E., Phinney, B. O., 1983: Evidence for the translocation of gibberellin A_3 and gibberellin-like substances in grafts between normal, dwarf $_1$ and dwarf $_5$ seedlings of *Zea mays* L. Plant Cell Physiol. **24,** 379—388.

Katsumi, M., Phinney, B. O., Jefferies, P. R., Hendrick, C. A., 1964: Growth response of the *d-5* and *an-1* mutants of maize to some kaurene derivatives. Science **144,** 849—850.

Keller, P. L., Coulter, M. W., 1982: The relationship of endogenous gibberellins to light-regulated stem elongation rates in dwarf and normal cultivars of *Pisum sativum* L. Plant Cell Physiol. **23,** 409—416.

King, W., Murfet, I. C., 1985: Flowering in *Pisum:* A sixth locus, *Dne.* Ann. Bot., **56,** 835—846.

Köhler, D., 1970: The effect of red light on the growth and gibberellin-content of pea seedlings. Z. Pflanzenphysiol. **62,** 426—435.

Koornneef, M., Jorna, M. L., Brinkhorst - van der Swan, D. L. C., Karssen, C. M., 1982: The isolation of abscisic acid (ABA) deficient mutants by selection of induced revertants in non-germinating gibberellin sensitive lines of *Arabidopsis thaliana* (L.). Heynh. Theor. Appl. Genet. **61,** 385—393.

Koornneef, M., van der Veen, J. H., 1980: Induction and analysis of gibberellin sensitive mutants in *Arabidopsis thaliana* (L.). Heynh. Theor. Appl. Genet. **58,** 257—263.

Koornneef, M., van der Veen, J. H., Spruit, C. J. P., Karssen, C. M., 1981: Isolation and use of mutants with an altered germination behaviour in *Arabidopsis thaliana* and tomato. In: Induced Mutants as a Tool for Crop Plant Improvement **251,** 227—232. Vienna: IAEA-SM.

Kuraishi, S., 1974: Biogenesis of auxin in the coleoptile of a semi-brachytic barley, uzu. Plant Cell Physiol. **15,** 295—306.

Kurogochi, S., Murofushi, N., Ota, Y., Takahashi, N., 1979: Identification of gibberellins in the rice plant and quantitative changes of gibberellin A_{19} throughout its life cycle. Planta **146,** 185—191.

Lockhart, J. A., 1956: Reversal of light inhibition of pea stem growth by gibberellins. Proc. Nat. Acad. Sci., U.S.A. **42,** 841—848.

Loy, J. B., Liu, P. B. W., 1974: Response of seedlings of a dwarf and a normal strain of watermelon to gibberellins. Plant Physiol. **53,** 325—330.

McComb, A. J., McComb, J. A., 1970: Growth substances and the relation between phenotype and genotype in *Pisum sativum.* Planta **91,** 235—245.

MacMillan, J., 1980: Hormonal Regulation of Development. I. Molecular Aspects of Plant Hormones (Encyclopedia of Plant Physiology, New Series, Vol. 9). Berlin - Heidelberg - New York: Springer-Verlag.

MacMillan, J., 1984: Analysis of plant hormones and metabolism of gibberellins. In: Crozier, A., Hillman, J. R. (eds.), The Biosynthesis and Metabolism of Plant Hormones (Society of Experimental Biology Seminar, Series 23), pp. 1—16. Cambridge: Cambridge University Press.

McVittie, J. A., Gale, M. D., Marshall, G. A., Westcott, B., 1978: The intra-chromosomal mapping of the Norin 10 and Tom Thumb dwarfing genes. Heredity **40,** 67—70.

Magara, J., 1963: Notes on the possible role of endogenous "gibberellins" in the determining of monofactorial dwarfism in dwarf sweet peas (*Lathyrus odoratus* L.) and in mutants d_1 and d_5 of maize. Ann. Physiol. Veg. **5,** 249—261.

Marx, G. A., 1968: Influence of genotype and environment on senescence in peas, *Pisum sativum* L. BioScience **18,** 505—506.

Metzger, J. D., Zeevaart, J. A. D., 1980a: Identification of six endogenous gibberellins in spinach shoots. Plant Physiol. **65,** 623—626.

Metzger, J. D., Zeevaart, J. A. D., 1980b: Effect of photoperiod on the levels of endogenous gibberellin in spinach as measured by combined gas chromatography-selected ion current monitoring. Plant Physiol. **66,** 844—846.

Moh, C. C., Alan, J. J., 1967: The response of a radiation-induced dwarf bean mutant to gibberellic acid. Turrialba **17,** 176—178.

Morris, R., Schmidt, J. W., Johnson, V. A., 1972: Chromosomal location of a dwarfing gene in "Tom Thumb" wheat derivative by monosomic analysis. Crop Sci. **12,** 247—249.

Murakami, Y., 1972: Dwarfing genes in rice and their relation to gibberellin biosynthesis. In: Carr, D. J. (ed.), Plant Growth Substances, 1970, pp. 166—174. Berlin - Heidelberg - New York: Springer-Verlag.

Murfet, I. C., 1971a: Flowering in *Pisum.* Three distinct phenotypic classes determined by the interaction of a dominant early and a dominant late gene. Heredity **26,** 243—257.

Murfet, I. C., 1971b: Flowering in *Pisum:* reciprocal grafts between known genotypes. Aust. J. Biol. Sci. **24,** 1089—1101.

Murfet, I. C., 1973: Flowering in *Pisum. Hr,* a gene for high response to photoperiod. Heredity **31,** 157—164.

Murfet, I. C., 1978: The flowering genes *Lf, E, Sn* and *Hr* in *Pisum:* their relationship with other genes, and their descriptions and type lines. *Pisum* Newsl. **10,** 48—52.

Murfet, I. C., 1982: Flowering in the garden pea: expression of gene *Sn* in the field and use of multiple characters to detect segregation. Crop Sci. **22,** 923—926.

Murfet, I. C., 1985: *Pisum sativum* L. In: Halevy, A. H. (ed.), Handbook of Flowering, Vol. IV, pp. 97—126. Boca Raton, Florida: CRC Press.

Murfet, I. C., Reid, J. B., 1973: Flowering in *Pisum:* evidence that gene *Sn* controls a graft-transmissible inhibitor. Aust. J. Biol. Sci. **26,** 675—677.

Murfet, I. C., Reid, J. B., 1985: The control of flowering and internode length in *Pisum.* In: Hebblethwaite, P. D., Heath, M. C., Dawkins, T. C. K. (eds.), The Pea Crop: a Basis for Improvement, pp 67—80. London: Butterworths.

Murofushi, N., 1983: Life cycle regulation in rice by endogenous plant hormones. In: Miyamoto, J. (ed.), Pesticide Chemistry. Human Welfare and the Environment (Proceedings of the 5th International Congress of Pesticide Chemistry, 1982, Kyoto), pp. 21—28. Oxford: Pergamon Press.

Musgrave, A., Kende, H., 1970: Radioactive gibberellin A_5 and its metabolism in dwarf peas. Plant Physiol. **45,** 53—55.

Ogawa, Y., 1962: Quantitative differences of gibberellin-like substances in normal and dwarf varieties of *Pharbitis nil.* Chois. Bot. Mag. Tokyo **75,** 449—450.

Perez, A. T., Marsh, H. V., Lachman, W. H., 1974: Physiology of the yellow-green 6 gene in tomato. Plant Physiol. **53,** 192—197.

Phinney, B. O., 1956: Growth response of single-gene dwarf mutants in maize to gibberellic acid. Proc. Nat. Acad. Sci., U.S.A. **42,** 185—189.

Phinney, B. O., 1961: Dwarfing genes in *Zea mays* and their relation to the gibberellins. In: Klein, R. M. (ed.), Plant Growth Regulation, pp. 489—501. Ames, Iowa: Iowa State College Press.

Phinney, B. O., 1984: Gibberellin A_1, dwarfism and the control of shoot elongation in higher plants. In: Crozier, A., Hillman, J. R. (eds.), The Biosynthesis and Metabolism of Plant Hormones (Society of Experimental Biology Seminar, Series 23), pp. 17—41. Cambridge: Cambridge University Press.

Phinney, B. O., Spray, C., 1982: Chemical genetics and the gibberellin pathway in *Zea mays* L. In: Wareing, P. F. (ed.), Plant Growth Substances 1982, pp. 101—110. London Academic Press.

Potts, W. C., 1982: The involvement of gibberellins with the internode length and flowering genotypes of *Pisum.* Ph. D. Thesis, University of Tasmania, Australia.

Potts, W. C., Reid, J. B., 1983: Internode length in *Pisum.* III. The effect and interaction of the *Na/na* and *Le/le* gene differences on endogenous gibberellin-like substances. Physiol. Plant. **57,** 448—454.

Potts, W. C., Reid, J. B., Murfet, I. C., 1982: Internode length in *Pisum.* I. The effect of the *Le/le* gene difference on endogenous gibberellin-like substances. Physiol. Plant. **55,** 323—328.

Potts, W. C., Reid, J. B., Murfet, I. C., 1985: Internode length in *Pisum.* Gibberellins and the slender phenotype. Physiol. Plant. **63,** 357—364.

Proano, V. A., Greene, G. L., 1968: Endogenous gibberellins of a radiation induced single gene dwarf mutant of bean. Plant Physiol. **43,** 613—618.

Proebsting, W. M., Davies, P. J., Marx, G. A., 1977: Evidence for a graft-transmissible substance which delays apical senescence in *Pisum sativum* L. Planta **135,** 93—94.

Proebsting, W. M., Davies, P. J., Marx, G. A., 1978: Photoperiod-induced changes in gibberellin metabolism in relation to apical growth and senescence in genetic lines of peas (*Pisum sativum* L.). Planta **141,** 231—238.

Proebsting, W. M., Heftman, E., 1980: The relationship of [^{3}H]GA_9 metabolism to photoperiod induced flowering in *Pisum sativum* L. Z. Pflanzenphysiol. **98,** 305—309.

Radley, M., 1970: Comparison of endogenous gibberellins and response to applied gibberellin of some dwarf and tall wheat cultivars. Planta **92,** 292—300.

Redei, G. P., 1975: Induction of auxotrophic mutations in plants. In: Ledoux, L. (ed.), Genetic Manipulations with Plant Material, pp. 329—349. New York - London: Plenum Press.

Reid, J. B., 1976: Regulation of Flowering in *Pisum*. Ph. D. Thesis, University of Tasmania, Australia.

Reid, J. B., 1979a: Red-far-red reversibility of flower development and apical senescence in *Pisum*. Z. Pflanzenphysiol. **93,** 297—301.

Reid, J. B., 1979b: Flowering in *Pisum:* the effect of age on the gene *Sn* and the site of action of gene *Hr*. Ann. Bot. **44,** 163—173.

Reid, J. B., 1980: Apical senescence in *Pisum:* a direct or indirect role for the flowering genes. Ann. Bot. **45,** 195—201.

Reid, J. B., 1983: Internode length genes in *Pisum*. Do the internode length genes affect growth in dark-grown plants? Plant Physiol. **72,** 759—763.

Reid, J. B., 1986: Internode length in *Pisum*. Three further loci, *lh, ls* and *lk*. Ann. Bot., **57,** in press.

Reid, J. B., Dalton, P. J., Murfet, I. C., 1977: Flowering in *Pisum:* does gibberellic acid directly influence the flowering process? Aust. J. Plant Physiol. **4,** 479—483.

Reid, J. B., Murfet, I. C., 1977: Flowering in *Pisum:* the effect of light quality on genotype *lf e Sn Hr*. J. Exp. Bot. **28,** 1357—1364.

Reid, J. B., Murfet, I. C., 1984: Flowering in *Pisum:* a fifth locus, *Veg*. Ann. Bot. **53,** 369—382.

Reid, J. B., Murfet, I. C., Potts, W. C., 1983: Internode length in *Pisum*. II. Additional information on the relationship and action of loci *Le, La, Cry, Na* and *Lm*. J. Exp. Bot. **34,** 349—364.

Reid, J. B., Potts, W. C., 1986: Internode length in *Pisum*. Two further mutants, *lh* and *ls,* with reduced gibberellin synthesis, and a gibberellin insensitive mutant, *lk*. Physiol. Plant., in press.

Sawhney, V. K., 1974: Morphogenesis of the stamenless-2 mutant in tomato. III. Relative levels of gibberellins in the normal and mutant plants. J. Exp. Bot. **25,** 1004—1009.

Sembdner, G., Schreiber, K., 1965: Activities of gibberellin A_3, A_5 and A_6 on the maize-dwarf mutants d_1, d_3 and d_5 in the dark. Flora, oder Allg. Bot. Z. *Abt.* A **156,** 359—363.

Shifriss, O., 1973: The drooping syndrome of *Ricinus*. J. Hered. **64,** 351—355.

Sidorova, K. K., 1970: The study of allelism in phenotypically identical mutants of pea in connection with the law of homologous series in hereditary variability. Genetika **6,** 23—35.

Smith, R. R., 1974: Inheritance of a gibberellin-responsive dwarf mutant in red clover. Euphytica **23,** 597—600.

Sponsel, V. M., 1982: Effects of applied gibberellins and napthylacetic acid on pod development in fruits of *Pisum sativum* L. cv. Progress No. 9. J. Plant Growth Regul. **1,** 147—152.

Sponsel, V. M., 1983 a: *In vivo* gibberellin metabolism in higher plants. In: Crozier, A. (ed.), The Biochemistry and Physiology of Gibberellins, Vol. 1, pp. 151—250. New York: Praeger.

Sponsel, V. M., 1983 b: The localisation, metabolism and biological activity of gibberellins in maturing and germinating seeds of *Pisum sativum* cv. Progress No. 9. Planta **159,** 454—468.

Spray, C., Phinney, B. O., Gaskin, P., Gilmour, S. J., MacMillan, J., 1984: Internode length in *Zea mays* L. The dwarf-1 mutant controls the 3β-hydroxylation of gibberellin A_{20} to gibberellin A_1. Planta **160,** 464—468.

Stoddart, J. L., 1983: Sites of gibberellin biosynthesis and action. In: Crozier, A. (ed.), The Biochemistry and Physiology of Gibberellins, Vol. 2, pp. 1—55. New York: Praeger.

Stoddart, J. L., 1984: Growth and gibberellin-A_1 metabolism in normal and gibberellin-insensitive *(Rht 3)* wheat (*Triticum aestivum* L.) seedlings. Planta **161,** 432—438.

Stoddart, J. L., Venis, M. A., 1980: Molecular and subcellular aspects of hormone action. In: MacMillan, J. (ed.), Hormonal Regulation of Development. I. Molecular Aspects of Plant Hormones (Encyclopaedia of Plant Physiology, New Series, Vol. 9), pp. 445—510. Berlin - Heidelberg - New York: Springer-Verlag.

Suge, H., 1972: Effect of uzu *(uz)* gene on the level of endogenous gibberellins in barley. Jpn. J. Genet **47,** 423—430.

Suge, H., 1978: The genetic control of gibberellin production in rice. Jpn. J. Genet. **53,** 199—207.

Suge, H., 1979: Gibberellin relationships in a dwarf mutant of sweet potato. Jpn. J. Genet. **54,** 35—42.

Suge, H., 1983: Gibberellin relationships in a dwarf mutant of barley: brachytic *(br_2)*. Jpn. J. Genet. **58,** 555—566.

Suge, H. Murakami, Y., 1968: Occurrence of a rice mutant deficient in gibberellin-like substances. Plant Cell Physiol. **9,** 411—414.

Suttle, J. C., Zeevaart, J. A. D., 1979: Stem growth, flower formation and endogenous gibberellins in a normal and a dwarf strain of *Silene armeria.* Planta **145,** 175—180.

Suzuki, Y., Kurogochi, S., Murofushi, N., Ota, Y., Takahashi, N., 1981: Seasonal changes of GA_1, GA_{19} and abscisic acid in three rice cultivars. Plant Cell Physiol. **22,** 1085—1093.

Trewavas, A., 1981: How do plant growth substances work? Plant Cell Environ. **4,** 203—228.

Vanderhoef, L. N., Kosuge, T., 1984: The molecular biology of plant hormone action: research directions of the future (Workshop summaries — II), p. 40. Rockville: Am. Soc. Plant Physiol.

Wareing, P. F., Seth, A. K., 1967: Aging and senescence in the whole plant. Symp. Soc. Exp. Biol. **21,** 543—558.

Weber, V. E., Gottschalk, W., 1973: Die Beziehungen zwischen Zellgröße und Internodienlänge bei strahleninduzierten *Pisum*-Mutanten. Beitr. Biol. Pflanz. **49,** 101—126.

Wellensiek, S. J., 1969: The physiological effects of flower forming genes in peas. Z. Pflanzenphysiol. **60,** 1388—1402.

Wellensiek, S. J., 1976: A genetical look at flower formation in *Silene armeria* L. In: Jacques, R. (ed.), Etudes de Biologie Végétale, Hommage au Professor Pierre Chouard, pp. 301—312. Paris: Louis-Jean.

Wylie, A., Ryugo, K., 1971: Diffusible and extractable growth regulators in normal and dwarf shoot apices of peach, *Prunus persica* Botsch. Plant Physiol. **48,** 91—93.

Zeevaart, J. A. D., 1971: Effects of photoperiod on growth rate and endogenous gibberellins in the long-day rosette plant spinach. Plant Physiol. **47,** 821—827.

Zeevaart, J. A. D., 1983: Gibberellins and flowering. In: Crozier, A. (ed.), The Biochemistry and Physiology of Gibberellins, Vol. 2, pp. 333—374. New York: Praeger.

Zeevaart, J. A. D., 1984: Gibberellins in single gene dwarf mutants of tomato. Plant Physiol. Suppl. **75,** 186.

Chapter 2

Genetic Aspects of Abscisic Acid

Maarten Koornneef

Department of Genetics, Agricultural University, 53, Generaal Foulkesweg, NL-6703 BM Wageningen, The Netherlands

With 3 Figures

Contents

I. Introduction

A. Abscisic Acid as a Plant Hormone

Abscisic acid (ABA) is a naturally occurring plant hormone, probably present in all higher plants. Its discovery in the sixties, and its chemical structure have been described in several reviews (Addicott and Carns, 1983; Milborrow, 1984). ABA may be involved, often as an inhibitor, in many physiological processes such as abscission, bud- and seed dormancy, elongation growth, stomatal opening, root growth, geotropism, fruit ripening and senescence (Walton, 1980; Addicott and Carns, 1983; Milborrow, 1984). In many cases, however, the role of ABA could not be established conclusively, mainly because of the inadequate experimental approaches. In addition, experiments examining the correlation between endogenous ABA levels and physiological effects, or those involving the exogenous application of ABA etc. never provide more than circumstantial evidence, and compartmentation as well as tissue- and time-specific differences in hormone sensitivity complicate the interpretation of results. Methods of manipulating endogenous hormone levels would provide more direct approaches (Karssen, 1982) but specific chemical inhibitors, such as there are for gibberellins, are not known for ABA. The use of isogenic genotypes differing in endogenous ABA content provides plant physiologists with an important tool to elucidate the regulatory function of this compound. This genetic approach has been successfully applied to study the relation of ABA with stomatal closure and seed dormancy. Mutants also contributed to research on the pathway by which ABA is synthesized in the plant.

B. Biosynthesis and Metabolism of ABA

ABA is a sesquiterpenoid derived from mevalonic acid (MVA). Two pathways have been suggested (Milborrow, 1983):

1. The *direct pathway,* where ABA is synthesized from a C-15 precursor, presumably farnesyl pyrophosphate.
2. The *indirect pathway,* where ABA is formed via cleavage of a C-40 precursor such as violaxanthin, a xanthophyll that is derived from carotenes. Xanthoxin is probably the ABA precursor derived from violaxanthin (Fig. 1).

Experiments by Creelman and Zeevaart (1984) showed that stress-induced ABA synthesized in an $^{18}O_2$ atmosphere contained a labelled oxygen atom only in the carboxyl group, providing strong evidence that at least stress-induced ABA is derived from a C-40 precursor. The arguments in favour of the direct pathway come from experiments where ^{14}C phytoene and ^{3}H-MVA were fed to avocado fruits; only ^{3}H was found in ABA (Milborrow, 1983). Creelman and Zeevaart (1984) have critically discussed these experiments.

ABA can be rapidly metabolized to phaseic acid (PA), for example upon rehydration of wilted leaves (Zeevaart, 1980) and PA thereafter is reduced to dihydrophaseic acid (DPA) (Fig. 1). ABA and its direct metabolites can be conjugated with glucose (Walton, 1980; Milborrow, 1983, 1984). Although in some systems PA has been reported to have hormone activity (e. g. Uknes and Ho, 1984), in general only ABA is found to be active in biological test systems (Walton, 1980).

C. Genetic Aspects of ABA

Genetic aspects of ABA have been reviewed by Quarrie (1983), who emphasizes the potential use of ABA in agriculture. The present review

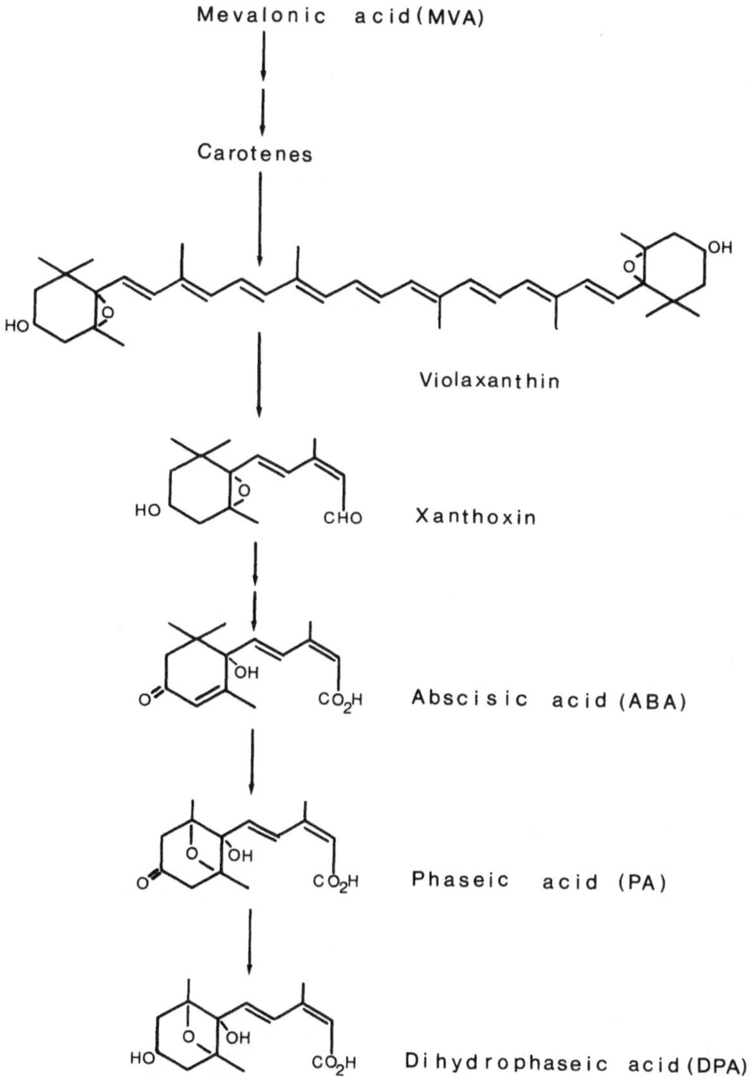

Mevalonic acid (MVA)

Carotenes

HO O O OH

Violaxanthin

HO O CHO

Xanthoxin

O OH CO_2H

Abscisic acid (ABA)

O O OH CO_2H

Phaseic acid (PA)

HO O OH CO_2H

Dihydrophaseic acid (DPA)

Fig. 1. The proposed "indirect" biosynthetic pathway for ABA and the major pathway for its catabolism

concentrates on the use of genetic variation for ABA in basic plant research. The genetic variation that is available derives from mutation induction experiments but may be found also among natural populations and domesticated cultivars of a specific species. In general, more extreme variation will be found among mutants than among "natural variants" as for the latter more severe limitations exist for the changed gene on survival and yield.

Table 1. Predicted Phenotypes of ABA Mutants

Process	Effect of mutation	Results in	*Effects compared to wild type on:* ABA-induced responses	ABA level	Sensitivity to ABA
Biosynthesis	decrease	deficiency	−	−	O
Biosynthesis	increase	overproduction	+	+	?
Catabolism	decrease	overproduction	+	+	?
Catabolism	increase	deficiency	−	−	−
Reception	decrease	hyposensitivity	−	O	−
Reception	increase	hypersensitivity	+	O	+
Uptake, transport	decrease	hyposensitivity	O	O	−
Uptake, transport	increase	hypersensitivity	+	O	+

O : comparable to wild type
\+ : increased compared to wild type
− : reduced compared to wild type

A crucial step in genetic studies is the ability to recognize genetic variation. Table 1 summarises the predicted phenotypes of possible ABA mutants. However, this scheme is compromised by the rapid changes in ABA levels due to stress. Thus, enhanced ABA levels may be a secondary effect of, for example, a mutation that reduces the root system and, therefore, increases stress and ABA levels. Furthermore, because ABA acts on several different physiological processes, extensive biochemical analysis may be required to establish the exact lesion. Phenotypic changes merely provide preliminary indications of possible mutations.

II. ABA-Deficient Mutants

A. Isolation, Genetic and General Phenotypic Characteristics

ABA-deficient mutants thus far have been recognized because they showed excessive wilting and/or reduced seed dormancy. These effects were shown to be reversed by exogenous ABA and correlated with greatly reduced endogenous ABA levels. In mutants of *Arabidopsis* (Koornneef *et al.,* 1982), potato (Quarrie, 1982) and tomato (Tal and Nevo, 1973; Koornneef *et al.,* 1982, 1985) both water relations and seed dormancy are affected. Effects of the wilty mutation in pea isolated by L. Cruger of the Del Monte Corporation, San Leandro, U. S. A. (see Marx, 1976) on seed dormancy have not been reported (Donkin *et al.,* 1983; Wang *et al.,* 1984). In the viviparous mutants of maize, which are seedling lethals, wilting is not obvious, although J. D. Smith (pers. comm.) mentions that they are hypersensitive to water stress.

All ABA-deficient mutants so far described are monogenic recessives. Tomato mutants with mutations at three different loci, *sitiens (sit), flacca (flc)* and *notabilis (not)* were isolated by Stubbe (1957, 1958, 1959) in the progeny of X-ray treated seeds. Another mutant at the *sit* locus induced by ethylmethanesulphonate was isolated by Van der Veen and Bosma (Koornneef *et al.,* 1985). *Droopy (dr)* mutants of potato *(Solanum tuberosum* L group *phureja)* were found in a cross of two clones (CPC 2862 × 2863) that were apparently heterozygous for *dr* (Simmonds, 1965).

All maize mutants with reduced ABA levels are carotenoid deficient and map to at least seven different loci viz.: *viviparous* 2, 5 and 9 *(vp2, vp5, vp9), pink scutellum (ps= vp7), albescent (al=y3), white seedling* 3 *(w3)* and *yellow (y9).* The mutants arose from different sources and many alleles have been independently isolated (Robertson, 1955, 1975). These seedling lethals are recognized on segregating ears (thus on heterozygous wild type mother plants) as white seeds, many of which germinate precociously on the ear. Mutant *vp8,* which is a viviparous green dwarf, may be an ABA-deficient mutant (Smith *et al.,* 1978), but no ABA determinations have been published.

In *Arabidopsis thaliana* (Koornneef and van der Veen, 1980) and in tomato (Koornneef, *et al.,* 1985), gibberellin-deficient mutants do not germinate without exogenous gibberellin (see also Chapter 1 in this volume). By selecting for germinating seeds amongst the progeny of mutagen-treated non-germinating *(ga-1) Arabidopsis* seeds, mutants were isolated that were reverted at least for the non-germination trait. Genetic analysis showed this reversion to be due to a mutation in a suppressor gene segregating independently from the *ga-1* gene.

Table 2 shows the scheme whereby this suppressor mutant was isolated. The fact that in tomato the double mutant sit^w/sit^w, *ga-1/ga-1* does not germinate without gibberellin indicates that the procedure used to isolate ABA mutants in *Arabidopsis* is not applicable to every species. However, the ability of ABA mutants (e. g. *sit* in tomato) to germinate in conditions that are adverse to the germination of the wild type, such as in a germination medium with high osmotic potential, or under continuous far-red irradiation, provides an alternative scheme for the direct selection of such non-dormant mutants (Koornneef *et al.,* 1985).

Mutants that have not yet been characterized with respect to ABA, but which have phenotypes similar to some of the ABA mutants mentioned above, include a non-dormant, carotenoid-free, white mutant of *Helianthus*

Table 2. Isolation of Isogenic, Recessive GA and ABA Mutants of *Arabidopsis thaliana*

Generation	*Phenotype*	*Genotype*
STAGE 1		
M_0	Wild type	*Ga-1/Ga-1, Aba/Aba*
↓ EMS[a]	↓	↓
M_1	Wild type	*Ga-1/ga-1, Aba/Aba*
↓ selfing	↓	↓
M_2	GA Mutant	*ga-1/ga-1, Aba/Aba*
	Germination: GA dependent	
	Plants: dwarfs	
STAGE 2		
M_0	GA mutant	*ga-1/ga-1, Aba/Aba*
↓ EMS[a]	↓	↓
M_1	GA mutant	*ga-1/ga-1, Aba/aba*
↓ selfing	↓	↓
M_2	Revertant	*ga-1/ga-1, aba/aba*
	Germination: GA independent	
	Plants: dwarfs	
STAGE 3		
	Revertant	*ga-1/ga-1, aba/aba*
	x	*x*
x	Wild type	*Ga-1/Ga-1, Aba/Aba*
	↓	↓
F_1	Wild type	*Ga-1/ga-1, Aba/aba*
↓ selfing	↓	↓
F_2	ABA Mutant	*Ga-1/Ga-1, aba/aba*
	Germination: GA independent	
	Plants: wilting, normal height	

[a]EMS = ethylmethanesulphonate; 10 mM applied for 24 hr in darkness at 24 °C.

annuus L. (Wallace and Habermann, 1958) and viviparous mutants in *Avena nuda* L. (Cummings *et al.*, 1978).

It should be emphasized that not all wilting mutants and mutants with reduced seed dormancy are ABA deficient. For example, the wilted *(wi)* mutant of maize (Postlethwait and Nelson, 1957) and wilting dwarf *(wd)* in tomato (Alldridge, 1964) wilt because anomalous vessel elements restrict water movement through the plants. Excessive stomatal opening, which is characteristic for all wilty ABA mutants, can result from the disturbed regulation of ion uptake mechanisms (Tal *et al.*, 1976) as found, for example, in the scabrous diminutive mutant of *Capsicum annuum* (Tal *et al.*, 1974).

A reduction in seed dormancy may also be caused by factors other than ABA deficiency, as shown by the germination behaviour of *Arabidopsis* seed coat mutants (Koornneef, 1981). The correlation between reduced ABA content and low sprouting resistance (reduced dormancy) in barley found by Goldbach and Michael (1976) is difficult to accept as a causal relationship as only two non-isogenic cultivars were compared and joint segregation of ABA levels and sprouting resistance in segregating progenies was not studied.

B. Pleiotropic Effects of ABA-Deficient Mutants

True pleiotropism is inferred from the joint occurrence of the same phenotypic characters in independently isolated mutations at the same locus and from the absolute association of the complex phenotype in segregating populations. Both the presence and the absence of particular pleiotropic effects in ABA mutants can provide useful information about the function of ABA. However, some pleiotropic effects are very indirect and have no causal relation to the hormone. For example, semi-dwarfness in ABA-deficient mutants is due to their disturbed water relations and should not be interpreted as a direct effect of ABA on cell division or cell elongation. Tomato ABA mutants show leaf epinasty, swelling of the upper stem and rooting along the stem and it is difficult to explain these characteristics as direct ABA effects. Tal *et al.* (1979) suggest that these traits are caused by auxin overproduction which subsequently enhances ethylene production. ABA deficiency is clearly the primary effect because all characters are reversed by exogenous ABA. In the *dr* mutant of potato (Quarrie, 1982) and *aba* mutants of *Arabidopsis,* such auxin-like effects have not been observed. An additional pleiotropic effect of *aba* mutants in *Arabidopsis* is the reduction of mucilage excretion (Karssen *et al.,* 1983).

All viviparous mutants of maize with reduced ABA levels are characterized by pale yellow endosperms in genotypes that otherwise would be yellow, and by white or almost white seedlings (Robertson, 1975). Defective carotenoid biosynthesis is the primary lesion in these mutants leading to the multiple defects described (Fong *et al.,* 1983). That this is a direct causal relationship is shown by the application of fluridone (a pyridone inhibitor of carotenoid biosynthesis) to the wild type, which results in phenocopies of the *vp5* mutants (Smith *et al.,* 1983).

C. Biochemistry of ABA-Deficient Mutants

The wilty mutants of tomato, potato, pea and *Arabidopsis* still contain low levels of ABA. For the three tomato mutants, ABA levels in non-stressed leaves were found to be 12—15 % of wild type for *sit,* 17—26 % for *flc* and 31—49 % for *not.* These percentages refer respectively to data published by Tal and Nevo (1973) and by Neill and Horgan (1985) and correlate approximately to the severity of the symptoms in the different mutants. Such a correlation has also been observed in the different alleles of the *aba* locus of *Arabidopsis*. The most viable mutant, *aba*^{3} (G4), contains 10—26 % of the wild-type value in immature and mature seeds (Koornneef *et al.*, 1982), whereas in *aba*^{1} (A26) ABA was below the detection level (Koornneef *et al.*, 1982, 1984). Still less vigorous *aba* alleles (e. g. *aba*^{4} = A73) are known which are almost lethal due to their extreme wilting and withering (Koornneef *et al.*, 1982). ABA levels in pea and potato are about 11 % of that of wild type in stressed leaves (Quarrie, 1982; Wang *et al.*, 1984).

In potato (Quarrie, 1982), *sit* and *flc* tomato (Neill and Horgan, 1985), pea (Wang *et al.*, 1984) and *aba*^{3} *Arabidopsis* (Zeevaart, pers. comm.) ABA does not accumulate in response to water stress as it does in the wild type. It may be that the wilty mutants are already stressed under normal conditions and have reached their maximal ABA level. Alternatively, stress-induced ABA cannot be produced in these mutants, suggesting different pathways in turgid and water-stressed leaves. However, because the mutations in tomato (Groot and Karssen, pers. comm.) and *Arabidopsis* (Koornneef *et al.*, 1982; Karssen *et al.*, 1983) affect ABA levels in seeds and leaves in a similar way, a leaf-specific pathway seems unlikely.

All carotenoid-deficient mutants of maize have reduced but significant levels of ABA in embryo tissue, the values being variable and age dependent (J. D. Smith, pers. comm.). Published data (Brenner *et al.*, 1977; Smith *et al.*, 1978; Robichaud *et al.*, 1980) range from 7—70 % of that of the wild type. However, ABA is not detectable in *w3, vp5* and *vp7* seedlings and roots (Moore and Smith, 1985). In view of these data and preliminary results of *in vitro* cultures of kernels (J. D. Smith, pers. comm.), the ABA found in mutant embryos produced on heterozygous wild-type mother plants is very probably of maternal origin. All deficient maize mutants accumulate specific carotenoid precursors but not zeaxanthin (Robertson, 1975; Fong *et al.*, 1983). The blocks in carotenogenesis suffered by these mutants are presented in Fig. 2 after Fong *et al.*, 1983). The relationships between the blocks in the carotenoid pathway and the effects on ABA levels, both in mutants (Moore and Smith, 1985) and after application of the carotenoid inhibitor fluridone (Moore and Smith, 1984), provide strong arguments for the C-40 pathway of ABA synthesis. As carotenoid deficiency indirectly affects many aspects of plant metabolism, for example leading to excessive photo-oxidation of chlorophylls, it may be argued that ABA deficiency is an indirect effect of carotenoid deficiency. ABA synthesis may need a functional chlorophyll system as suggested by Quarrie and Lister (1984) who found no ABA in leaves of the barley plastid (albino) mutant *albostrians.* All dicot ABA-deficient mutants are green and contain carotenoids. However, it seems unlikely that there is a different ABA pathway in monocots and dicots. The occurrence of non-dormancy and carotenoid deficiency in a sunflower mutant (Wallace and Habermann, 1958) indicates that this relationship at least is not unique for monocots. Most probably the absence of carotenoid mutants in dicots is due to the fact that non-dormancy in dicots rarely expresses itself as vivipary. When it does, it is less obvious because the viviparous seeds are covered by the fruit tissue. Furthermore, lethal albinos in general are not maintained in mutant collections.

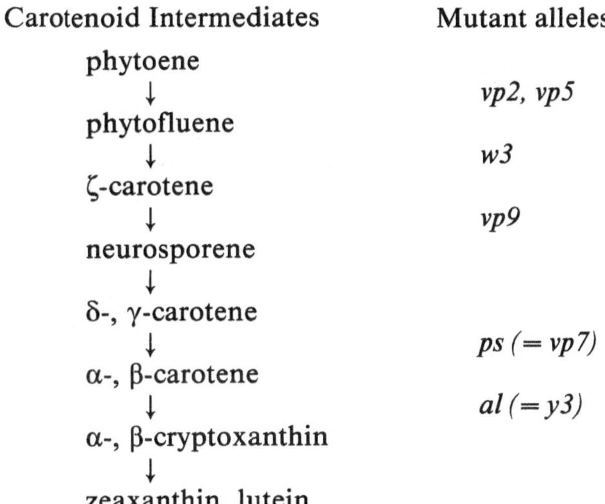

Fig. 2. Carotenoid biosynthesis in maize and the metabolic blocks of different viviparous mutants (after Fong *et al.*, 1983)

The ghost *(gh)* mutant of tomato, which is a developmentally unstable chlorophyll and fruit carotenoid mutation (Rick *et al.*, 1959), deserves more attention with respect to ABA effects. In their initial report neither disturbed water relations nor deviating seed germination were mentioned.

The green wilty mutants apparently represent, if the C-40 pathway operates, mutations in the last part of the ABA pathway . Nevo and Tal (1973) showed that *flacca* contained slightly higher levels of xanthophylls (including violaxanthin) than its wild type, Rheinlands Ruhm. Recently Bowman *et al.* (1984) observed the accumulation of 2,7-dimethyl-2,4-octadienedioic acid (ODA) in *flacca,* which they suggest derives from oxidative cleavage of violaxanthin (Taylor, 1984). However, as pointed out by Neill and Horgan (1985), there are problems in explaining why this compound accumulates when the ABA pathway is blocked between violaxanthin and ABA. Although not conclusive, the biochemistry of both the viviparous maize and *flacca* tomato mutants suggests the involvement of the indirect C-40 pathway. No detailed biochemistry has been applied to the other mutants mentioned above.

The interrelations between the three tomato mutants were studied genetically by Taylor and Tarr (1984). When studying double mutants, they found that the double recessive *not/not, flc/flc* and *not/not, sit/sit* had a much more severe phenotype than the most extreme mutant *sit.* However, in *flc/flc, sit/sit* the effects were not additive. Taylor and Tarr give two possible explanations for these effects:

1. *Flc* and *sit* are mutated in a pathway different but parallel to a pathway in which *not* regulates one step.

2. *Flc* and *sit* code for different subunits of the same enzyme and *not* for a different enzyme in the same pathway.

D. The Use of ABA Mutants in Water Relations Research

The rapid wilting of ABA-deficient mutants in several species is due to greater stomatal conductance, which can be normalized by exogeneously applied ABA (Tal, 1966; Imber and Tal, 1970; Tal and Nevo, 1973; Bradford, 1983; Neill and Horgan, 1985). This results in excessive transpiration and wilting because stomata of the mutant remain partially open during periods of water stress and in darkness. On the other hand, the stomatal conductance of the *flc* tomato (Bradford *et al.*, 1983) and wilted pea (Donkin *et al.*, 1983) is reduced by increased CO_2 pressure, increased leaf-to-air vapour pressure gradient and decreased light intensity, showing that the regulation of stomatal closure still functions. Since these environmental factors do not act by inducing ABA synthesis, it appears that ABA overrides the factors mentioned above when water deficit threatens (Bradford *et al.*, 1983), thus enabling stomata to close despite the strong promotion of stomatal opening by, for example, light. Apart from effects on stomatal opening, hydraulic conductance (osmotic potential of the root exudate) is also reduced in the mutant and this effect is ABA restorable (Tal and Nevo, 1973; Bradford, 1983). Although not studied in as much detail as tomato and pea, the overall picture for the potato (Waggoner and Simmonds, 1966; Quarrie, 1982) and *Arabidopsis* (Koornneef *et al.*, 1982) mutants seems similar.

E. The Use of ABA Mutants in Seed Physiology

Seeds as they mature enter a period of arrest when development ceases and dehydration ensues. The dehydrated seed may be either dormant or quiescent. A quiescent seed will immediately start to germinate when it is rehydrated. Dormancy is a physiological or morphological state that restricts germination completely or to a limited range of environmental conditions. Dormancy must be relieved before germination can occur. Based on correlative evidence only, a regulatory role for ABA has been suggested in the termination of seed growth, the start of dehydration, the prevention of precocious germination, the stimulation of food reserves, the induction of the developmental arrest and the induction of primary dormancy (Karssen, 1982; Black, 1983). In *Arabidopsis* it was shown that the induction of dormancy during seed development depends on endogenous ABA. Immature seeds of the wild type germinate if they are isolated half-way through development whereas mature seeds will not germinate. In ABA-deficient mutants the germination potential increases instead of decreases towards maturity (Fig. 3, after Karssen *et al.*, 1983). The ABA content during seed development in both genotypes is also shown in Fig. 3.

Reciprocal crosses of wild type and ABA mutants point to a dual origin of endogenous ABA in seeds (Table 3). The genotype of the mother plant regulated a sharp rise in ABA content with a maximum at 10 days after pollination (maternal ABA). The genotype of the embryo was responsible for a second ABA fraction (embryonic ABA) which reached far lower levels than

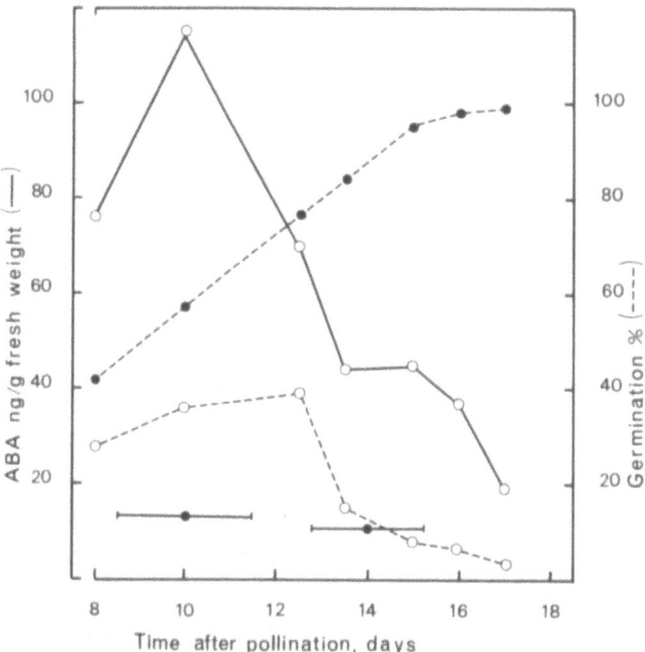

Fig. 3. Changes with time after pollination in ABA content (—) and in germination (- - -) of isolated seeds of the *Arabidopsis* wild type (open symbols) and the *aba*[3] mutant (closed symbols). Modified after Karssen *et al.* (1983), with permission

maternal ABA, but persisted for some time during maturation after the maternal ABA had decreased. Induction of dormancy only occurred if embryonic ABA was present and was independent of the presence of maternal ABA.

Applied ABA mimicked maternal ABA in this respect: dormancy was not induced in ABA-deficient lines sprayed with a 100 μM ABA solution every four days during seed development (Karssen, 1982; Karssen *et al.*, 1983). Thus dormancy induction occurs entirely within the embryo. To induce dormancy ABA must be synthesized close to its site of action or must be localized in a specific subcellular compartment not accessible to ABA transported from maternal tissues or supplied exogenously. These results add further weight to the criticism of the classic hormone concept which includes transport of hormones to the sites of action (Karssen, 1982).

That ABA in seeds partially originates in the mother plant has been shown in the work on maize discussed in section B above. Using translocations with heterochromatic B chromosomes, Robertson (1955) showed that the viviparous character in the *vp1, vp8, vp5* and *ps* mutants, of which the latter two are ABA deficient, is determined by the genotype of the embryo and not the endosperm.

As previously mentioned, lack of endogenous ABA does not always result in viviparous germination in the fruit. In *Arabidopsis* the dehydrated state of the seeds prevents germination of the ABA mutants (Karssen *et al.*,

Table 3. ABA content and germination of immature (10 and 16 days) and mature (26 days) F_1 seeds of different crosses with Arabidopsis wild type and *aba* mutant

		Days after pollination				
		10		16		ripe
♀ × ♂	F_1	ABA ng/g	Germ. %	ABA ng/g	Germ. %	Germ. %
Aba/Aba × *aba/aba*	*Aba/aba*	598	39	59	29	0
aba/aba × *Aba/Aba*	*Aba/aba*	35	44	36	49	0
aba/aba × *aba/aba*	*aba/aba*	8	64	1	99	95

1983), as do osmotic constituents of the pericarp tissue in tomato (Groot *et al.*, pers. comm.). Robichaud *et al.* (1980) consider precocious germination in maize to be due to the absence of developmental arrest in the mutant because embryo development is already completed. Thus it is not only dormancy that is under ABA control.

The role of ABA on food reserve accumulation has not been studied using mutants. It is interesting to note, however, that in spite of the generally poor vegetative growth of the mutants, seed development is comparable to wild type.

F. ABA Deficiency in Relation to Other Physiological Effects

None of the processes such as fruit development, senescence and especially root geotropism (Pilet, 1982; Audus, 1983) which have been associated with ABA are clearly affected in ABA mutants (Koornneef, unpublished; Moore and Smith, 1985). This suggests either that ABA plays no part or that the residual levels of ABA in the mutants (below detection level in maize) are sufficient for these processes.

III. Mutants Affecting ABA Sensitivity

Differences in ABA sensitivity for growth inhibition (Sloger and Caldwell, 1970) and the inhibition of seed germination (Rudnicki *et al.*, 1973) are found between soybean cultivars. Stable variants with an up to 10-fold decrease in ABA sensitivity (resistance to 0.1 mM ABA) have been selected in suspension cultures of haploid *Nicotiana sylvestris* (Wong and Sussex, 1980). Preliminary experiments suggest enhanced ABA catabolism in some of these variants (J. R. Wong, pers. comm.). In carrot suspension cultures ABA at O.1 mgl^{-1} does not inhibit cell growth, but particular stages of somatic embryo development are affected, for example, the formation of plantlets from torpedo stage embryos is inhibited (Sung *et al.*, 1984). Variants that developed into plantlets at this ABA concentration were selected and have been characterized as putative uptake mutants (C. Borkird and Z. M. Sung, pers. comm.).

An elegant selection scheme aiming at the selection of tissue-specific alterations in ABA sensitivity was developed by Ho *et al.* (1980) for barley endosperm. Endosperm half seeds are placed on top of starch- and gibberellin-containing agar with the cut edge in contact with the agar surface. Gibberellin induced α-amylase synthesis is inhibited by ABA. After incubation the agar plates are stained with an I_2-KI solution to visualize α-amylase activity as a colourless halo around enzyme-producing half seeds. ABA-insensitive mutants were identified as seeds that produced enzyme on media containing both gibberellin and ABA.

Selecting amongst the progeny of mutagen-treated seeds for seeds that germinated in ABA concentrations inhibitory to the germination of the parental variety, Hayter and Allison (1976) selected barley mutants with a higher diastatic power (a measure of the activity of amylase enzymes). The authors suggest that the increased enzyme activity is due to higher gibberellin levels which overcome ABA inhibition. This is not, therefore, an example of selection for true ABA insensitivity.

In the fern *Ceratopteris,* where ABA retards growth and inhibits sexual differentiation in gametophytes, there has been a preliminary report of several types of ABA resistant mutants (Hickok, 1984).

Mutants of *Arabidopsis* were selected on 10 μM ABA that were 5—20 times less sensitive to ABA at the germination and seedling growth level than the wild type (Koornneef *et al.,* 1984). The ABA-insensitive mutants have mutations at three different loci and show similarities in phenotype (Table 4) with the ABA-deficient mutant. The altered sensitivity to ABA is possibly due to reduced availability of a "receptor", or reduced binding-capacity of a "receptor" for ABA. The "receptor" may be an ABA-binding protein (Hornberg and Weiler, 1984), a membrane fraction (Hocking *et al.,* 1978) or the ABA target itself, i. e. molecules that react with

Table 4. Some characteristics of ABA-insensitive mutants of *Arabidopsis (abi)* as compared to the wild type and the ABA-deficient mutant *(aba)*

Genotype	Sensitivity to exogenous ABA	Seed dormancy	Transpiration	ABA content
Wild type	+	+	+	+
abi-3	–	–	+	+
abi-1/abi-2	–	–	–	+
aba	+	–	–	–

Character:	+	–
ABA sensitivity	normal	reduced
Seed dormancy	normal	reduced
Transpiration	normal	excessive
ABA content	normal	reduced

ABA or the ABA-receptor complex to give a physiological response. The decreased sensitivity of these *Arabidopsis* mutants to exogenous ABA could be also due to reduced ABA uptake (which would not explain the changes in water relations and seed dormancy) or an enchanced ABA metabolism. Enhanced metabolism is also unlikely since ABA levels in developing seeds of the mutants were comparable or somewhat higher than in the wild type (Koornneef *et al.*, 1984).

The *vp1 (= vp)* mutant of maize is also insensitive to the germination inhibiting effect of ABA when grown *in vitro* (Smith *et al.*, 1978; Robichaud *et al.*, 1980). The mutant is green, has normal maize carotenoids and normal ABA levels. It has been suggested that *vp1* is defective in a seed-specific ABA receptor (Smith *et al.*, 1978). However, the absence of flavenoid pigments in its endosperm, which is not a characteristic of other viviparous, ABA-deficient maize mutants, is difficult to explain by this hypothesis. Dooner (1985) noticed that in addition to three enzymes of flavenoid synthesis viz. phenylalanine ammonia lyase, chalcone synthetase and UDPG-flavenoid glucosyl transferase, the enzymes glucose-6-phosphate dehydrogenase, catalase and alcohol dehydrogenase, which in the wild type show a similar pronounced increase in the later stages of aleurone development, were also deficient in *vp1* aleurone layers. These data suggest that *vp1* may be a major regulatory gene responsible for turning on a battery of structural genes at a certain time in development (Dooner, 1985). The relationship between ABA insensitivity and vivipary remains unexplained by this hypothesis, unless an ABA receptor gene is one of the genes regulated by *vp1.*

IV. Genetic Differences in ABA Accumulation

Reduction of leaf water potential, ABA accumulation and stomatal closure result in the avoidance of excessive water loss in periods of water shortage. Genetic variation in ABA accumulation and in the effect of ABA upon the induction and relief of drought stress has therefore attracted the interest of plant breeders. The subject has been reviewed recently by Quarrie (1983; 1984).

The levels to which ABA accumulated after stress induction and drought resistance have been compared in cultivars of wheat (Quarrie and Jones, 1979), maize (Larqué-Saavedra and Wain, 1976; Ilahi and Dörffling, 1982), millet (Henson *et al.*, 1981) and sorghum (Durley *et al.*, 1983). The reationship between ABA accumulation and drought resistance differed between species, probably due mainly to the complexity of the interaction of ABA accumulation with drought resistance, which is often measured as crop yield. In contrast to beneficial effects on water use, high ABA levels may also have detrimental effects on photosynthesis (by decreased stomatal conductance), pollen fertility, etc. Quarrie (1984) mentions three factors that determine how endogenous ABA modifies yield: 1) the amount of ABA produced in response to drought and the organs to which it is distributed, 2) the extent to which individual processes are regulated by a particular endogenous ABA level, and 3) the cumulative effects of interactions between these processes and other factors that determine crop yield. In any case, differences in experimental procedures (eg. how stress was applied) and lack of isogenenicity make interpretation of the results difficult. The near-isogenic material differing in ABA accumulation which has now been obtained in wheat (Quarrie, 1981; Innes *et al.*, 1984) and rice (Henson *et al.*, 1985) will provide much better material for studying ABA effects on yield.

Apart from genetic differences in the extent to which external stress will stimulate ABA production, genetic differences in ABA accumulation have been reported for cotton (Ibragimov *et al.*, 1978), maize (Larqué-Saavedra and Wain, 1976), millet (Henson, 1984), rice (Henson, 1983) and wheat (Quarrie, 1981). The differences were shown to be heritable in millet, rice and wheat. Homozygosity was obtained for this character in millet and wheat after limited inbreeding, which, together with the high heritability, suggests the segregation of only a small number of genes in the crosses studied. In neither millet nor wheat (Quarrie and Henson, 1982) was there evidence for dominance or a maternal influence on ABA accumulation. However, in maize one example of maternal inheritance has been reported (Larqué-Saavedra and Rodrigues, 1979; Quarrie, 1983), suggesting the genetic involvement of organelles (most probably plastids).

The biochemical basis of the differences in ABA accumulation reported above has not been extensively studied. Durley *et al.* (1983) consider the lower levels of ABA and higher levels of PA and total ABA metabolite concentration in particular genotypes to be the result of a higher efficiency of conversion of ABA to its metabolites. Mutants that overproduce ABA constitutively have not been described. The tomato mutants blind *(bl)* and lateral suppressor *(ls)*, characterized by increased apical dominance, were initially reported to contain high levels of ABA (Tucker, 1976), but Taylor and Rossall (1982), using reverse-phase, high-performance, liquid chromatography instead of a bioassay, found no increased ABA levels in the *ls* mutant. When ABA, PA and DPA levels of a similar *torosa-2* mutant were compared to wild type (Mapelli and Rocchi, 1985) no clear differences were detected.

Differences in ABA levels have also been associated with dwarfness (Quarrie, 1983). In comparisons between different non-isogenic apple genotypes, an association of dwarfness with low ABA levels was found in two studies (Lee and Looney, 1977; Yadava and Lockard, 1977), and with higher levels in a third study (Robitaille and Carlson, 1976). The differences were not large and levels of other hormones were also affected, therefore a causal relationship is doubtful. The association observed between ABA accumulation and the presence of *Rht* genes in wheat cultivars suggested a causal relationship, but comparison of near-isogenic lines differing in *Rht* genes showed no such effect (Quarrie, 1983).

V. Conclusions

ABA-deficient mutants have been identified in a number of species. Common to all the mutants is an excessive water loss due to a lower stomatal conductance which results in a wilty phenotype. Most (probably all) mutants also have reduced seed dormancy. These characteristics, especially the effects on seed dormancy, allow direct selection for ABA-deficient mutants. Mutations resulting in ABA insensitivity can be selected directly on the basis of ABA resistance. Several different mechanisms can give rise to ABA insensitivity.

Genetic variation in ABA accumulation has been found among cultivars, especially in cereals. No direct mutant selection schemes have yet been developed for ABA overproducers. For this type of mutation, which may be based upon reduced catabolism of ABA, direct screening using a rapid ABA assay, like an immunoassay, may be applicable.

Although potentially of great importance, there has so far been only limited use of mutants in ABA biochemistry, although in physiological studies mutants have been used successfully to analyse the role of ABA in stomatal opening and seed dormancy.

Acknowledgements

I thank many colleagues for providing preprints and other unpublished information and discussions. In particular I would like to mention Drs. R. Horgan, C. M. Karssen, J. M. Mottley, J. D. Smith, S. A. Quarrie and J. A. D. Zeevaart.

VI. References

Addicott, F. T., Carns, H. R., 1983: History and introduction. In: Addicott, F. T. (ed.), Abscisic Acid, pp. 1—21. New York: Praeger Scientific.

Alldridge, N. A., 1964: Anomalous vessel elements in wilty dwarf tomato. Bot. Gaz. **125,** 138—142.

Audus, L. J., 1983: Abscisic acid in root growth and geotropism. In: Addicott, F. T. (ed.), Abscisic Acid, pp. 421—477. New York: Praeger Scientific.

Black, M., 1983: Abscisic acid in seed germination and dormancy. In: Addicott, F. T. (ed.), Abscisic Acid, pp. 331—363. New York: Praeger Scientific.

Bowman, W. R., Linforth, R. S. T., Rossall, S., Taylor, I. B., 1984: Accumulation of an ABA analogue in the wilty tomato mutant, *flacca.* Biochem. Genet. **22,** 369—377.

Bradford, K. J., 1983: Water relations and growth of the *flacca* tomato mutant in relation to abscisic acid. Plant Physiol. **72,** 251—255.

Bradford, K. J., Sharkey, T. D., Farquhar, G. D., 1983: Gas exchange, stomatal behaviour and $\delta^{13}C$ values of the *flacca* tomato mutant in relation to abscisic acid. Plant Physiol. **72,** 245—250.

Brenner, M. L., Burr, B., Burr, F., 1977: Correlation of genetic vivipary in corn with abscisic acid concentration. Plant Physiol. Suppl. **63,** 36.

Creelman, R. A., Zeevaart, J. A. D., 1984: Incorporation of oxygen into abscisic acid and phaseic acid from molecular oxygen. Plant Physiol. **75,** 166—169.

Cummings, D. P., Stuthman, D. D., Green, C. E., 1978: Morphological mutations induced with ethyl methanesulphonate in oats. J. Hered. **69,** 3—7.

Donkin, M. E., Wang, T. L., Martin, E. S., 1983: An investigation into the stomatal behaviour of a wilty mutant of *Pisum sativum.* J. Exp. Bot. **34,** 825—834.

Dooner, H. K., 1985: Viviparous-1 mutation in maize conditions pleiotropic enzyme deficiencies in the aleurone. Plant Physiol. **77,** 486—488.

Durley, R. C., Kanangara, T., Seetharama, N., Simpson, G. M., 1983: Drought resistance of *Sorghum bicolor* 5. Genotypic differences in the concentrations of free and conjugated abscisic, phaseic and indole-3-acetic acids in leaves of field-grown drought-stressed plants. Can. J. Plant Sci. **63,** 131—145.

Fong, F., Koehler, D. E., Smith, J. D., 1983: Fluridone induction of vivipary during maize seed development. In: Krueger, J. E., La Berge, D. E. (eds.), III. Int. Symp. on Pre-harvest Sprouting in Cereals, pp. 188—196. Boulder, Colo.: Westview Press.

Goldbach, H., Michael, G., 1976: Abscisic acid content of barley grains during ripening as affected by temperature and variety. Crop Sci. **16,** 787—800.

Hayter, A. M., Allison, M. J., 1976: Breeding for high diastatic power. In: Gaul, H. (ed.), Barley Genetics III Proc. of the 3rd Int. Barley Genet. Symp., Garching 1975, pp. 612—619. München: Verlag Karl Thiemig.

Henson, I. E., 1983: Abscisic acid accumulation in detached leaves of rice (*Oryza sativa* L.) in response to water stress: a correlation with leaf size. Ann. Bot. **52,** 385—398.

Henson, I. E., 1984: The heritability of abscisic acid accumulation in water-stressed leaves of pearl millet (*Pennisetum americanum* [L.] Leeke). Ann. Bot. **53,** 1—11.

Henson, I. E., Mahalakshmi, V., Bidinger, F. R., Alagarswamy, G., 1981: Genotypic variation in pearl millet (*Pennisetum americanum* [L.] Leeke) in the ability to accumulate abscisic acid in response to water stress. J. Exp. Bot. **32,** 899—910.

Henson, I. E., Loresto, G. C., Chang, T. T., 1985: Drought tolerance: Production of closely-related lines of rice differing in drought-induced abscisic acid accumulation. Int. Rice Res. Newsl., **10,** 12—13.

Hickok, L. G., 1984: Selection and analysis of ABA-resistant mutants in *Ceratopteris.* Am. J. Bot. **71,** 141.

Ho, T. H. D., Shih, S. C., Kleinhofs, A., 1980: Screening for barley mutants with altered hormone sensitivity in their aleurone layers. Plant Physiol. **66,** 153—157.

Hocking, A., Clapham, J., Catsell, K. J., 1978: Abscisic acid binding to subcellular fractions from leaves of *Vicia faba.* Planta **138,** 303—304.

Hornberg, C., Weiler, E. W., 1984: High-affinity binding sites for abscisic acid on the plasmalemma of *Vicia faba* guard cells. Nature (London) **310,** 321—324.

Ibragimov, A. P., Igamberdyeva, Z. I., Saidova, S. A., 1978: Effect of moisture stress on the level of abscisic acid in cotton leaves. Uzb. Biol. Zh. **4,** 11—14.

Ilahi, I., Dörffling, K., 1982: Changes in abscisic acid and proline levels in maize varieties of different drought resistance. Physiol. Plant. **55,** 129—135.

Imber, D., Tal, M., 1970: Phenotypic reversion of *flacca,* a wilty mutant of tomato, by abscisic acid. Science **169,** 592—593.

Innes, P., Blackwell, R. D., Quarrie, S. A., 1984: Some effects of genetic variation in drought-induced abscisic acid accumulation on the yield and water use of spring wheat. J. Agric. Sci. **102,** 341—351.

Karssen, C. M., 1982: Role of endogenous hormones during seed development and the onset of primary dormancy. In: Wareing, P. F. (ed.), Plant Growth Substances 1982, pp. 623—632. London: Academic Press.

Karssen, C. M., Brinkhorst-van der Swan, D. L. C., Breekland, A. E., Koornneef, M., 1983: Induction of dormancy during seed development by endogenous abscisic acid: studies on abscisic acid deficient genotypes of *Arabidopsis thaliana* (L.) Heynh. Planta **157,** 158—165.

Koornneef, M., 1981: The complex syndrome of *ttg* mutants. Arabidopsis. Inf. Serv. **18,** 45—51.

Koornneef, M., van der Veen, J. H., 1980: Induction and analysis of gibberellin-sensitive mutants in *Arabidopsis thaliana* (L.) Heynh. Theor. Appl. Genet. **58,** 257—263.

Koornneef, M., Jorna, M. L., Brinkhorst-van der Swan, D. L. C., Karssen, C. M., 1982: The isolation of abscisic acid (ABA)-deficient mutants by selection of induced revertants in non-germinating gibberellin-sensitive lines of *Arabidopsis thaliana* (L.) Heynh. Theor. Appl. Genet. **61,** 385—393.

Koornneef, M., Reuling, G., Karssen, C. M., 1984: The isolation and characterization of abscisic acid insensitive mutants of *Arabidopsis thaliana.* Physiol. Plant. **61,** 377—383.

Koornneef, M., Cone, J. W., Karssen, C. M., Kendrick, R. E., van der Veen, J. H., Zeevaart, J. A. D., 1985: Plant hormone and photoreceptor mutants in *Arabidopsis* and tomato. In: Freeling, M. (ed.), Plant Genetics (UCLA Symposia on Molecular and Cellular Biology, New Series, Vol. 35), pp. 103—114. New York, NY: Alan R. Liss. Inc.

Larqué-Saavedra, A., Rodriguez, G. M. T., 1979: Maternal inheritance of abscisic acid (ABA) in *Zea mays* L. In: Abstracts of the 10th Intern. Conf. on Plant Growth Subst. Madison, Wisconsin, p. 23.

Larqué-Saavedra, A., Wain, R. L., 1976: Studies on plant growth-regulating substances. XLII. Abscisic acid as a genetic character related to drought tolerance. Ann. Appl. Biol. **83,** 291—297.

Lee, J. M., Looney, N. E., 1977: Abscisic acid levels and genetic compaction in apple seedlings. Can. J. Plant Sci. **57,** 81—85.

Mapelli, S., Rocchi, P., 1985: Endogenous abscisic acid in *torosa-2* mutant tomato plant. Plant Cell Physiol. **26,** 371—374.

Marx, G. A., 1976: "Wilty": a new gene of *Pisum.* Pisum Newsl. **8,** 40—41.

Milborrow, B. V., 1983: Pathways to and from abscisic acid. In: Addicott, F. T. (ed.), Abscisic Acid, pp. 79—111. New York: Praeger Scientific.

Milborrow, B. V., 1984: Inhibitors. In: Wilkins (ed.), Advanced Plant Physiology, pp. 76—110. London: Pitman.

Moore, R., Smith, J. D., 1984: Growth, graviresponsiveness and abscisic acid content of *Zea mays* seedlings treated with fluridone. Planta **162,** 342—344.

Moore, R., Smith, J. D., 1985: Graviresponsiveness and abscisic-acid content of roots of carotenoid-deficient mutants of *Zea mays* L. Planta **164,** 126—128.

Neill, S. J., Horgan, R., 1985: Abscisic acid production and water relations in wilty tomato mutants subjected to water deficiency. J. Exp. Bot. **36,** 1222—1231.

Nevo, Y., Tal, M., 1973: The metabolism of abscisic acid in *flacca,* a wilty mutant of tomato. Biochem. Genet. **10,** 79—90.

Pilet, P. E., 1982: Abscisic acid, one of the endogenous growth inhibitors regulating root gravireaction. In: Wareing, P. F. (ed.), Plant Growth Substances 1982, pp. 529—536. London: Academic Press.

Postlethwait, S. N., Nelson, O. E., 1957: A chronically wilted mutant of maize. Am. J. Bot. **44,** 628—633.

Quarrie, S. A., 1981: Genetic variability and heritability of drought-induced abscisic acid accumulation in spring wheat. Plant Cell Environ. **4,** 147—151.

Quarrie, S. A., 1982: Droopy: a wilty mutant of potato deficient in abscisic acid. Plant Cell Environ. **5,** 23—26.

Quarrie, S. A., 1983: Genetic differences in abscisic acid physiology and their potential uses in agriculture. In: Addicott, F. T. (ed.), Abscisic Acid, pp. 365—419. New York: Praeger Scientific.

Quarrie, S. A., 1984: Abscisic acid and drought resistance in crop plants. News Bull. Br. Plant Growth Regulator Group **7,** 1—15.

Quarrie, S. A., Henson, I. E., 1982: Biparental inheritance of drought-induced accumulation of abscisic acid in wheat and pearl millet. Ann. Bot. **49,** 265—268.

Quarrie, S. A., Jones, H. G., 1979: Genotypic variation in leaf water potential, stomatal conductance and abscisic acid concentration in spring wheat subjected to artificial drought stress. Ann. Bot. **44,** 323—332.

Quarrie, S. A., Lister, P. G., 1984: Evidence of plastid control of abscisic acid accumulation in barley (*Hordeum vulgare* L.). Z. Pflanzenphysiol. **114,** 295—308.

Rick, C. M., Thompson, A. E., Brauer, O., 1959: Genetics and development of an instable chlorophyll deficiency in *Lycopersicon esculentum.* Am. J. Bot. **46,** 1—11.

Robertson, D. S., 1955: The genetics of vivipary in maize. Genetics **40,** 745—760.

Robertson, D. S., 1975: Survey of the albino and white-endosperm mutants of maize. J. Hered. **66,** 67—74.

Robichaud, C. S., Wong, J., Sussex, I. M., 1980: Control of *in vitro* growth of viviparous embryo mutants of maize by abscisic acid. Dev. Genet. **1,** 325—330.

Robitaille, H. A., Carlson, R. F., 1976: Gibberellin and abscisic acid-like substances and the regulation of apple shoot extension. J. Am. Soc. Hortic. Sci. **101,** 388—392.

Rudnicki, R., Blumenfeld, A., Bukovac, M. J., 1973: *Glycine* seed germination: Differential response to abscisic acid. Experientia **29,** 231.

Simmonds, N. W., 1965: Mutant expression in diploid potatoes. Heredity **20,** 65—72.

Sloger, C., Caldwell, B. E., 1970: Response of cultivars of soybean to synthetic abscisic acid. Plant Physiol. **46,** 634—635.

Smith, J. D., McDaniel, S., Lively, S., 1978: Regulation of embryo growth by abscisic acid *in vitro.* Maize Genet. Coop. Newsl. **52,** 107—108.

Smith, J. D., Fong, F., Magill, C. W., Herlick, S., 1983: Fluridone-induced phenocopies of *vp-5* in *Zea mays* seed. Genetics Suppl. **104,** 66.

Stubbe, H., 1957: Mutanten der Kulturtomate, *Lycopersicon esculentum,* Miller I. Kulturpflanze **5,** 190—220.

Stubbe, H., 1958: Mutanten der Kulturtomate, *Lycopersicon esculentum,* Miller II. Kulturpflanze **6,** 89—115.

Stubbe, H., 1959: Mutanten der Kulturtomate, *Lycopersicon esculentum,* Miller III. Kulturpflanze **7,** 82—112.

Sung, Z. M., Fienberg, A., Chorneau, R., Borkird, C., Furner, I., Smith, J., Terzi, M., Loschiavo, F., Giuliano, G., Pitto, L., Nuti-Ronchi, V., 1984: Developmental biology of embryogenesis from carrot culture. Plant Mol. Biol. Rep. **2,** 3—14.

Tal, M., 1966: Abnormal stomatal behaviour in wilty mutants of tomato. Plant Physiol. **41,** 1387—1391.

Tal, M., Nevo, Y., 1973: Abnormal stomatal behaviour and root resistance, and hormonal imbalance in three wilty mutants of tomato. Biochem. Genet. **8,** 291—300.

Tal, M., Eshel, A., Witztum, A., 1976: Abnormal stomatal behaviour and ion imbalance in *Capsicum* scabrous diminutive. J. Exp. Bot. **27,** 953—960.

Tal, M., Imber, D., Erez, A., Epstein, E., 1979: Abnormal stomatal behaviour and hormonal imbalance in *flacca,* a wilty mutant of tomato. Plant Physiol. **63,** 1044—1048.

Tal, M., Witztum, A., Shifriss, C., 1974: Abnormal stomatal behaviour and leaf anatomy in *Capsicum annuum,* scabrous diminutive, a wilty mutant of pepper. Ann. Bot. **38,** 983—988.

Taylor, I. B., 1984: Abnormalities of abscisic acid accumulation in tomato mutants. In: Menhennet, R., Lawrence, D. K. (eds.), Biochemical Aspects of Synthetic and Naturally Occurring Plant Growth Regulators, Monograph 11, pp. 73—90. Wantage: British Plant Growth Regulator Group.

Taylor, I. B., Rossall, S., 1982: The genetic relationship between the tomato mutants, *flacca* and lateral suppressor, with reference to abscisic acid accumulation. Planta **154,** 1—5.

Taylor, I. B., Tarr, A. R., 1984: Phenotypic interactions between abscisic acid deficient tomato mutants. Theor. Appl. Genet. **68,** 115—119.

Tucker, D. J., 1976: Endogenous growth regulators in relation to side shoot development in the tomato. New Phytol. **77,** 561—568.

Uknes, S. J., Ho, T. H. D., 1984: Mode of action of abscisic acid in barley aleurone layers. Plant Physiol. **75,** 1126—1132.

Waggoner, P. E., Simmonds, N. W., 1966: Stomata and transpiration of droopy potatoes. Plant Physiol. **41,** 1268—1271.

Wallace, R. H., Habermann, H. M., 1958: Absence of seed dormancy in a white mutant strain of *Helianthus annuus* L. Plant Physiol. **33,** 252—254.

Walton, D. C., 1980: Biochemistry and physiology of abscisic acid. Annu. Rev. Plant Physiol. **31,** 453—489.

Wang, T. L., Donkin, M. E., Martin, E. S., 1984: The physiology of a wilty pea: abscisic acid production under water stress. J. Exp. Bot. **35,** 1222—1232.

Wong, J. R., Sussex, I. M., 1980: Isolation of abscisic acid-resistant variants from tobacco cell cultures. II. Selection and characterization of variants. Planta **148,** 103—107.

Yavada, U. L., Lockard, R. G., 1977: Abscisic acid and gibberellin in three ungrafted apple *(Malus sylvestris)* rootstock clones. Physiol. Plant. **40,** 225—229.

Zeevaart, J. A. D., 1980: Changes in the levels of abscisic acid and its metabolites in excised leaf blades of *Xanthium strumarium* during and after water stress. Plant Physiol. **66,** 672—678.

Chapter 3

Mutants as Tools for the Elucidation of Photosynthetic Processes

Christa Critchley and Warwick Bottomley

Botany Department, The Faculties, Australian National University, GPO Box 4, Canberra ACT 2601, and CSIRO Division of Plant Industry, GPO Box 1600, Canberra ACT 2601, Australia

With 4 Figures

Contents

I. Introduction

Genetic, biochemical and physiological analyses of photosynthetic mutants have been used for many years in attempts to broaden our understanding of the structure-function relationships of the photosynthetic apparatus. The recent development of molecular biology has given us some insight into the primary structure of plant genes and is providing the techniques for the investigation of the molecular basis of mutations which were previously only detectable through their phenotypic expression. In spite of these advances we still have little understanding of the mechanisms regulating the expression of these genes. The use of recombinant DNA techniques in directly probing and manipulating the molecular pathways responsible for particular aspects of plant performance, such as photosynthesis, may lead to significant advances in our knowledge and understanding of these processes.

In this chapter we will neither deal with the genetic analysis of "photosynthetic mutants" as such nor will we provide a comprehensive review of the recent literature on "photosynthetic mutants". We will instead attempt a critical evaluation of the usefulness of these mutants to assist in the solution of some of the physiological and biochemical problems concerning the structure, organization and function of the photosynthetic apparatus. We will draw on results of selected investigations which, we believe, illustrate the power or otherwise of this particular approach. We will also explore the potential for recombinant DNA technology to aid in this approach by delineating the limitations of each of the disciplines involved, i. e. genetics, molecular biology, biophysics, biochemistry and physiology, stressing the importance of a multidisciplinary approach.

Chloroplasts, the organelles in the plant cells responsible for photosynthesis, are bounded by a double membrane, the envelope, and contain a highly organized and complex internal membrane system, the thylakoids, in which the initial energy conversion and electron transport processes occur. The fixation of carbon dioxide and the ensuing metabolic reactions which lead to carbohydrate synthesis take place in the stroma, which is the non-membranous part of the chloroplast. Photorespiration, part of which

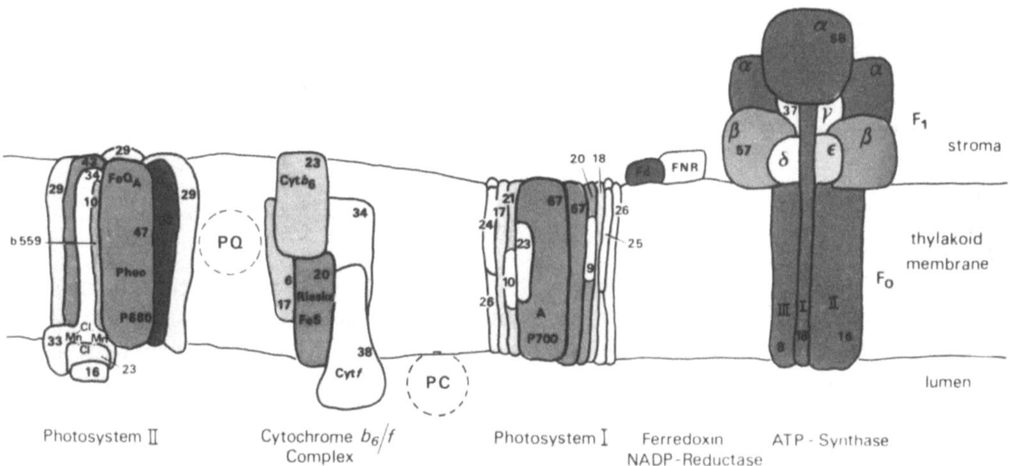

Fig. 1. Model of the thylakoid membrane with the four major supramolecular complexes designated Photosystem II, Photosystem I, Cytochrome b6/f complex and ATP synthase. The numbers indicating molecular weights of individual polypeptides are those most commonly obtained from sodium dodecyl sulfatepolyacrylamide gel electrophoresis, which differ somewhat in most instances from those obtained by deduction from gene sequences (for comparison see Table I)

also takes place in the stroma, is a process by which ribulose bisphosphate, the acceptor for CO_2 in photosynthesis, is oxidized by molecular oxygen. The carbon reduction and oxidation cycles are interlocked and are both initiated by the bifunctional enzyme ribulose-1,5-bisphosphate carboxylase/oxygenase. The enzymatic reactions which constitute the two cycles are well characterized biochemically and use the reducing power of the $NADPH_2$ and the ATP generated in the membrane dependent light reactions. Although most of the soluble enzymes involved in the carbon reduction and oxidation cycles are well characterized, little is known of the genetic factors which govern the regulation of expression and activity of these enzymes. Much research effort has recently been concentrated on the structure-function relationships of the constituents of the thylakoid membrane and their organization.

Photosynthetic mutants have the potential to yield several categories

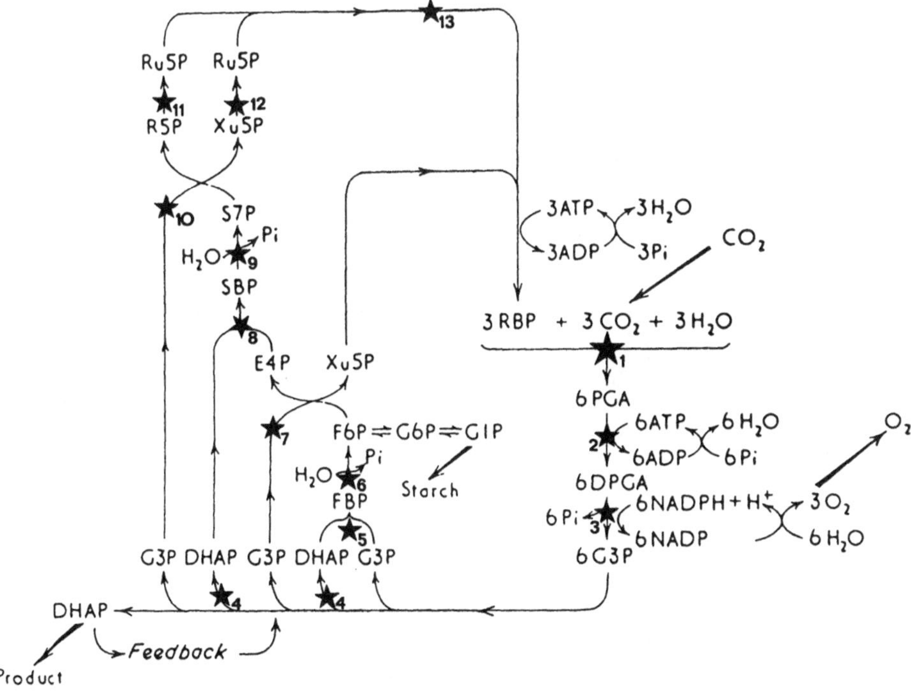

Fig. 2. The Reductive Pentose Phosphate Pathway (or Benson-Calvin Cycle) with numbered stars indicating the enzymes involved: 1, Ribulose Bisphosphate Carboxylase Oxygenase; 2, Phosphoglycerate Kinase; 3, Triosephosphate Dehydrogenase; 4, Triosephosphate Isomerase; 5, Aldolase (FBP Aldolase); 6, Fructose Bisphosphatase (FBPase); 7, Transketolase; 8, Aldolase (SBP Aldolase); 9, Sedoheptulose Bisphosphatase (SBPase); 10, Transketolase; 11, Ribose Phosphate Isomerase; 12, Ribulose Phosphate 3-Epimerase; 13, Phosphoribulokinase.
(With permission from Walker, D. A. and Edwards, G. C3, C4: mechanisms, and cellular and environmental regulation of photosynthesis. Blackwell Scientific Publications, Oxford 1983)

of data. These include the identification of specific proteins which have roles in photosynthetic processes, such as electron transport and carbon metabolism, as well as the localization of such proteins in the supramolecular complexes of the thylakoid membrane (Fig. 1) or the positions of enzymes in a metabolic pathway (Fig. 2). They can also be used to facilitate biochemical and biophysical measurements when, for instance, the two photosystems (Fig. 3) interfere with one another, as is the case when studying ESR signals from photosystem II. Here, the use of a mutant deficient in photosystem I components and function can eliminate interference from this complex without the need to use detergents which may remove other parts of the electron transport chain (see also Section IV).

In most of the past work on photosynthetic mutants neither the molecular basis for the mutation(s) nor their location, i. e. whether chromosomal or extrachromosomal, was known because only in plants with which formal genetic analysis was possible could it be established whether Mendelian or maternal inheritance patterns prevailed. Therefore mutant work was restricted to the detection of phenotypic expression of a lesion at the biochemical level. Identification of a molecular defect or mutation, i. e. a DNA-based alteration in gene structure, has only been possible since the development of recombinant DNA techniques which enable positive identification of genes and their products, and the determination of their

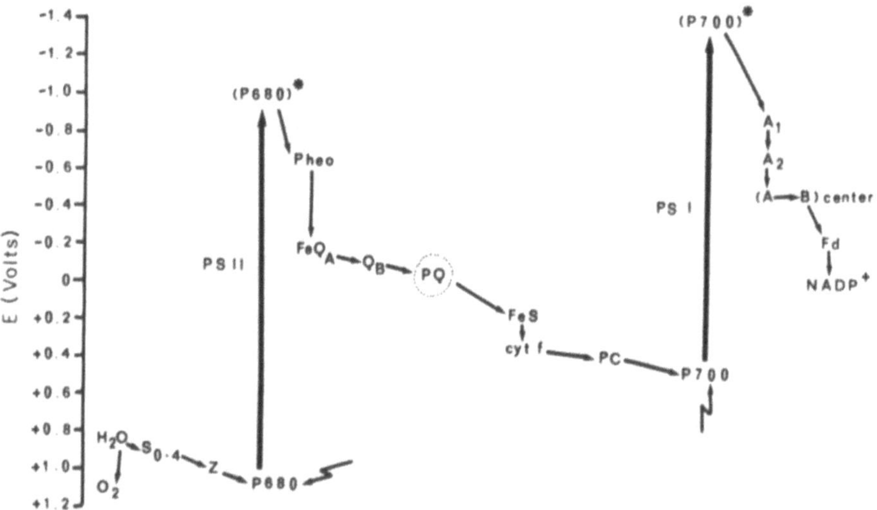

Fig. 3. The Z-Scheme of photosynthetic electron transport indicating the redox components operating within and between the two photosystems. S_{0-4}, S-states of the oxygen evolving system; Z, secondary donor to PS II; P680, reaction centre of PS II; Pheo, pheophytin, primary electron acceptor of PS II; Fe QA, quinone, first stable electron acceptor of PS II; QB, quinone, secondary acceptor of PS II; PQ, plastoquinone pool; FeS, Rieske iron-sulphur centre; Cyt f, cytochrome f; PC, plastocyanin; P700, reaction centre of PS I; A1, A2, primary and secondary electron acceptors of PS I; Fd, ferredoxin

primary structure through DNA sequencing. Such a molecular defect or change, however, needs to be correlated with, or shown to bear a causal relationship to, the phenotypic characteristic in question. This causal relationship has to be extended further to the change in gene structure and the physical manifestation of the mutation, such as alterations in the primary, secondary or tertiary structure of the protein which may, for instance, affect its binding affinity for substrates or cofactors or alter its temperature dependence.

As long as the pleiotropic effects of a mutation are not ubiquitous and the mutant characteristics can be described adequately and comprehensively by comparison with the wild type, mutants should be useful in solving some of the biochemical problems in metabolism or membrane structure. The question really is whether we can find examples showing that the mutant approach has clearly contributed to our knowledge of photosynthetic membrane structure and function or metabolic pathways which straightforward biochemical or biophysical analysis could not provide.

II. Two Genomes Code for the Structural and Regulatory Elements of the Photosynthetic Apparatus

It has long been known that chloroplasts contain their own DNA as well as possessing the ability to express the genetic information encoded in that DNA. They contain their own RNA polymerase and also ribosomes which

Table 1. Genes known to be located on the chloroplast genome

Gene	Gene Product	Complex	Molecular Weight
atpA	CF1-alpha	ATP-synthase	56,000
atpB	CF1-beta	ATP-synthase	54,000
atpE	CF1-epsilon	ATP-synthase	14,700
atpF	CFo-I	ATP-synthase	24,000
atpH	CFo-III	ATP-synthase	8,000
rbcL	large subunit	RuBPC'ase	54,000
psaA	reaction centre	PS I	68,000
psbA	QB-protein	PS II	39,000
psbB	reaction centre	PS II	56,000
psbC	reaction centre	PS II	51,000
psbD	D2	PS II	39,000
psbE	cytochrome b 559	PS II	9,000
petA	cytochrome f	cytb 6/f	33,000
petB	cytochrome b 6	cytb 6/f	23,000
petD	polypeptide IV	cytb 6/f	17,000

are distinct from those of the cytoplasm (for reviews see Bohnert *et al.,* 1982; Bottomley and Bohnert, 1982; Whitfeld and Bottomley, 1983). It is also well documented that the genes coding for the chloroplast components are not all contained in chloroplast DNA but, indeed, that most of these genes are localized in the nucleus and are transcribed and translated in the cytoplasm. Only a relatively small number of genes is coded for by chloroplast DNA. Table 1 lists the genes currently known to be contained in chloroplast DNA. Fig. 4 shows the physical map of the spinach chloroplast genome denoting the approximate locations of the genes identified to date.

Most of the chloroplast polypeptides whose genes have been localized occur in macromolecular complexes and it appears that the genetic information for the constituents of each of these complexes is distributed between the two genomes. That is, part of each complex is coded for by the nuclear genome and part by the chloroplast DNA.

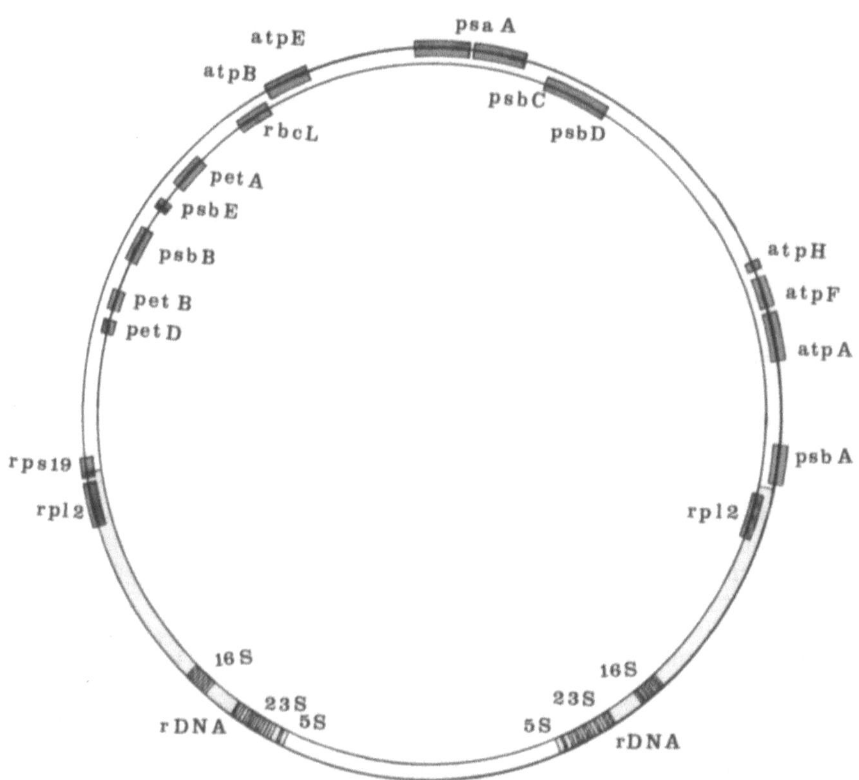

Fig. 4. A representation of the circular spinach chloroplast genome showing the location of the known genes. The genes coding for constituents of the photosynthetic system are as shown in Table 1. Other genes are rps 19 and rpl 2 — the genes for the ribosomal proteins S 19 and L 2; rRNA — the genes for the 23 S, 16 S and 5 S ribosomal RNAs

In addition, each chloroplast has, on average, about 25 copies of its DNA and, since there are around 100 chloroplasts in each cell, it follows that there are, on average, over 2000 copies of chloroplast DNA in each green cell. While the mechanism of chloroplast DNA replication is largely unknown at present, it is obvious that the means by which chloroplast mutants arise is unlikely to be simple. A further complication is that the nuclear encoded chloroplast genes that have been studied so far, those for the small subunit of ribulose-1,5-bisphosphate carboxylase (Dunsmuir *et al.*, 1983a; Coruzzi *et al.*, 1983; Smith *et al.*, 1983) and the chlorophyll a/b binding proteins of the light-harvesting complex of photosystem II (Dunsmuir *et al.*, 1983b; Dunsmuir, 1985; Cashmore, 1984), occur in multiple copies in each genome. These copies are not all identical but can be divided into families according to the hybridizability of the untranslated regions of the genes. These factors not only increase the complexity of the interpretation of results derived from the use of mutants in the elucidation of structural and functional aspects of photosynthesis but also of the isolation of mutants with particular genotypic alterations.

III. Identification of Thylakoid Membrane Proteins and Their Functions

Mutants can facilitate biochemical and biophysical analysis of specific photosynthetic functions (Moller *et al.*, 1980; Nugent *et al.*, 1980). Criteria for assessment of photosynthetic properties are important because some structural or functional characteristics may not be useful in mutant characterization (Simpson and von Wettstein, 1980). Characterization of a photosynthetic mutant based on the presence or absence of components of a particular complex or the measurement of functional or biophysical properties does not necessarily constitute an accurate characterization of the mutation since they measure only phenotypic manifestations. An example of this is the barley chlorina *f2* mutant where the thylakoid membranes have been shown to be deficient in chlorophyll-binding proteins of the photosystem II light-harvesting complex. However, it has also been shown that not only are these proteins synthesized *in vivo*, but also that the chloroplasts of both the wild type and the mutants are capable of transporting them across the chloroplast envelope (Bellemare *et al.*, 1982; Ryrie, 1983). This indicates that the lesion affects assembly of the complex rather than synthesis of the polypeptide. The mutant is known also to be deficient in chlorophyll b and it has been proposed by Bellemare *et al.* (1982) that normally the polypeptides are stabilized by binding to chlorophyll and the absence of this binding results in an increased turnover of the proteins.

A good deal of work has been directed towards the identification of polypeptides which are absent from mutant thylakoids. However, without direct evidence concerning the synthesis of these peptides in the mutant little can be concluded regarding the nature of the lesion.

Metz *et al.* (1980) have recently identified a 34 kD protein in PS II of *Scenedesmus* as a component of the water splitting system, possibly involved in manganese binding and maintenance of cytochrome b 559 HP. It appears that this protein, in contrast to the 16 and 23 kD polypeptides of the water splitting system (Honberg, 1984), is encoded in the chloroplast DNA and it may well be identical with the recently sequenced so-called D2 polypeptide (Rasmussen *et al.*, 1984; Holschuh *et al.*, 1984).

The *Scenedesmus* mutant LF-1, blocked on the oxidizing side of PS II and shown to be deficient in a 34 kD protein, contained instead a 36 kD protein (Metz and Seibert, 1984). These authors also confirmed that this 34 kD protein is not identical with the 33 kD "Murata" protein involved in oxygen evolution (Miyao und Murata, 1984; Kuwabara *et al.*, 1985).

Recently Browse *et al.* (1985) and McCourt *et al.* (1985) described a mutant of *Arabidopsis thaliana* which is deficient in the unusual fatty acid trans-3-hexadecenoic acid. This acyl group exclusively esterifies the 2-position of phosphatidylglycerol (PG), one of the phospholipid constituents of normal thylakoid membranes, and has been thought to play an important role in the structure and/or function of these membranes. Significantly, the complete lack of this fatty acid does not alter the structural or functional characteristics of the mutant chloroplasts when compared to those of the wild type, either by electron microscopy of grana structure or by measurement of photosynthetic activities for PS I and PS II partial reactions *in vitro*. Electrophoretic analysis of the chlorophyll-protein complexes and properties of chlorophyll fluorescence under various conditions showed no significant effect on energy transfer efficiency or related properties in the mutant (McCourt *et al.*, 1985). This is an example of a useful photosynthetic mutant which has assisted in the elucidation of photosynthetic structure and function. Indeed it has been shown very elegantly that this unusual fatty acid, the subject of much controversy concerning its role in the membrane, is not required for either structure or function in the thylakoids.

IV. From Phenotype to Gene Structure

A. Herbicide Resistance

In 1984, the first reports appeared of a mutation affecting a protein involved in photosynthesis for which it is believed the molecular alteration in the primary gene structure is known. This is a mutation of the rapidly turned over QB membrane protein that is an intrinsic component of PS II and is thought to bind the herbicides atrazine and diuron. Interestingly, the gene for this protein is encoded in the chloroplast DNA and is extremely highly conserved between a number of diverse plant species. The only difference in the amino-acid sequences, as deduced from the nucleotide sequences of the genes of the sensitive and resistant biotypes, is a substitution for the serine at position 264 in the protein sequence as a result of a single base mutation. In *Solanum nigrum* (Hirschberg *et al.*, 1983; Goloubinoff and Edelman, 1984) and *Amaranthus hybridus* (Hirschberg and

McIntosh, 1983) glycine is the substitute while in *Chlamydomonas* it is alanine (Erickson *et al.*, 1984). The coincidence of triazine resistance and a single amino-acid substitution in the triazine binding protein provides only circumstantial evidence, although quite compelling and appealing evidence nonetheless, that the substitution is the direct cause of the alteration in the affinity of the polypeptide for triazine. The future challenge is to elucidate the nature of the alteration in the secondary or tertiary structure of the protein which will explain the change in triazine-binding properties leading to resistance, and perhaps to introduce the suspected specific mutation into the genome of a suseptible species and thereby confer triazine resistance on its photosynthetic electron transport function.

B. Barley

A great deal of painstaking work has been carried out, mainly by the Carlsberg Laboratories in Copenhagen, on a series of barley mutants with alterations resulting in various differences in their photosynthetic capabilities. The mutations have mainly been characterized in terms of fluorescence properties, photosystem activities and ultramicroscopic structure.

One such mutant which has a potential to localize genes for chloroplast constituents in either the nuclear or the chloroplast DNA has been described by Hoyer-Hansen and Casadoro (1982). This mutant is a conditional temperature sensitive mutant of barley which, if grown at the permissive temperature of 30° C, contains complete sets of functional ribosomes whereas, if grown at 23° C, appears to lack both 70 S ribosomes and chloroplast ribosomal proteins. Examination of the polypeptide composition of the chloroplast thylakoids showed that nine polypeptides known to be synthesized on chloroplast ribosomes were missing.

However, it has also been shown that the mutation is not in any of the genes associated with ribosome assembly or function but rather in the synthesis of carotenoids. The destruction of ribosomes at the lower temperature is a secondary effect of the mutation and can be overcome by selection of the wavelength of light in which the plants are grown. At long wavelengths even at 23° C there is a normal complement of chloroplast ribosomes. This mutant has been used to show that the gene for a 23 kD component of the water-splitting complex associated with photosystem II is located in the nucleus, since the protein is present in the mutant when grown under the conditions where the chloroplast ribosomes are absent (Honberg, 1984).

This 23 kD component of the water-splitting complex is also present in the membranes of mutants which are deficient in photosystem II. Thus it appears that the water-splitting complex is capable of assembling in the thylakoid membrane in the absence of a complete photosystem II complex.

Hoyer-Hansen *et al.* (1979) found that three mutants deficient in stroma thylakoids also have decreased amounts of all five subunits of the CF_1 complex of ATP-synthase. This finding supports the contention that the ATP-synthase complex is localized exclusively on the stroma thylakoids and does not occur within the granal stacks.

While some of the mutants appear to be deficient in a single protein, others, such as some of the photosystem I mutants, have deficiencies in at least four polypeptides. This is probably due to pleiotropic effects such as occur in the chlorina *f2* mutant which contains decreased amounts of four photosystem I proteins and three proteins of the photosystem II light-harvesting complex. However, it has been demonstrated by Bellemare *et al.* (1982) that all of these proteins are synthesized in the plants and that they are transportable across the chloroplast membrane. Although a number of mutants in PS II which are deficient in only one polypeptide are known, all PS I mutants so far isolated are deficient in at least three polypeptides and it would be useful if PS I mutants lacking individual polypeptides were available (Hoyer-Hansen *et al.*, 1982).

Moller *et al.* (1980), Nugent *et al.* (1980) and Moller *et al.* (1981) describe some of the biochemical properties of photosystem I mutant viridis-*n 34*. The thylakoid membranes of this mutant are deficient in four polypeptides, one of which is the chlorophyll-protein-1 of PS I and the other three polypeptides are believed to be iron-sulphur proteins. All four proteins are found in PS I particles isolated from the wild type.

EPR measurements on PS I particles from this barley mutant show that the characteristic signals for PS I are present in those particles which contain only three polypeptides of 110, 18.3 and 15.2 kD (Moller *et al.*, 1981).

EPR measurements of PS II signals are facilitated by the use of PS I mutant viridis-*b 63*, because of the lack of interference from PS I signals. These studies supplement and support those with PS II particles and whole thylakoids.

Haworth *et al.* (1982) claim that barley mutant chlorina *f2* lacks the LHC II polypeptide which is phosphorylated *in vivo* and *in vitro* and which, through phosphorylation and dephosphorylation, regulates the excitation energy distribution between PS II and PS I. They claim that this lack of protein and phosphorylation make state 1 — state 2 transitions impossible.

By contrast, Ryrie (1983) showed that three of the four LHC II polypeptides are present in chlorina *f2*, although in greatly reduced quantities, and that phosphorylation also takes place. This would appear to be an instance where mutant work has confused rather than clarified the issue in question and we do not seem to have learned anything about either the role of LHC II in excitation energy distribution or that of its phosphorylation. The lack of mutants with specific and definable lesions together with the difficulties associated with pleiotropic effects, complicate the interpretation of results obtained with these mutants.

C. Chlamydomonas

The work on a variety of *Chlamydomonas* mutants designed to unravel the relationship between photosynthesis, bicarbonate transport and carbonic anhydrase activity, for example, has generated a lot of data that are quite difficult to interpret. It has, however, unambiguously shown that the coordinated regulation of all three processes is necessary, and has provided a direct demonstration that a component other than carbonic anhydrase is involved in the carbon dioxide concentrating mechanism (Spalding *et al.*, 1983; Ogren *et al.*, 1984; Spalding *et al.*, 1985).

D. Arabidopsis

Somerville (1984) has recently published an excellent review article on the subject of carbon fixation and photorespiration mutants. We will therefore only deal with one more recent example. The dicarboxylate transporter mutant of *Arabidopsis thaliana* is an example of how biochemical and physiological analysis of a mutant helped to determine the specificity of the transporter and its primary *in vivo* role, which were either previously not known or disputed (Somerville and Ogren, 1983). Subsequently Somerville and Somerville (1985) described the comparison of chloroplast envelope polypeptides of wild type and transporter mutant and found that a 42 kD protein was missing from the polypeptide pattern of the mutant envelope. This protein is suggested to be a component of the dicarboxylate transporter, about whose structural identity nothing was known previously.

E. Oenothera

The chloroplast encoded genes of most plants used in the study of photosynthesis, indeed of most higher plants, exhibit maternal inheritance. That is, the total genome of the F1 plastids is derived from the female parent. Segregation is in the ratio of 0:4 instead of the usual ratio of 1:2:1, which is the Mendelian ratio found in nuclear inheritance, where equal contributions to the genetic make-up come from each parent. One genus in which chloroplasts can be derived from either parent is *Oenothera.* In this genus the chloroplast genomes of either parent can be transferred to the progeny. An extensive study of the inheritance patterns of the genus has been made by Stubbe and Herrmann (1982) who have described five types of chloroplast genomes and three types of nuclear genomes. Stubbe and his coworkers described the phenotypic effects of transferring the various chloroplast types into the three types of nuclear background, particularly with regard to the chlorophyll content (Kutzelnigg and Stubbe, 1974). They isolated more than 40 of these mutants varying from green to white and described some combinations which were lethal.

Unfortunately, only a relatively small amount of biochemical work has been carried out on these mutants to date and only one instance of a location of a mutation has been reported. The mutant, named *lsigma,* has been shown to lack the large subunit of ribulose-1,5-bisphosphate carboxylase (Hallier *et al.,* 1978). More recent work (Hildebrandt *et al.,* 1984), using immunoblotting of polyacrylamide gels, has revealed that the mutant produces a smaller polypeptide reacting with antiserum to the large subunit, suggesting that the gene contains a mutation which causes the premature termination of transcription or translation.

This mutant has also been used in an effort to shed some light on the mechanism of coordination of the expression of chloroplast and nuclear genes. Since the gene for the large subunit of ribulose-1,5-bisphosphate carboxylase is located in the chloroplast DNA, while that for the small subunit is nuclear encoded, and they are required in equal numbers in the enzyme complex, the possibility exists that the presence or absence of one of the subunits affects the synthesis of the other. For instance Ellis *et al.* (1978) have suggested that the synthesis of the small subunit regulates the synthesis of the large subunit. One way to test this hypothesis would be to determine whether the inhibition of the synthesis of one subunit affected the production of the other. If the small subunit was controlling the synthesis of the large subunit then, under conditions where synthesis of the large subunit was inhibited by some means other than by inhibiting the synthesis of the small subunit, it could be expected that the small subunit would still be present or even would accumulate in amounts in excess of normal. A mutant in which one subunit is not produced is obviously ideal material for such an experiment. Hallier *et al.* (1978) used this *lsigma Oenothera* mutant in such an experiment and came to the conclusion that, on the basis of polyacrylamide gels and immunodiffusion, no small subunit was produced by the mutant. This was in contrast with the results of Feierabend (1978) who used a temperature-sensitive mutant of rye which contained no chloroplast ribosomes and therefore synthesized no large subunit when grown at the nonpermissive temperature. This author found that the small subunit was still produced in the absence of the large subunit. More recently the *lsigma* mutant has been re-examined by Hildebrandt *et al.* (1984), who found that the small subunit was indeed produced in the mutant. However, since they also showed that a truncated large subunit was produced by the mutant the situation is by no means clear and the possibility remains that the presence of the small subunit is affecting the synthesis of the large subunit.

Sears and Herrmann (1985) have examined an *Oenothera* mutant deficient in two of the eight subunits of the ATP-synthase complex. Although the results of the *in vivo* experiments suggested that the two subunits, which in all plants so far examined are juxtaposed on the genome and cotranscribed (Whitfeld and Bottomley, 1983), are fused to one another perhaps by a deletion of the adjacent ends, translation of the mRNA from this mutant yielded polypeptides which were indistinguishable from those of the wild type. Sears and Herrmann (1985), therefore, suggest that the mutation is in some component which affects a post-transcriptional event. This then can presumably also be classed as a pleiotropic effect which results in an apparent photosynthetic mutation.

F. Summary

Genes which code for proteins involved in the photosynthetic process are distributed between the nuclear and the chloroplast genomes. Of the genes coding for proteins which have so far been shown to occur in chloroplast

DNA, the majority have been genes for proteins that are involved directly in the photosynthetic process. That is, with the exception of genes for tRNAs, ribosomal proteins (Schmidt *et al.,* 1983), elongation factor, ribosomal RNA and RNA polymerase, the genes so far located are either constituents of the photosynthetic membranes or the large subunit of ribulose-1,5-bisphosphate carboxylase (see Table 1 and Fig.4). The number of structural proteins whose genes are located on the chloroplast genome is small compared to the total number of genes required for the development and function of the chloroplast. The significance of the occurrence of those specific genes may become more apparent when one or several complete chloroplast genomes have been sequenced and analyzed. Another remarkable aspect of the distribution of genes for proteins associated with photosynthesis is that, to date, no incidence has been reported of differences in the location of a gene for a specific protein; in other words at present there is an absolute commonality of the distribution of genes between the nucleus and plastid among and across plant species from *Euglena* and *Chlamydomonas* to higher plants. It is most interesting in this context to note that the genes which code for the small and the large subunits of ribulose bisphosphate carboxylase, which in all higher plants and algae with chloroplasts are genetically separated, are linked in the blue-green alga *Anabaena,* which has only one genome and no separate plastid-bound thylakoid membranes. The two subunits are cotranscribed so that the requirement for coordinated expression would appear to be less of a problem in these primitive organisms (Nierzwicki-Bauer *et al.,* 1984).

It seems to be fairly well established now that there is no mRNA exchange between the two sites of protein synthesis: this is to say that all cytoplasmic mRNA appears to be translated on 80S cytoplasmic ribosomes, whereas all chloroplastic mRNA seems to be translated on 70S chloroplast ribosomes. Chloroplast mRNA is generally believed not to be polyadenylated and this structural feature may serve to distinguish between sites of protein synthesis, and may in fact be related to structural differences between 70S and 80S ribosomes. Another feature of the distinction between the two genomes is the finding that most nuclear-encoded proteins which are destined for the chloroplast are synthesized as precursors, whereas most chloroplast-coded proteins are not. The leader sequences on the nuclear-encoded proteins are involved in their transport across the chloroplast membrane, since these sequences are cleaved off during passage into the organelle. As pointed out before, the complexity associated with the dual genomes clearly increases the complexities and the ambiguity associated with interpretation of the results obtained with mutants.

It was highlighted above that, considering the number of chloroplasts per cell and the number of DNA copies per chloroplast, each cell should contain around 2000 copies of chloroplast DNA. To date, only one organism has been reported to contain heterogeneous chloroplast DNA molecules (Jenni *et al.,* 1981). It was found that the multiple copies of the *Euglena gracilis* chloroplast genome are not uniform in size and that the difference between the extremes amounted to about 800 basepairs or somewhat less than 0.6% of the average genome length. These authors could not distinguish whether the differences were intra- or inter-chloroplastic or -cellular. It is, however, quite feasible that such differences, if they were intrachloroplastic for example, could be the equivalent of the multigene families found for nuclear genes (Dunsmuir *et al.*, 1983a, b; Cashmore, 1984; Broglie *et al.*, 1983). Both phenomena may have a role in the differential expression of a particular gene in response to environmental or other regulatory factors.

The greater majority of plant chloroplast genomes so far investigated contain an inverted repeat sequence of around 20 kbp, which suggests the possibility that the DNA circle can exist in two forms as a result of intramolecular recombination at the ends of these repeats. This was indeed shown to be so for the two cases in which a restriction enzyme could be found which did not cut within the repeats, thus allowing the detection of two different sized fragments containing the inverted repeats. These cases were *Phaseolus* (Palmer, 1983) and also the cyanelle DNA from *Cyanophora paradoxa* (Bohnert and Loeffelhardt, 1982).

V. Conclusions

The photosynthetic process is a complex one carried out in a specialized organelle with the aid of a number of protein complexes in a highly structured membrane system and a large amount of soluble enzymes in the stroma. In addition, the genetic information coding for these components is contained in at least two distinct genetic systems. The expression of these two genomes is carried out by two different transcription and translation systems, one of which, that of the chloroplast, is largely procaryotic in nature while that of the cytosol is eucaryotic. In view of the complexity it is obvious that mutants should be extremely useful tools for the elucidation of the mechanisms of the photosynthetic processes as well as that of the regulation and coordination of the two systems. There is no doubt that mutants have played, and are continuing to play, an important role in furthering our understanding of these processes. In particular they can provide material in which the complexity is reduced and interpretation is simplified by the removal of parts of the systems. However, the overall processes are so complex that the interpretation of the results is still difficult, particularly because of the possibility of phenotypic expression being the result of indirect interactions of the constituents, as has been demonstrated a number of times.

To date, the contributions of recombinant DNA technology to the solution of the problems of photosynthetic mechanisms are only relatively minor. The potential for the application of such techniques as site directed mutagenesis is just beginning to be tapped. For example, the elegant work of Gruissem *et al.* (1983) who have used this technique, combined with *in vitro* transcription in a chloroplast cell-free system, to localize the regulatory regions of the gene for $tRNA_{met}$ and the work of Mullet *et al.* (1985), who applied the same approach to the genes for the large subunit of ribulose bisphosphate carboxylase and the beta subunit of ATP-synthase, give an indication of the advances in our understanding of the subject that should be made in the near future.

The application of the knowledge gained from the understanding of the photosynthetic processes to the solution of problems in plant science will ultimately depend on the availability of a transformation system for chloroplasts. While there are indications that this will soon be achieved, the problem remains as to the possible complications caused by the existence of the large number of copies of chloroplast DNA in each cell.

Some recent research that may have implications on the interpretation of results using mutants are the reports that copies of parts of the chloroplast DNA are present in other genomes in the cell. It was been reported by Lonsdale *et al.* (1983) that maize mitochondrial DNA contains regions homologous with both chloroplast ribosomal cistrons as well as the gene for the large subunit of ribulose bisphosphate carboxylase. In addition, Scott and collaborators (Scott and Timmis, 1984; Whisson and Scott, 1985) have reported that the nuclear DNA of spinach contains regions homologous with chloroplast DNA. Further substantiation and extension of these results, together with the possible demonstration that these extra chloroplast DNA genes are expressed *in vitro,* would not only complicate the interpretation of some results but would also raise interesting questions regarding the mechanism of such duplications and their possible role in the overall regulation of photosynthesis.

VI. References

Bellemare, G., Bartlett, S. G., Chua, N.-H., 1982: Biosynthesis of chlorophyll a/b-binding polypeptides in wild type and the chlorina f2 mutant of barley. J. Biol. Chem. **257,** 7762—7767.

Bohnert, H.-J., Crouse, E. J., Schmitt, J. M., 1982: Organization and expression of plastid genomes. In: Parthier, B., Boulter, D. (eds.), Nucleic Acids and Proteins in Plants II. Encyclopedia of Plant Physiology, New Series, Vol. 14B, pp. 475—530. New York — Berlin — Heidelberg: Springer-Verlag.

Bohnert, H.-J., Lieffelhardt, W., 1982: Cyanelle DNA from Cyanophora paradoxa exists in two forms due to intramolecular recombination. FEBS Lett. **150,** 403—406.

Bottomley, W., Bohnert, H.-J., 1982: The biosynthesis of chloroplast proteins. In: Parthier, B., Boulter, D. (eds.), Nucleic Acids and Proteins in Plants II. Encyclopedia of Plant Physiology, New Series, Vol. 14B, pp. 531—596. New York - Berlin — Heidelberg: Springer-Verlag.

Broglie, R., Coruzzi, G., Lamppa, G., Keith, B., Chua, N.-H., 1983: Structural analysis of nuclear genes coding for the precursor to the small subunit of wheat ribulose-1,5-bisphosphate carboxylase. Bio/Technology **1,** 55—61.

Browse, J., McCourt, C. R., Somerville, C. R. 1985: A mutant of *Arabidopsis* lacking a chloroplast-specific lipid. Science **227,** 763—765.

Cashmore, A. R., 1984: Structure and expression of a pea nuclear gene encoding a chlorophyll a/b-binding polypeptide. Proc. Nat. Acad. Sci., U.S.A. **81,** 2960—2964.

Coruzzi, G., Broglie, R., Cashmore, A., Chua, N.-H., 1983: Nucleotide sequence of two pea cDNA clones encoding the small subunit of ribulose 1,5-bisphosphate carboxylase and the major chlorophyll a/b-binding thylakoid polypeptide. J. Biol. Chem. **258,** 1399—1402.

Dunsmuir, P., 1985: The petunia chlorophyll a/b-binding protein genes: a comparison of Cab genes from different gene families. Nuc. Acids Res. **13,** 2503—2519.

Dunsmuir, P., Smith, S. M., Bedbrook, J., 1983a: A number of different nuclear genes for the small subunit of RuBCase are transcribed in petunia. Nuc. Acids Res. **11,** 4177—4183.

Dunsmuir, P., Smith, S. M., Bedbrook, J., 1983b: The major chlorophyll a/b-binding protein of *Petunia* is composed of several polypeptides encoded by a number of distinct nuclear genes. J. Mol. Appl. Genet. **2,** 285—300.

Ellis, R. J., Highfield, P. E., Silverthorne, J., 1978: The synthesis of chloroplast proteins by subcellular systems. In: Hall, D. O., Coombs, J., Goodwin, T. W. (eds.), Photosynthesis 1977, pp. 497—506.

Erickson, J. M., Rahire, M., Bennoun, P., Delepelaire, P., Diner, B., Rochaix, J.-D., 1984: Herbicide resitance in *Chlamydomonas reinhardii* results from a mutation in the chloroplast gene for the 32-kilodalton protein of photosystem II. Proc. Nat. Acad. Sci., U.S.A. **81,** 3617—3621.

Feierabend, J., 1978: Cooperation of cytoplasmic and plastidic protein synthesis in rye leaves. In: Akoyunoglou, G., Argyroudi-Akoyunoglou, J. H. (eds.), Chloroplast Development, pp. 207—213. Amsterdam: Elsevier.

Goloubinoff, P., and Edelman, M., 1984: Chloroplast-coded atrazine resistance in *Solanum nigrum:* psbA loci from susceptible and resistant biotypes are isogenic except for a single codon change. Nuc. Acids Res. **12,** 9489—9496.

Gruissem, W., Greenberg, B. M., Zurawski, G., Prescott, D. M., Hallick, R. B., 1983: Biosynthesis of chloroplast transfer RNA in a spinach chloroplast transcription system. Cell **35,** 815—828.

Hallier, U. W., Schmitt, J. M., Heber, U., Chaianova, S. S., Volodsarsky, A. D., 1978: Ribulose-5-bisphosphate carboxylase-deficient plastome mutants of *Oenothera.* Biochim. Biophys. Acta **504,** 67—83.

Haworth, P., Kyle, D. J., Arntzen, C. J., 1982: Protein phosphorylation and excitation energy distribution in normal, intermittent-light-grown, and a chlorophyll b-less mutant of barley. Arch. Biochem. Biophys. **218,** 199—206.

Hildebrandt, J., Bottomley, W., Moser, J., Herrmann, R. G., 1984: A plastome mutant of *Oenothera hookeri* has a lesion in the gene for the large subunit of Ribulose-1,5-bisphosphate carboxylase/oxygenase. Biochim. Biophys. Acta **783,** 67—73.

Hirschberg, J., Bleecker, A., Kyle, D. J., McIntosh, L., 1983: The molecular basis of triazine-herbicide resistance in higher-plant chloroplasts. Z. Naturforsch. **39c,** 412—420.

Hirschberg, J., McIntosh, L., 1983: Molecular basis of herbicide resistance in *Amaranthus hybridus.* Science **222,** 1346—1349.

Holschuh, K., Bottomley, W., Whitfeld, P. R., 1984: Structure of the spinach chloroplast genes for the D2 and 44 kD reaction-centre proteins of photosystem II and for tRNASer (UGA). Nuc. Acids Res. **12,** 8819—8834.

Honberg, L. S., 1984: Probing barley mutants with a monoclonal antibody to a polypeptide involved in photosynthetic oxygen evolution. Carlsberg Res. Commun. **49,** 703—719.

Hoyer-Hansen, G., Moller, B. L., Pan, L. C., 1979: Identification of coupling factor subunits in thylakoid polypeptide patterns of wild-type and mutant barley thylakoids using crossed immunoelectrophoresis. Carlsberg Res. Commun. **44,** 337—351.

Hoyer-Hansen, G., Casadoro, G., 1982: Unstable chloroplast ribosomes in the cold-sensitive barley mutant tigrina-0^{34}. Carlsberg Res. Commun. **47,** 103—118.

Hoyer-Hansen, G., Moller, B. L., Henrey, L. E. A., Casadoro, G., 1982: Thylakoid polypeptide synthesis and assembly in wild-type and mutant barley. Cell Function and Differentiation, Part **B,** 111—125.

Jenni, B., Fasnacht, M., Stutz, E., 1981: The multiple copies of the *Euglena gracilis* chloroplast genome are not uniform in size. FEBS Lett. **125,** 175—179.

Kutzelnigg, H., Stubbe, W., 1974: Investigations on plastome mutants in *Oenothera.* 1. General considerations. Sub-Cell. Biochim. **3,** 73—89.

Kuwabara, T., Miyao, M., Murata, T., Murata, N., 1985: The function of 33-kDa protein in the photosynthetic oxygen-evolution system studied by reconstitution experiments. Biochem. Biophys. Acta **806,** 283—289.

Lonsdale, D. M., Hodge, T. P., Howe, C. J., Stern, D. B., 1983: Maize mitochondrial DNA contains a sequence homologous to the ribulose-1,5-bisphophate carboxylase large subunit gene of chloroplast DNA. Cell **34,** 1007—1014.

McCourt, P., Browse, J., Watson, J., Arntzen, C. J., Somerville, C. R., 1985: Analysis of photosynthetic antenna function in a mutant of *Arabidopsis thaliana* (L.) lacking trans-hexadecenoic acid. Plant Physiol. **78,** 853—858.

Metz, J. G., Seibert, M., 1984: Presence in photosystem II core complexes of a 34-kilodalton polypeptide required for water photolysis. Plant Physiol. **76,** 829—832.

Metz, J. G., Wong, J., Bishop, N. I., 1980: Changes in electrophoretic mobility of chloroplast membrane polypeptide associated with the loss of the oxidizing side of photosystem II in low fluorescent mutants of *Scenedesmus.* FEBS Lett. **144,** 61—66.

Miyao, M., Murata, N., 1984: Effect of urea on photosystem II particles: Evidence for an essential role of the 33 kilodalton polypeptide in photosynthetic oxygen evolution. Biochim. Biophys. Acta **765,** 253—257.

Moller, B. L., Nugent, J. H. A., Evans, M. C. W., 1981: Electron paramagnetic resonance spectrometry of photosystem I mutants in barley. Carlsberg Res. Commun. **46,** 373—382.

Moller, B. L., Smillie, R. M., Hoyer-Hansen, G., 1980: A photosystem I mutant in barley (*Hordeum vulgare* L.). Carlsberg Res. Commun. **45,** 87—99.

Mullet, J. E., Orozco, E. M., Chua, N.-H., 1985: Multiple transcripts for higher plant rbcL and atpB genes and localization of the transcription initiation site of the rbcL gene. Plant Mol. Biol. **4,** 39—54.

Nierzwicki-Bauer, S. A., Curtis, S. E., Haselkorn, R., 1984: Cotranscription of genes encoding the small and large subunits of ribulose-1,5-bisphosphate carboxylase in the cyanobacterium *Anabaena* 7120. Proc. Nat. Acid Sci., U.S.A. **81,** 5962—5965.

Nugent, J. H. A., Moller, B. L., Evans, M. C. W., 1980: EPR detection of the primary photochemistry of photosystem II in a barley mutant lacking photosystem I activity. FEBS Lett. **121,** 355—357.

Ogren, W. L., Somerville, C. R., Somerville, S. C., Spreitzer, R. J., Spalding, M. H., Jordan, D. B., 1984: Genetic analysis of photosynthetic carbon pathways. In:

Sybesma, C. (ed.), Advances in Photosynthesis Research, Vol. III, pp. 429—435. The Hague: Nijhoff/Junk.

Palmer, J. D., 1983: Chloroplast DNA exists in two orientations. Nature (London) **301,** 92—93.

Rasmussen, O. F., Bookjans, G., Stummann, B. M., Henningsen, K. W., 1984: Localization and nucleotide sequence of the gene for the membrane polypeptide D2 from pea chloroplast DNA. Plant Mol. Biol. **3,** 191—199.

Ryrie, I. J., 1983: Immunological evidence for apoproteins of the light-harvesting chlorophyll-protein complex in a mutant of barley lacking chlorophyll b. Eur. J. Biochem. **131,** 149—155.

Schmidt, R. J., Richardson, C. B., Gillham, N. W., Boynton, J. E., 1983: Sites of synthesis of chloroplast ribosomal proteins in *Chlamydomonas.* J. Cell Biol. **96,** 1451—1463.

Scott, N. S., Timmis, J. N., 1984: Homologies between nuclear and plastid DNA in spinach. Theor. Appl. Genet. **67,** 279—288.

Sears, B. B., Herrmann, R. G., 1985: Plastome mutation affecting the chloroplast ATP synthase involves a post-transcriptional defect. Current Genet. **9,** 521—528.

Simpson, D. J., von Wettstein, D., 1980: Macromolecular physiology of plastids. XIV. Viridis mutants in barley: genetic, fluoroscopic and ultrastructural characterization. Carlsberg Res. Commun. **45,** 283—314.

Smith, S. M., Bedbrook, J., Spiers, J., 1983: Characterisation of three cDNA clones encoding different mRNAs for the precursor to the small subunit of wheat ribulosebisphosphate carboxylase. Nuc. Acids Res. **11,** 8719—8734.

Somerville, C. R., 1984: The analysis of photosynthetic carbon dioxide fixation and photorespiration by mutant selection. Oxford Surveys of Plant Molecular & Cell Biology **1,** 103—131.

Somerville, S. C., Ogren, W. L., 1983: An *Arabidopsis thaliana* mutant defective in chloroplast dicarboxylate transport. Proc. Nat. Acad. Sci., U.S.A. **80,** 1290—1294.

Somerville, S. C. Somerville, C. R., 1985: A mutant of *Arabidopsis* deficient in chloroplast dicarboxylate transport is missing an envelope protein. Plant Sci. Lett. **37,** 217—220.

Spalding, M. H., Spreitzer, R. J., Ogren, W. L., 1983: Genetic and physiological analysis of the CO_2-concentrating system of *Chlamydomonas reinhardii.* Planta **159,** 261—266.

Spalding, M. H., Spreitzer, R. H. Ogren, W. L. 1985: Use of mutants in analysis of the CO_2-concentrating pathway of *Chlamydomonas reinhardii.* In: Lucas, W. J., Berry, J. A., (eds.), Inorganic carbon uptake by aquatic photosynthetic organisms. American Society of Plant Physiology, Rockville, MD, pp. 361—375.

Stubbe, W., Herrmann, R. G., 1982: Selection and maintenance of plastome mutants and interspecific genome/plastome hybrids from *Oenothera.* In: Edelman, M., Hallick, R. B., Chua, N.-H. (eds.), Methods in Chloroplast Molecular Biology, pp. 149—165. Amsterdam: Elsevier.

Whisson, D. L., Scott, N. S., 1985: Nuclear and mitochondrial DNA have sequence homology with a chloroplast gene. Plant Mol. Biol. **4,** 267—273.

Whitfeld, P. R., Bottomley, W., 1983: Organization and structure of chloroplast genes. Annu. Rev. Plant Physiol. **34,** 279—310.

Chapter 4

Maize Alcohol Dehydrogenase: A Molecular Perspective

Wayne L. Gerlach, Martin M. Sachs*, Danny Llewellyn, E. Jean Finnegan and Elizabeth S. Dennis

Division of Plant Industry, CSIRO, GPO Box 1600, Canberra, ACT 2601, Australia
*Dept. of Biology, Washington University, St. Louis, MO 63130, U.S.A.

With 5 Figures

Contents

I. Introduction

During the 1960's a number of advances were made toward understanding the regulation of prokaryotic and fungal gene systems. From there interest turned to gene regulation in higher eukaryotes and, amongst plant gene systems, the genetics and biochemistry of the maize alcohol dehydrogenases (ADH, EC 1.1.1.1) received immediate attention from two groups (Scandalios, 1966; Schwartz, 1966; Schwartz and Endo, 1966).

There were several reasons for this interest in the maize ADH system. Maize itself was readily amenable to genetic analysis and a strong background knowledge of maize genetics already existed. ADH was found to be very stable in extracts and could be isolated to homogeneity (Felder *et al.*, 1973; Kelly and Freeling, 1980). Analysis of ADH activity was fairly straightforward using a spectrophotometric assay (Hageman and Flesher, 1960; Freeling, 1973). An activity-specific stain was also developed and this could be applied to both gel electrophoresis and *in situ* studies of cells and tissues (Schwartz and Endo, 1966). Furthermore, studies had shown that ADH activity could be induced by flooding maize seedlings (Hageman and Flesher, 1960). If, as was thought, this induction of activity acted at the gene level then the ADH system seemed to provide an opportunity to approach the study of gene expression and its regulation in a plant system.

The alcohol dehydrogenase system of maize has now been extensively characterized in genetic, physiological and, more recently, molecular terms. This chapter will review this work from the characterization of the system in terms of enzyme activity and how this led to the isolation of the genes, through the molecular organization of the *Adh* genes and on to the insights this has provided about this and other genetic systems of maize. Because of the large body of knowledge which has accumulated on the *Adh* genes of maize, we will primarily confine this review to the maize ADH system, but will also draw upon information about *Adh* genes of other plants whenever relevant to the concepts under consideration.

II. Genetics and Expression of ADH Enzymes in Maize

A. Two Genes Encode ADH in Maize

Gel electrophoresis of inbred maize lines reveals three ADH isozymes. Genetic analysis using different inbred lines (Schwartz and Endo, 1966) and EMS-induced mutant lines (Schwartz, 1969 a, 1969 b) which exhibit simultaneous polymorphisms in electrophoretic migration or disappearance of two of the three isozymes showed that ADH activity is encoded by two unlinked genes (Freeling and Schwartz, 1973). These are *Adh 1* on chromosome 1 (Schwartz, 1971) and *Adh 2* on chromosome 4 (Dlouhy, 1980). Additionally, the ADH enzyme is active as a dimer and the three isozymes are explained as ADH1.ADH1 and ADH2.ADH2 homodimers and the ADH1.ADH2 heterodimer (Freeling, 1974).

B. Organ Specificities of ADH1 and ADH2 Activities

Adh1 and *Adh2* were found to be expressed differently relative to each other in different maize tissues and organs. Neither ADH1 nor ADH2 is found to any great extent in seedling tissues, but both activities are induced by flooding and by treatment with the synthetic plant hormone 2,4-dichlorophenoxyacetic acid, (2,4-D) (Freeling, 1973). There are high levels of ADH1 and ADH2 in the naturally developing endosperm but little or no activity in mature endosperm. ADH1 activity is predominant in the scutellum, aleurone, embryo and anther, while ADH2 predominates in tassel nodes and the peduncle. ADH1 is found in pollen but there is no detectable ADH2 activity.

C. Mutations of Adh Genes of Maize

This topic has been extensively reviewed by Freeling and Birchler (1981) and more recently by Freeling and Bennett (1985) and this section will, therefore, give only a brief indication of the mutational analyses of maize *Adh* genes.

Numerous natural variants of *Adh1* have been characterized. These include variants that affect electrophoretic mobility (e. g. *S, F, C;* Schwartz and Endo, 1966), specific activity (e. g. C^{m} and natural null alleles; Schwartz and Laughner, 1969; Stuber and Goodman, 1983), ability or inability of alleles to intragenically recombine with each other (Freeling, 1976, 1978), relative organ specific expression (Woodman and Freeling, 1981), transcript length variants (Dennis *et al.,* 1984; Sachs *et al.,* 1986), polymorphisms in regions flanking the *Adh1* gene (Johns *et al.,* 1983; Sachs *et al.,* 1985) and a natural, tightly linked duplication of *Adh1* (Fc^{m}; Schwartz and Endo, 1966).

A number of variants of *Adh2* have been characterized by Dlouhy (1980; described also in Freeling and Birchler, 1981). These include electrophoretic variants (R, P, N, L) and naturally occurring null alleles. *Adh2* alleles also exhibit differences in relative expression in the presence of ethylene (Schwartz, 1978), and extensive restriction site polymorphisms (Dennis *et al.,* 1985).

Three methods have been developed to screen for induced *Adh1* mutants. The allozyme screen, developed by Schwartz (see Freeling and Birchler, 1981) involves mutagenesis of one *Adh1* electrophoretic allele and crossing to a different electrophoretic variant. M1 seeds are then analysed for mutation. Allyl alcohol selection of pollen (Schwartz and Osterman, 1976; Freeling and Cheng, 1978) takes advantage of the huge number of pollen grains from a single plant and the relative insensitivity of pollen to allyl alcohol unless it has normal levels of ADH which convert the alcohol to the highly toxic aldehyde, acrolein. A third method can be used to select in both directions, for and against ADH1 activity. When kernels are flooded with aerated water, seeds with normal levels of ADH1 will germinate. Those with lower levels will not germinate but can be subsequently rescued by germination in air (Chen, pers. comm; see also Freeling and Bennett, 1985).

Several *Adh1* mutants have been isolated after EMS treatment, mostly by Schwartz (see Freeling and Birchler, 1981; Freeling and Bennett, 1985). Many, perhaps all, appear to be point mutations most easily explained as base substitutions. They include mutants with altered electrophoretic mobility, specific activity, heat lability, dimerization ability, as well as CRM+ and CRM− nulls. Most appear to have normal levels of normal sized mRNAs and none appear to affect gene regulation. One exception is *S664* which appears to have an unusually long transcript in addition to one of the normal size and seems to be affected in transcription or 3′ processing (Hake *et al.*, 1984). Most of the remaining *Adh1* mutants have been isolated by insertion of transposable elements. These will be discussed in more detail below.

D. The Maize Anaerobic Response

Hageman and Flesher (1960) were the first to show that ADH activity increases as a result of flooding maize seedlings. Freeling (1973) later showed that ADH activity increased at a zero order rate between 5 hr and 72 hr of anaerobic treatment reflecting a simultaneous expression of two unlinked genes, *Adh1* and *Adh2*. Schwartz (1969 a) showed that ADH activity is required to allow maize seeds and seedlings to survive anaerobic treatment and presumably flooding under natural conditions. Since this anaerobic response formed the basis of the isolation of the *Adh1* and *Adh2* genes we will consider it here in more detail.

In many ways the maize anaerobic response is analogous to the heat-shock response described in many animals and plants. However, except for one possible overlap, it involves a different set of proteins (Sachs *et al.*, 1980; Cooper and Ho, 1983; Kelley and Freeling, 1982). There is an immediate repression of pre-existing protein synthesis and the first five hours of anaerobic treatment are a transition period during which there is a rapid increase in the synthesis of a class of polypeptides with an approximate molecular weight of 33 kd (Fig. 1). These have been referred to as the transition polypeptides (TPs). After approximately 90 minutes, the synthesis of an additional group of 20 novel "anaerobic" polypeptides (ANPs) is induced, amongst which are ADH1 and ADH2. This group of 20 ANPs represents more than 70 percent of the total label incorporation after five hours of anaerobic treatment. By this time synthesis of the TPs is at a minimal level; however, pulse-chase experiments have shown that the TPs are very stable. The synthesis of the ANPs continues in a quantitatively stable ratio for up to 72 hours of anaerobic treatment, at which time protein-synthesis decreases as the seedlings begin to die (Sachs *et al.*, 1980).

The rapid repression of pre-existing protein synthesis, as seen also in anaerobically treated soybean seedlings (Lin and Key, 1967), is correlated with a near complete dissociation of polysomes in anaerobically treated maize tissue (Dennis and Pryor, unpublished). It is not due to degradation

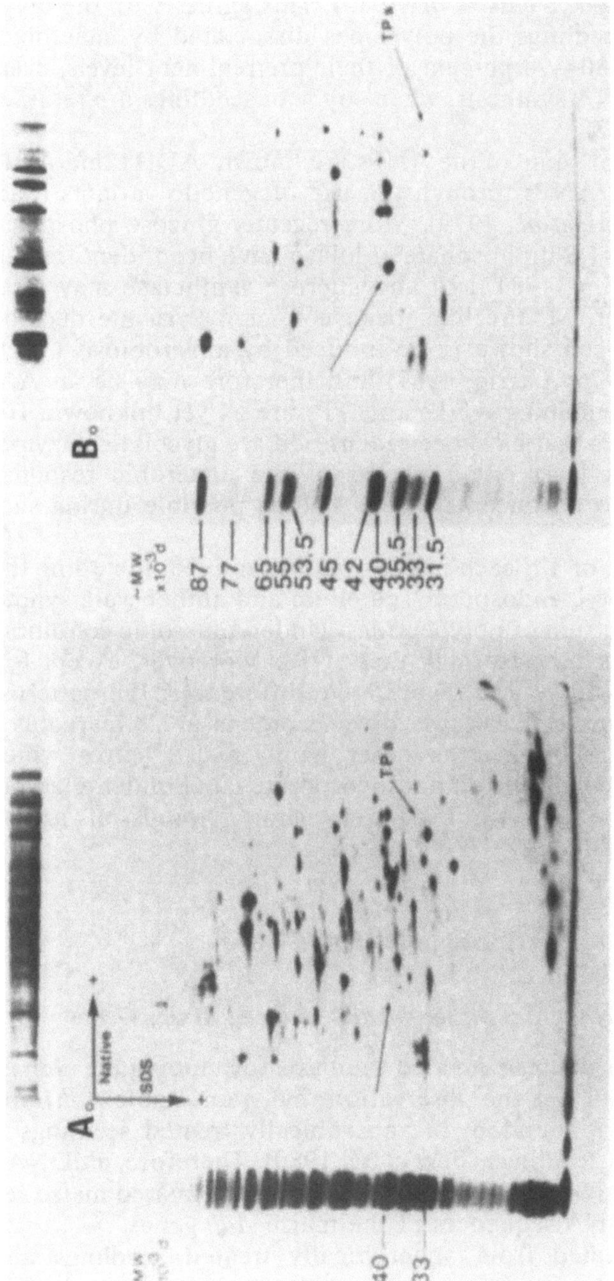

Fig. 1. Proteins synthesized in maize primary roots during aerobic (A) and anaerobic (B; 12—17 hours anaerobiosis) growth. Proteins being synthesized were radioactively pulse labelled and visualized by autoradiography after two dimensional electrophoresis. The unlabelled arrow points to ADH 1 location. TP arrow points to position of transition polypeptides (Sachs *et al.*, 1980)

of "aerobic" mRNAs, since the mRNAs encoding the "aerobic" proteins remain translatable in an *in vitro* system at least five hours after anaerobic treatment is initiated (Sachs *et al.*, 1980). This agrees with the observation that in soybean seedlings the polysomes dissociated by anaerobiosis can rapidly reform to 80—90 percent of their pretreatment levels, even in the absence of new RNA synthesis, when soybean seedlings are returned to air (Lin and Key, 1967).

The functions of some of the ANPs are known. ADH1 and ADH2 have been identified as ANPs through the use of genetic variants (Sachs and Freeling, 1978; Ferl *et al.*, 1979). More recently glucose phosphate isomerase and fructose-1,6-diphosphate aldolase have been identified as ANPs (Kelley and Freeling, 1984a, b) and sucrose synthetase may also be an ANP (P. Starlinger, M. Freeling, pers. comms.). Pyruvate decarboxylase activity has also been shown to be induced by anaerobiosis (Wignarajah and Greenway, 1976; Laszlo, 1981) and therefore may be an ANP. The functions of the remaining ANPs and TPs are as yet unknown. However, all five of the ANPs that have been identified are glycolytic enzymes.Thus, it appears that at least one function of the anaerobic response is to enable the plant to produce as much ATP as possible during short-term flooding.

In the presence of air, each maize organ examined, including the roots, coleoptile, mesocotyl, endosperm, scutellum and anther wall, synthesizes a tissue-specific spectrum of polypeptides. Under anaerobic conditions all of the above organs synthesize only the ANPs. Moreover, except for a few characteristic qualitative and quantitative differences, the patterns of anaerobic protein synthesis in these diverse organs are remarkably similar (Okimoto *et al.*, 1980). On the other hand, maize leaves which have emerged from the coleoptile do not incorporate label under anaerobic conditions and do not survive even very short periods of anaerobiosis (Okimoto *et al.*, 1980).

III. Isolation of *Adh* Genes

A. cDNAs from Anaerobically Induced Maize Genes

In vitro translation studies provided the basis for cloning the *Adh* genes of maize. Fundamental was the observation that translatable *Adh1* and *Adh2* mRNAs were more abundant in anaerobically treated seedlings than in aerobically treated seedlings (Ferl *et al.*, 1980). Therefore, a cDNA library prepared from mRNA isolated from anaerobically treated maize seedlings should contain cDNA sequences of the maize *Adh* genes.

RNA was isolated from anaerobically treated seedlings and fractionated according to size by sucrose-gradient centrifugation. An RNA size class which produced ADH-sized polypeptides in *in vitro* translation experiments was chosen for cloning reactions. A cDNA library was prepared from this RNA and transformant colonies were hybridized in parallel with two different radioactive probes prepared from RNA of anaerobically or aerobically grown plants. Colonies were chosen which showed greater hybridization with the anaerobic probe than with the aerobic probe (Fig. 2A). These were presumed to contain cDNA sequences complementary to induced transcripts in the anaerobic RNA population, amongst which would likely be cDNAs from ADH mRNAs (Gerlach *et al.*, 1982).

B. Isolation and Identification of Adh1 cDNA Clones

The chimaeric plasmids from individual clones were used to hybrid-select mRNAs from an mRNA population prepared from anaerobic seedlings. The selected mRNAs were then translated *in vitro*. One clone, pZML84, selected an mRNA which translated to a 40 kd polypeptide, the same size as ADH1 and ADH2. Further tests, which drew upon the wealth of background information on maize *Adh* genetics, indicated that it was a cDNA clone for *Adh1* of maize.

First of all, RNAs from genetic variants were used in hybrid-selection experiments. One experiment involved RNA from a line which produced a variant of ADH with altered electrophoretic mobility (the *Adh1-U725* allele). It was found that the clone pZML84 selected an mRNA which translated *in vitro* to produce an ADH polypeptide with the expected alteration in electrophoretic migration. When RNA from a heterozygote for alleles encoding ADHs with these different electrophoretic mobilities (genotype: *Adh1*-F/*Adh1*-U725; Ferl *et al.*, 1979) was used, the hybrid selection experiment produced the two products expected.

Physical properties of the maize ADH1 polypeptide were also used to verify that the pZML84 hybrid selection product was ADH1 polypeptide and that the clone therefore contained an *Adh1* cDNA sequence. It was found that the HRT product co-migrated with authentic ADH1 (Ferl *et al.*, 1979) on two dimensional isoelectric focussing/polyacrylamide gel electrophoresis. Previous techniques of purification of RNA had also led to production of antibody against ADH1. A polyclonal antibody prepared against ADH1 also recognized and precipitated the HRT product. Conclusive evidence came from a comparison of amino-acid sequences of ADH1 protein and the amino-acid sequence predicted from the cDNA nucleotide sequence (Dennis *et al.*, 1984). Three fragments were isolated from purified ADH1 protein after partial proteolytic digestion and subjected to sequence analysis from their N-termini. The amino-acid sequences were found in the inferred sequence from the cDNA clone, as were other peptide sequences which had been determined in M. Freeling's laboratory (Kelly and Freeling, unpublished).

In later experiments (Dennis *et al.*, 1984) other *Adh1* cDNA clones were isolated using the original pZML84 cDNA insert sequence as a probe. The cDNA sequences in these newer clones had been size-fractionated prior to cloning to ensure that only long cDNA sequences were included. Among these, one clone designated pZML793 included the entire 3′ non-coding

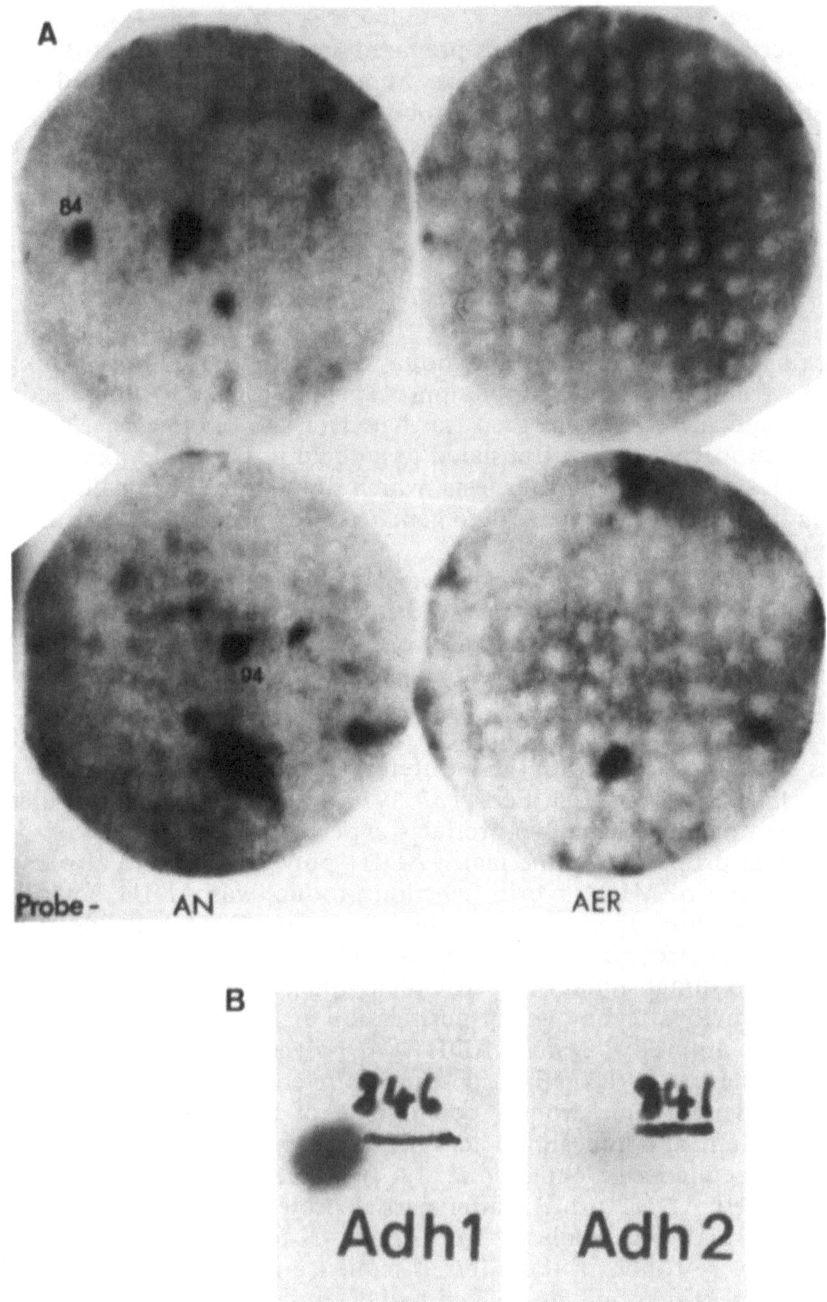

Fig. 2. Colony hybridizations. (A) Signals seen when duplicate filters containing arrays of different anaerobic cDNA clones were hybridized with radioactive probes prepared from anaerobic "An" or aerobic "Aer" mRNAs. Colony *84* contains sequences corresponding to an anaerobically induced RNA and proved to be *Adh 1* cDNA. (B) Different signal strengths seen when *Adh 1* cDNA is used as probe on colonies containing either homologous *Adh 1* cDNA plasmid or heterologous *Adh 2* cDNA (80 % sequence homology, see text)

region, the translated region and extended into the 5′ non-coding region i. e. a near full-length *Adh1* cDNA clone.

C. Isolation and Identification of Adh2 cDNA Clones

The *Adh1* cDNA clone was used in the isolation of *Adh2* cDNA sequences. During experiments in which the *Adh1* cDNA sequence was used as a probe to isolate further cDNA clones in colony hybridization experiments, colonies were observed which hybridized to the probe under moderate stringency, albeit at a lower level than colonies containing authentic *Adh1* cDNA (Fig. 2 B). Because ADH1 and ADH2 can form functional heterodimers and crossreact immunologically, it was considered likely that they would be related in amino-acid sequence and, hence, that their genes might have related nucleotide sequences. Thus, it seemed probable that these weakly cross-hybridizing colonies contained *Adh2* cDNA sequences.

Isolation and sequencing of these cDNA sequences showed that they probably did encode *Adh2* (Dennis *et al.*, 1985). The cDNA sequence was 82 % homologous with the *Adh1* sequence and derived amino-acid sequences had 87 % identity. Verification that the clone contained *Adh2* cDNA again used the existing genetic information on the maize ADH system. It was known that lines containing the *Adh2-33* null allele did not accumulate translatable *Adh2* mRNA as measured by *in vitro* translation (Ferl *et al.*, 1980) and the putative *Adh2* cDNA did not hybridize to mRNA from any line containing this allele, whereas it did hybridize to all lines containing active *Adh2* alleles.

Genomic clones containing gene sequences and flanking regions of *Adh1* and *Adh2* from maize have been isolated from libraries using these cDNA probes (e. g. see Dennis *et al.*, 1984, 1985).

D. Isolation of Adh Genes from Other Plant Species

The cloned *Adh* sequences from maize can be used as probes to detect and isolate cross-hybridizing sequences in cloned libraries from other plants. Isolation and subsequent characterization, particularly sequence analysis, of such clones will determine whether they do in fact contain *Adh* sequences from the plant under consideration.

This approach has been used in our laboratory to isolate three *Adh* cDNA clones from pea (E. J. F and D. L., manuscript in preparation). They were detected by sequence homology with the maize *Adh1* cDNA sequence. Sequence data have been obtained for all three clones and they appear to be from two distinct but very similar *Adh1*—like genes with 96 % homology over 386 bp of coding region and 89 % over 150 bp of 3′-untranslated region. When compared with the nucleotide and predicted amino-acid sequences of maize *Adh1* cDNA, the degree of homology is 78 % and 85 % respectively. When compared with maize *Adh2*, the values are 73 % and 84 % respectively.

Cross hybridization with the maize *Adh1* cDNA was also used to isolate an *Adh* genomic clone from barley (M. Trick and K. Edwards, pers. comm.), and a 2.3 kb Hind III restriction fragment from within the *Adh1* genomic sequence of maize has been used to isolate a cross-hybridizing genomic clone from *Arabidopsis thaliana* (C. Chang and E. Meyerowitz, pers. comm.). In both cases sequencing of the isolated clones indicates that they encode ADH polypeptides.

The use of prokaryotic expression vectors has enabled the isolation of an *Adh* cDNA clone from tomato cell suspension cultures (B. Williams, pers. comm.) in which tomato ADH2 is found at high levels. A library of cDNA clones inserted into the β-galactosidase gene of a phage lambda vector was prepared. Colonies which produced ADH polypeptide as a β-gal fusion protein were detected by immunological cross reaction with polyclonal and monoclonal antibodies to tomato ADH. The cloned tomato cDNA sequences have homology with the maize *Adh* genes and are, therefore, from an *Adh* gene. Again, the ability to isolate pure ADH for the production of antibodies was central to the success of this approach.

IV. Structure of Plant *Adh* Genes

Genomic clones of maize *Adh1-1S* and *Adh2-N* alleles have been fully sequenced and permit a comparison of the two genes (Fig. 3; and Dennis *et al.*, 1984, 1985). The coding regions appear to be identical in length and are more than 80% homologous at both the amino-acid and nucleotide sequence level. Most nucleotide sequence differences occur in the redundant third base positions of codons. Both genes are interrupted by nine introns in identical positions within the gene. These introns account for up to half of the length of the genes. Except for the short regions at intron-exon boundaries which contain splice signals, there is very little length or sequence resemblance between corresponding introns in *Adh1* and *Adh2*. The splice signals are essentially the same as those found in other plant and animal genes, reflecting the probable similarity in RNA splicing mechanism between animals and plants. The close structural homology and similarity in number and location of introns in the genes is evidence that they arose by duplication and subsequent divergence from a single ancestral gene.

One interesting region of difference between *Adh1* and *Adh2* lies in exon 5. This is shown in Fig. 4A. One interpretation is that compensating frame-shift mutations have occurred in this region in one of the *Adh* genes during their evolution. Presumably the non-essential nature of the *Adh* genes (lines containing *Adh*-null mutations are viable under normal aerobic growth conditions) has allowed the original frameshift null mutation to be tolerated until a compensating frameshift mutation occurred nearby to restore activity.

Another noteworthy structural rearrangement which has occurred during evolution of these genes is found at the 3′ end of intron 6 of the *Adh1* gene (Dennis *et al.*, 1984). Here a region of approximately 100 bp is

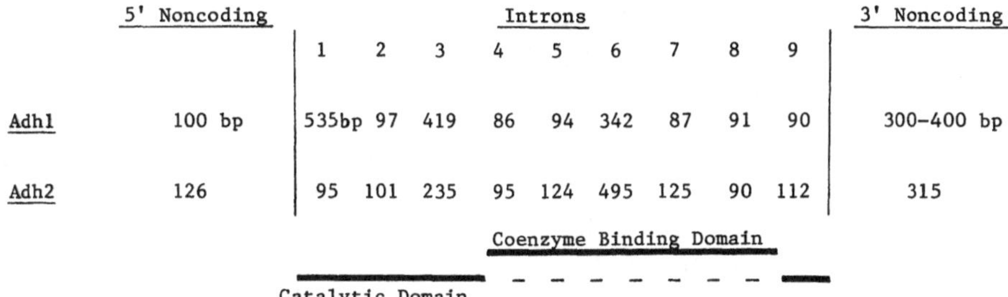

	5' Noncoding	Introns 1	2	3	4	5	6	7	8	9	3' Noncoding
Adhl	100 bp	535bp	97	419	86	94	342	87	91	90	300–400 bp
Adh2	126	95	101	235	95	124	495	125	90	112	315

Fig. 3. Structural features of maize *Adh 1-S* and *Adh 2-N* alleles. Each gene contains 9 introns at identical locations in the coding sequence. Intron sizes as well as 5′ leader and 3′ non-coding region lengths are shown. Relationship of coding regions to protein structure are shown on the lower lines

repeated three times (Fig. 4B). The first two repeats are entirely within intron 6. Both contain in-frame stop codons following the putative splice acceptor sites, and so could only code for a truncated protein if they were incorrectly used as the intron exit during splicing of an RNA transcript. The third copy of the repeat spans the junction of intron 6 and its adjacent exon region. The sequence chosen for splicing *in vivo* is that in the third repeat. This sequence has high homology with the corresponding regions in the other two repeats within intron 6, suggesting that sequence *per se* of the intron-exon boundary is not the sole determinant of a functional splice junction. This situation provides a natural anology to the observation that when duplications are artificially introduced into a rabbit β-globin gene the effective acceptor site is the most 3′ copy of the acceptor sequence (Wieringa *et al.*, 1983). Sequence divergence between *Adhl* and *Adh2* is so high in intron 6 that we cannot determine whether *Adh2* once also had this triplication. We consider it unlikely since the mutation rate which has allowed divergence of the non-functional intron 6 sequences between *Adhl*

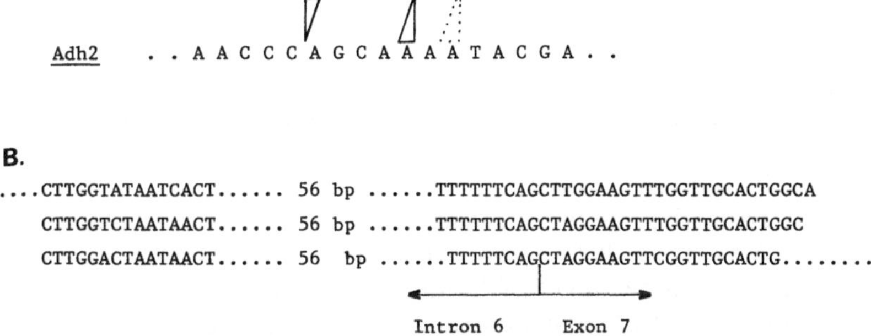

Fig. 4. A. Comparison of *Adh 1* and *Adh 2* sequence homologies in a portion of exon 5 where the genes differ by a likely double-frameshift (initial mutation followed by compensating mutation). B. Intron 6/exon 7 junction of *Adh 1* showing triplicate nature of base sequence

and *Adh2* should also have obscured the triplication in *Adh1,* unless some rectification mechanisms were operating.

Except for an 11 bp region which includes the TATA box and some regions of 8 bp or less, the 5′ and 3′ regions of the *Adh1* and *Adh2* genes show little or no homology. The 5′ non-translated region is 99 bp for *Adh1* and 126 bp for *Adh2.* The 3′ non-translated regions are approximately 300 to 400 bp long for both genes, depending on the allele under consideration. Some transcription signals can be recognized in these regions (discussed below) but in general it appears that while the overall structure and protein coding sequences of *Adh1* and *Adh2* are highly conserved, the non-translated sequences have almost completely diverged since the duplication event occurred.

Structural information on gene organization is available for *Adh* genes from two other species. Two genomic clones have been isolated from barley by their cross hybridization with a maize *Adh1* probe (M. Trick and K. Edwards, pers. comm.). Although neither of them contains a complete gene, sufficient data is available to make some comparisons. One clone obviously contains an *Adh* sequence, having high homology to both the maize *Adh1* gene (80 % identity over 222 nucleotides) and maize *Adh2* (79 %). It also has 78 % homology with another barley *Adh* gene. This second barley clone contains an *Adh2*-like sequence with 89 % homology to maize *Adh2* but only 80 % homology to *Adh1.* In both barley genes the introns which have been sequenced are located at identical positions to those seen in the maize *Adh* genes.

An *Arabidopsis Adh* gene has also been fully sequenced (C. Chang and E. Meyerowitz, pers. comm.). The coding sequence is 73 % and 72 % homologous with those of maize *Adh1* and *Adh2* respectively. The predicted amino-acid sequence is 81 % and 79 % conserved with the maize sequences. Six of the nine introns found in maize are also present in the *Arabidopsis* gene but differ in size and sequence, generally being smaller. Three introns (4, 5 and 7) which are present in the maize genes are entirely absent from the *Arabidopsis* gene. Presumably this reflects a mechanism by which *Arabidopsis,* one of the lowest nuclear C-value plants, is able to streamline its genome.

V. Three-Dimensional Structure of ADH Enzymes

The amino acid sequences of the maize ADH1 and ADH2 enzymes have been derived from the nucleotide sequences of the genes (Dennis *et al.,* 1984, 1985). They have 20 % amino-acid sequence homology with yeast ADH and an even higher level of homology, 50 %, with horse liver ADH. The three dimensional (3D) structure of the horse-liver enzyme has been determined by X-ray diffraction and, considering its homology to the maize enzymes, it can be used as a basis for 3D structuring of the maize enzymes (Branden *et al.,* 1984). Computer modelling and physical model building have enabled the complete 3D maize ADH structure to be predicted and the functional domains of the enzyme to be identified (Fig. 3, Branden *et al.,* 1984). The catalytic domain extends from residue 1 to 174 and 318 to 374 and the coenzyme binding domain roughly comprises residues 175 to 315. The coenzyme binding domain is separated from the main part of the catalytic domain by intron 4 and from the carboxy terminal fragment of the catalytic domain by intron 9. The coenzyme binding domain is further separated by intron 7 into two structurally similar $(\alpha\beta)_3$ domains. There is evidence that the two $(\alpha\beta)_3$ units are themselves structural modules; one of them is subdivided by introns 5 and 6 into three similar supersecondary structural units, while the other is separated by intron 8 into one $\alpha\beta$ and one $(\alpha\beta)_2$ units. In the ancestor of the maize gene a tenth intron may have been located between the now adjacent units in $(\alpha\beta)_2$ region and its absence could be due to loss during evolution of the gene. Certainly loss of introns can occur, as evidenced by the loss of three introns during evolution of an *Arabidopsis Adh* gene (see above).

Branden *et al.* (1984) suggest that, as with most eukaryotic dehydrogenases, maize *Adh* may have evolved as two separate domains, the catalytic- and coenzyme-binding domains, which later fused. The basic building block of the coenzyme-binding site was a unit of about twenty amino acids. Three such blocks were fused together to form a mononucleotide-binding domain which later duplicated to form the basic structure of the NAD-binding domain present in all characterized dehydrogenases.

From a consideration of the 3D structures of the proteins and a comparison of maize and horse-liver ADHs the following observations can be made (H. Eklund and C.-I. Branden, pers. comm.):

1. The inter-subunit contact area is very different in the maize enzymes to that in the horse-liver enzyme. However, it is conserved between maize ADH1 and ADH2. This is compatible with the fact that association can occur to produce ADH1/ADH2 heterodimers.

2. Residues that are essential for coenzyme or catalytic activity can be identified from an analysis of which amino acids are conserved between maize ADH1 and ADH2 and horse-liver enzyme. The position of these regions in the molecule can be determined from the 3D model. Thus, ADH provides a system in which fundamental studies on structure-function relationships of proteins can be made.

3. The different substrate specificities and inhibitor kinetics of the maize and horse-liver enzymes can be accounted for on the basis of the 3D structure. At position 48 there is a substitution of a threonine in the maize ADHs for a serine in the horse, and this probably accounts for most of these effects since it is located in the cleft of the catalytic domain close to important functional groups.

4. The substitution of amino acids between the products of *S* and *F* alleles of maize *Adh1* (ala at 126 for gly, and asp at 360 for asn) has no effect on the catalytic activity as both these residues are on the surface of the catalytic domain. The ten fold difference in activity between ADH1 and ADH2 is probably explained by the substitution at position 228 of lys by arg.

VI. Genetic Change Around and Within *Adh* Genes in Maize

One of the strengths of the *Adh1* and *Adh2* genetic system of maize is the number of mutants and variants available. These have been provided largely by Drew Schwartz of Indiana University and his students and associates. Knowledge of the *Adh1* gene structure together with the molecular characterization of a number of these mutants has led to an understanding of the structural nature of transposable elements, allelic variation, gene duplication and somaclonal variation in maize.

A. Allelic Variation

Two naturally occurring variants of *Adh1, Adh1-1S* and *Adh1-1F* (Schwartz, 1966), encode, respectively, slow and fast anodally migrating electrophoretic allozymes of *Adh1.* The altered electrophoretic migration is due to a single charge change (Kelly and Freeling, 1980) which we have determined from the derived amino-acid sequence of the gene sequence to be due to a change from aspartate to asparagine (Sachs *et al.,* 1985).

In addition to the difference in electrophoretic mobility, *Adh1-1S* and *Adh1-1F* differ from each other in their tissue specific expression (Schwartz, 1971; Woodman and Freeling, 1981). Expression of the ADH1-1S polypeptide is higher than that of *F* in the scutellum of the mature kernel. However, in both root and shoot tissue of anaerobically induced seedlings, ADH-1F is found at approximately twice the level of ADH1-1S. Another difference between the two alleles is that *Adh1* null mutants derived from the *1S* allele recombine among themselves and so do a series of null mutants derived from the *1F* allele. However, intragenic recombination between the *Adh1-1S* and *Adh1-1F* alleles has not been detected (Freeling, 1978). A further difference between the two alleles is that, at least in the standard *Adh1-F* line (BKF), there are two different length messenger RNAs but in the standard *Adh1-S* line (BKS) there is only one (Gerlach *et al.,* 1982).

Cloning and sequence comparison of these two alleles has shown them to differ mainly at the 3′ end. The coding regions and the introns are very highly conserved as is the 5′ nontranscribed region up to about 1200 bases 5′ of the coding region. There are differences at the 3′ end of the gene, even in the transcribed region. This accounts for the two mRNAs in *Adh1-1F* and one in *1S.* One of the main differences between the two alleles is the presence of an insert with the form of a transposable element remnant having terminal inverted repeats flanked by a short direct duplication of a genome segment in *Adh1-1F* but not in *Adh1-1S.* This extra DNA contributes extra sites for polyA addition in the *Adh1-1F* transcript. The other differences between the 3′ ends of the *1F* and the *1S* allele are a tandem duplication of 86 bp in *1F* and one of 14 bp in *1S.* There are also 108 bp present in *1S* that are absent from *1F* and, at the 3′ end of the gene, the sequences of both alleles diverge completely.

From sequence analysis we know that the coding and immediate upstream regions (1 kb) remain fairly constant, but that the 3′ ends diverge close to the coding region. Further from the conserved region, Southern analysis suggests extensive sequence variation; restriction enzyme polymorphism is the rule when the two alleles are compared in these regions (Johns *et al.,* 1983; Sachs *et al.,* 1986). We do not know whether these flanking sequences are sufficiently different that a synaptonemal complex cannot be formed and, therefore, that the alleles cannot recombine. Since the main difference between the two alleles is at the 3′ end, one suggestion is that tissue specific expression may involve sequences in the 3′ region. The differences in the 5′ region of the gene are minimal and we cannot point to any sequence difference that appears likely to produce the different expression, although single base changes or small additions/deletions could be significant. Thus, from the analysis of the *1S* and *1F* alleles we have a picture of the sequence differences between two naturally occurring allelic variants in maize.

B. Somaclonal Variation

Plants which have been regenerated following a cycle of tissue culture show variation (somaclonal variation) and such variation has generally been described for morphological characters (Larkin and Scowcroft, 1981). In order to investigate the molecular basis of somaclonal variation, mutations arising in the *Adh1* gene of maize in tissue culture have been sought and one has been isolated (Brettell *et al.,* 1986). This mutant has an altered electrophoretic mobility for the *Adh1* locus but retains full enzymatic activity. Both Southern blot analysis and mapping of the mutant allele show no obvious size difference between the gene and its progenitor. This holds for all of the restriction fragments examined. It is likely that the change in charge in the mutant is due to a point mutation in the coding region. The difference between the mutant and the progenitor will be investigated at the nucleotide level to determine the mutational event. A meaningful comparison can be made because we know the complete nucleotide sequence of the progenitor *Adh1* gene.

C. Gene Duplication

There is a naturally occurring duplication of the *Adh1-FC*^{m} locus (Schwartz, 1973), two electrophoretic products being produced from the same gene. The *F* polypeptide has full activity while *C*^{m} has much lower activity but is more heat stable. The coding regions specifying the two polypeptides cannot be separated by recombination, reflecting either the closeness of the two alleles or else a block in recombination similar to the block in intragenic recombination seen in the *Adh1-F/Adh1-S* comparison. Mutation studies definitely show two coding regions since single mutations affect either *C*^{m} or *F* but not both. Southern blot analysis of the *FC*^{m} allele following BamH1 digestion shows a single fragment of 9 kb different to that seen from either the standard *S* (11 kb) or the standard *F* (7 kb) alleles.

When this 9 kb Bam fragment was cloned and mapped it was shown to carry only a single copy of the *Adh1* gene. C^m also resides on a 9 kb Bam fragment. Further Southern analysis showed that the duplicated region extends for more than 14 kb. Thus, although FC^m is genetically a duplication, the two copies of the structural gene must be separated by more than 14 kb. This separate nature of the *F* and C^m portions of the duplication had been suggested already by the mutational analysis of Birchler and Schwartz (1979). More extensive studies will be needed to show the exact limits of the duplication and how it occurred.

D. Ds Element Mutations in Adh1

Knowing the sequence of the *Adh1* locus has enabled also the isolation and sequencing of *Ds* controlling elements. A *Ds* controlling element mutation of the *Adh1* gene (allele designation *Adh1-Fm335*) was isolated by Osterman and Schwartz (1981). Comparison of the progenitor, mutant and revertant alleles identified the *Ds1* element and led to the first sequencing of a controlling element (Peacock *et al.*, 1983; Sutton *et al.*, 1984). In *Fm335* the *Ds1* element was inserted into the 5′ untranslated region of the gene and caused about a 90 % decrease in ADH1 activity. Analysis of the 405 bp *Ds* element showed 11 bp inverted repeats at the termini and that it had caused an 8 bp direct duplication of the genomic DNA at the site of insertion. Between the 11 bp inverted repeats the internal sequence of the *Ds* element is AT rich (80 % AT) and does not have any recognizable coding capacity. S_1 mapping and Northern analysis suggest that much of the *Ds* element behaves as an intron which is processed out of the transcript. It is possible that the unprocessed RNA transcript is rapidly degraded since the mRNA level is much lower than normal (1 %) and there is no obvious high level of precursor transcripts. Examination of a number of revertants of this mutation showed that the *Ds* element does not excise precisely and much or all of the 8 bp target site duplication remains, although in altered form (Sachs *et al.*, 1985; Dennis *et al.*, in preparation). During excision the bases bordering the site of insertion are either deleted or undergo a complementary transversion. The study of a number of these revertants provides the rules and models for the excision of *Ac/Ds* controlling elements (Peacock *et al.*, 1984; Saedler and Nevers, 1985).

Starting with the same original line used by Osterman and Schwartz (1981), another *Ds* element mutation has been isolated at the *Adh1* locus (Doring *et al.*, 1984). Molecular characterization showed it to be a 1.5 kb element derivative of the *Ac9* family of *Ac/Ds* controlling elements which has inserted into the coding region of the gene (see Doring and Starlinger, 1984).

E. Robertson's Mutator

Having an isolated and characterized *Adh1* gene has enabled characterization of Robertson's mutator elements. "Robertson's mutator" is a phenomenon in which lines carrying the mutator activity give rise to mutants at defined genetic loci at high frequency (Robertson, 1978). Mutations of *Adh1* have been isolated and characterized (Strommer *et al.*, 1982). One such mutant allele, *Adh1-S3034*, was selected phenotypically as an allyl alcohol resistant pollen grain from a stock containing Robertson's mutator. S3034 has about 40 % normal ADH1 activity and reverts to full activity at about one hundred times the normal mutation rate. Mutant derivatives that express 13 % and 0 % *Adh1* (*S3034* A) protein were further selected. Comparison of the *Adh1-1S* progenitor and the *S3034* mutant and its derivative showed the presence of a 1.5 kb insert which was identified as a Robertson's mutator element inserted into the first intron of the *Adh1* gene (Bennetzen *et al.*, 1984). The *S3034* A derivative with zero *Adh1* activity results from a deletion extending from the element into the first exon and removing some of the coding region. The Robertson's mutator element has been sequenced and is being used for transposon mutagenesis in attempts to clone other genes.

VII. Approaches to the Mechanism of *Adh* Gene Regulation

As with any other plant or animal gene, there are three paths towards a better understanding of *Adh* gene regulation. First, direct comparisons of DNA sequences of different but functionally or physiologically related genes give an indication of the gene regions conserved during evolution, and hence of some possible functional significance. Second, correlations between mutant phenotypes and the physical lesions found in natural or induced mutations of genes can suggest a function for certain sequences. Finally, this process can be refined by the specific *in vitro* mutation of cloned genes and their re-insertion into the genome to test the effect of the mutation on *in vivo* expression. Each of these paths complements the other and will provide a picture of the molecular nature of the controls on the expression of the *Adh* genes.

A. DNA Sequence Comparisons

Adh1 and *Adh2* share many biochemical and genetic characteristics that suggest their origin from some ancestral gene duplication and subsequent evolutionary divergence (Dennis *et al.*, 1985). A direct comparison at the sequence level should reveal those parts of the gene responsible for similarities in expression and regulation, since they are more likely to be conserved. As already indicated, the homologies within the coding regions of *Adh1* and *Adh2* are very high but elsewhere there is very little overall homology between the genes. The corresponding introns bear little resemblance to one another except, as expected, around the intron-exon junctions where specific splicing signals are to be found. It is in the non-conserved 5′ and 3′ region that we should first look for sequence features which might be responsible for transcription and translation rates as well as for 3′ polyadenylation signals. S1 mapping and cDNA cloning have indicated that there are multiple polyadenylation sites in *Adh1*: four for *Adh1-1S* and seven for the *Adh1-1F* allele (Sachs *et al.,* 1985). In contrast, *Adh2* has a single polyadenylation site. Analysis of the sequence near the polyadenylation sites in *Adh1-1F* and *Adh1-1S* alleles reveals no conserved sequence that could serve as a polyadenylation signal. The sequence AATAAT, on the other hand, is located 20 bp upstream of the single polyadenylation site of *Adh2* mRNA. This is similar to the AATAAA signal found 15—30 bp upstream of the polyadenylation site of animal genes (Proudfoot and Brownlee, 1976).

In the regions 5′ to the start of translation, several oligonucleotide homologies between *Adh1* and *Adh2* have been identified by computer analysis of the DNA sequences. The translation start point itself is within a short region of homology (Fig. 5). This sequence, GGAATGGC, corresponds to the conserved sequence for translation start sites in animal genes (Kozak, 1981). Nearby, within the 5′ untranslated region, there is a region of homology located about 45 bp upstream of the translation start of each gene. The role of this sequence is unknown as yet but it may serve as a recognition signal to the translation machinery of the cell, since anaerobic mRNAs are translated selectively and preferentially during anaerobiosis despite the persistence of non-stress mRNA. Alternatively, it may be important for *Adh* mRNA stability, since mutations in this region severely affect mRNA levels and transcription rates (see below).

There is a cluster of three short stretches of homology associated with the transcription start for the two *Adh* genes of maize which are probably important for the precise location of the beginning of the mRNA (Fig. 5). Transcription is initiated at an A within a pyrimidine rich tract, which resembles the beginning of animal genes (Corden *et al.,* 1980). The longest stretch of homology outside of the coding region of the *Adh* genes occurs about 35 bp upstream of the transcription start (Dennis *et al.,* 1985). This sequence, . . . ACCACTATATAAATCAG . . ., contains a classical TATA

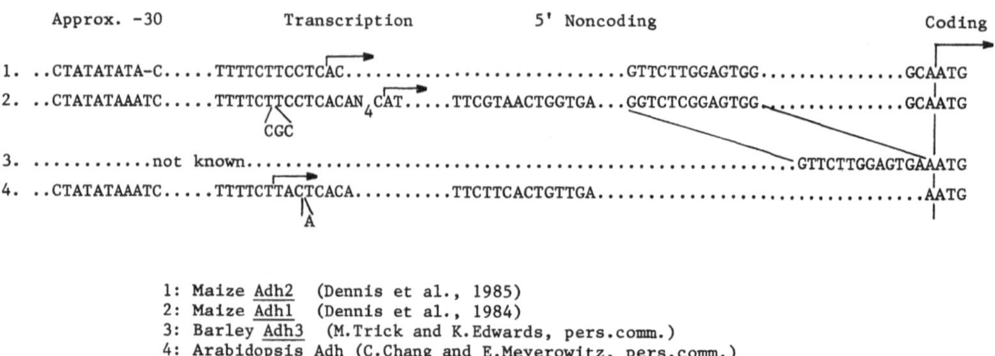

Fig. 5. Some of the sequence homologies detectable in comparisons of the 5′ leader and upstream sequences of Adh genes from maize, barley and *Arabidopsis*

box characteristic of animal viral and cellular genes (Breathnach and Chambon, 1981). In animal systems at least, the TATA box is primarily involved in the fixation of the site of transcription initiation to within the narrow pyrimidine-rich region already described (Corden *et al.,* 1980; Benoist and Chambon, 1981; Grosveld *et al.,* 1982). A TATA-like sequence has been found in a similar position in all plant genes so far examined, and its strong conservation, together with some flanking sequences, between *Adh1* and *Adh2* confirms its importance in plant cells for transcription by RNA polymerase II.

In some animal genes other sequences upstream from the TATA box have been also implicated in efficient transcription *in vivo.* In the globin genes both the TATA box and the so-called CCAAT box at —80 are essential for high levels of transcription. However, the animal gene consensus CCAAT sequence can be rarely fitted with any confidence to sequences in similar places in plant genes (Messing *et al.,* 1983), and although we had previously assigned the sequences GGCCAAACC at —90 in the *Adh1* gene as a putative CCAAT sequence (Dennis *et al.,* 1984) there is no homologous sequence in the *Adh2* gene (Dennis *et al.,* 1985). Unless such a sequence is involved in tissue-specific expression, which differs between *Adh1* and *Adh2,* it should have been conserved like the TATA box region. In genes where the CCAAT box is functionally significant, e. g. the beta and alpha-globin genes, both it and the TATA box have been conserved during evolution (Efstratiadis *et al.,* 1980).

Regions further upstream in *Adh1* and *Adh2* have been scanned in an attempt to locate any homologous sequence(s) that may be responsible for the anaerobic induction of these genes. Genes such as the metallothionens (Mayo *et al.,* 1982) and the heat-shock genes (Pelham, 1982), which are coordinately induced in response to a given stimulus, have been found to include a common regulatory sequence 5′ to the coding sequence. When placed upstream of any other gene, these sequences confer on that gene the inducible response of the gene from which it was derived. Comparisons of quite extensive regions of the 5′ ends of *Adh1* and *Adh2* revealed very little sequence homology. A short eight base sequence, CACCTCCC, at —170 in *Adh2* and —200 in *Adh1* is a candidate for an anaerobic control signal (Dennis *et al.,* 1984b). If the homology criteria are relaxed, two more extensive regions of imperfect homology, each extending about thirty bases, can be seen at about —80 and —160 in both genes.

Although these sequence comparisons reveal regions of homology between the maize *Adh1* and *Adh2* genes, their importance in the regulation of *Adh* gene expression can only be determined by functional assays. Some indication of importance may come from comparisons with *Adh* genes of other plant species as they are characterized, but it is important to note that different nucleotide signals may have evolved to perform the same function in other species.

Some sequence data of the 5′ flanking region has been obtained for the barley *Adh2*-like gene (M. Trick, unpublished) and an *Arabidopsis Adh* gene (C. Chang, pers. comm.). For the barley gene, the sequence information includes the translation start and some 80 bp of untranslated region. There is a 12 bp perfect match to the conserved sequence seen in the 5′ leader of *Adh1* and *Adh2* of maize although the location is somewhat different (Fig. 5). S1 mapping of this barley gene has defined the start of transcription at 70 bp upstream from the start of translation at the sequence TCATCAGCAA which shows considerable homology to the *Adh2* transcription start (TCCTCACCAA), although the untranslated region is much shorter (126 bp for *Adh2*).

For the *Arabidopsis* gene, the homology again involves the conserved sequence found in the 5′ non-translated sequence of the maize genes and, like barley, this sequence is found very close to the translation start point (Fig. 5). A sequence ACATCACAA, which has some homology with the maize and barley transcription start regions, is found approximately 50 bp upstream of the translation start. However, it is not yet known where transcription begins in the *Arabidopsis Adh* gene.

B. Mutations in Adh Which Affect Expression

Preliminary results show that the control of induction of the *Adh1* gene is at the level of transcription. This has been done by isolating nuclei and measuring "runoff" transcription (L. Beach, pers. comm.). Under anaerobic conditions there is an increased amount of incorporation of radioactive precursors into *Adh*-RNA. This is reflected in a 50-fold increased steady-state level of *Adh1* mRNA. *Adh2* mRNA steady-state levels also increase 50-fold upon anaerobiosis but we have not done the experiments to discover whether this is a transcriptional response.

Some of the mutants of *Adh1* have both a decreased mRNA level and a decreased transcription rate. For example, in the *Ds* mutant allele *Adh1-Fm335* there is a very low steady-state level of *Adh1* specific mRNA. The *Ds1* element has inserted into the 5′ untranslated region of the gene, and nuclease S1 mapping shows that transcription still starts at the same point as *Adh1*. These results show that insertion of 402 bp about 45 bp upstream from the translation start in the 5′ untranslated region does not affect the point of transcription initiation, so the 402 bp is not providing a new promoter. Although there is a marked effect on the amount of mRNA, preliminary results from "runoff" transcription studies (L. Beach, unpublished) show that the transcription rate is maintained. Therefore, it is not the amount of RNA made but its stability that is affected. S1 nuclease data and sequence analysis suggest that the *Ds* element acts in the *Adh1* gene as an extra intron which may not be excised at the normal rate, thus possibly affecting the stability of the RNA. In revertants of the *Adh1-Fm335* allele which have full activity we found that the extra 6—8 bp present in the same position in the 5′ untranslated region (the remnant of the duplication formed upon insertion) and present in the mRNA have no effect on either the level of RNA or of ADH enzyme.

A null derivative of the *Adh1-Fm335* allele has been isolated with a deletion of 70 bp from the insertion of the *Ds* element at position 45 into the first exon of the gene. As expected, this mutant produces no enzyme because it is missing a portion of the coding region including the start of translation, but, more important, there is also very little mRNA. Again this could be due to instability of the RNA or a decreased rate of transcription. The notable feature of these two mutations, the *Adh1-Fm335* Ds insertion and the null derivative with a deletion, is that the 5′ non-coding region affected is close to a sequence conserved between *Adh1* and *Adh2.*

In another case, the insertion of a Robertson's mutator element into the first intron of *Adh1* (Strommer *et al.,* 1982) lowers the level of mRNA but the cause of this decrease is not known.

Overall, we see that some insertions into the 5′ untranslated region or the introns of the maize *Adh1* gene do have an effect on expression, altering the rate of transcription or the stability of the message, while certain smaller insertions have no effect. Many more mutants and variants of the *Adh1* locus are still to be examined and molecular analysis of these should assist in our understanding of gene regulation.

For *Adh2,* the only mutant which has been analysed is the *Adh2-33* allele found as a naturally occurring null mutation in the line Knobless Wilbur Flint (Dlouhy, 1980). This gene has a 200 bp insertion in the 3′ region downstream of the polyadenylation site (J. Ellis, unpublished) and this may be the reason for the lack of detectable message for this mutation (Dennis *et al.,* 1985). However, recent results have shown that this 3′ insertion does not affect polyadenylation activity in tobacco when tested in T-DNA transformation experiments (S. Ben-Tahar, pers. comm.), so there may be some other reason for the null phenotype of the *Adh2-33* allele.

A transformation system is needed for studying expression of natural and *in vitro* mutated gene segments of *Adh1* genes and for defining those elements important for regulation and expression.

VIII. *In Vivo* Expression of *Adh*

Sequence comparisons and mutant studies have given some insights into the possible nature and location of regulatory signals within and adjacent to the *Adh1* structural gene, but regulatory functions for these sequences must, at best, be considered tentative. Confirmation of an *in vivo* function for any specific sequence will only come from *in vivo* expression studies on genetically-engineered promoters or genes. Such analyses using rearrangements of DNA segments adjacent to genes in animal cells have identified signal sequences responsible for regulating expression, e. g. during heavy metal induction (Mayo *et al.,* 1982) and heat shock induction (Pelham, 1982) of specific genes or sets of genes. However, in plants it has been difficult to put *in vitro* modified genes back into cells to assess the effects of the alterations on gene expression. The difficulties are now being resolved and allow the study of *Adh* gene regulation on three broad fronts: first, and possibly most important, in a completely homologous system involving maize protoplasts, pollen or endosperm as recipients for modified *Adh* genes; second, in an heterologous system with, for example, tobacco, and finally, the introduction of *Adh* genes into animal cells.

A. Attempts at Expression in Cereal Tissue Culture Cells

It is now possible to prepare protoplasts from callus or suspension cultures of a limited number of cell lines of maize, wheat and rice which retain the potential for cell wall regeneration and continued cell division to reform a callus but will not, unfortunately, regenerate into plants. Transformation of cereal protoplasts may lead to either short-term (transient) expression or stable integration and expression of introduced genes. Both require the uptake of vector DNAs into protoplasts but only the latter demands the survival and growth of protoplasts for extended periods of time under a specific selection regime. The transient expression assays will be useful in the optimization of vector entry into protoplasts (since many variations can be tested in a short time) and in the study of genes expressed in protoplasts. Stable transformation will be essential for any studies of gene regulation and genetic engineering of plants.

A transient expression system is now being used to examine *Adh1* expression. A chimeric gene has been constructed with a maize *Adh1* promoter fragment, extending from −1 (relative to the start of translation) to approximately 1.5 kb upstream, fused with a bacterial chloramphenicol resistance (CAT) gene. We are using protoplasts of the maize line of Chourey and Zurawski (1981) in which we have already shown transient expression of the CAT gene with the Ti plasmid nopaline synthase promoter. As yet we have no data on the expression of this *Adh* construct.

Stable transformation has not as yet been achieved with maize, but has been reported recently for *Triticum monococcum* (Lorz *et al.,* 1985). Wheat protoplasts were treated with polyethylene glycol and a vector containing a chimeric gene (pNOS-Kan) known to express a bacterial aminoglycoside phosphotransferase (NPT-II) activity when introduced into dicot plants (Herrera-Estrella *et al.,* 1984). In tobacco this gene confers resistance to as much as 500 μg/ml kanamycin, a dose twenty times greater than the minimum lethal concentration. Experiments are under way to establish an effective selection system in maize using kanamycin and G418 (K. Danna, unpublished), attempting to select for resistant cells after treatment of protoplasts with DEAE-Dextran and vector DNAs containing various combinations of pNOS-Kan, pAdh-Kan, with *Ac* and *Ds* controlling element sequences.

The most interesting properties of *Adh1* and *Adh2* (and most other genes) are their spatial and temporal regulation in the whole plant. There is thus considerable interest in the development of a transformation system in which tissue-specific and developmental aspects of *Adh* expression can be studied at the molecular level. There has been some suggestion of the transfer of *Tripsacum* genetic characters into maize after the fertilization of maize ovules with pollen treated with isolated *Tripsacum* genomic DNA (D. Jewell, pers. comm.) but no molecular proof of actual gene transfer is available. The same approach can be used with sequences such as the cloned *Adh* genes as donor DNAs and defined mutants at the *Adh* loci as recipients. Pollen from an *Adh1⁻Adh2⁻* plant treated with vectors containing either the *Adh1-F* of *Adh1-S* genomic clones was used to pollinate another *Adh1⁻Adh2⁻* plant or cultured ovules (J. Waldron, unpublished). Three out of more than 1000 seeds set by this procedure and screened for ADH activity on starch gels showed a low level of the appropriate *Adh* allele. Unfortunately all three lacked vigour and died shortly after germination making confirmation of transformation impossible.

B. Testing Maize Adh Gene Activity in Dicotyledonous Plant Cells

The transformation of dicot plants provides a different perspective by studying maize *Adh* gene regulation in a phylogenetically unrelated plant species. Gene transfer based on the Ti-plasmid of *Agrobacterium tumefaciens* is the most established technique for dicots and we have used this to introduce maize *Adh* constructs into tobacco species.

Initially, the entire *Adh1-S* genomic clone, including considerable 5′ and 3′ flanking sequences, was inserted into two different types of Ti-plasmid and the resultant bacterial strains used to produce tumours on *Nicotiana tabacum* and *N. plumbaginifolia*. Sterile tumour tissues contained the intact maize gene integrated into tobacco DNA but the gene was not expressed at the mRNA level (Llewellyn *et al.*, 1985). Lack of expression may arise from a defect at any step in the transcription or processing of RNA, so other chimeric constructs, including the pAdh-CAT gene (described above) and new constructs with slightly shorter maize promoter fragments coupled to the octopine synthase gene (pAdh-OCS), have been used to try to define the problem. In both cases 3′ signals were those of the nopaline synthase gene. No CAT or OCS activity was detected in tumours or regenerated plants containing these chimeric genes (J. Ellis, unpublished). The 3′ end of *Adh1-S* tested in tobacco by substitution for the NOS 3′ end in pNOS-CAT functioned as well or better than the NOS 3′ end (S. Ben-Tahar, pers. comm.). Polyadenylation occurs at the same polyadenylation site as in the Adh2 gene in maize.

Although we cannot discount the possibility that a low level of RNA is transcribed using the maize promoter but is rapidly turned over, the data strongly suggest that the *Adh1* promoter is not functional in tobacco. An identical problem has been encountered with other maize and wheat genes introduced into tobacco and *Petunia* and this may be a more general phenomenon.

C. Expression of Maize Adh in Animal Cells

The *Adh1* gene has been intoduced into monkey cells using SV40-based vectors but was not transcribed at a detectable level (E. S. Dennis, unpublished). A chimeric gene containing an *Adh1* promoter attached to a kanamycin resistance gene was expressed, and at increased levels in the presence of a viral enhancer sequence. Transcription was shown to start in the same place as in maize but was not anaerobically inducible. The 3′ end of *Adh1* tested in monkey cells in a chimeric SV40 early gene promoter-maize *Adh1-F* cDNA gene was functional but used a polyadenylation site upstream from the true site. This new site was just beyond a sequence that mimicked the animal AATAAA polyadenylation signal, but which is not used in maize. Clearly, plant genes do not function in the same way when taken out of their normal environment.

IX. Conclusions

It is obvious that the wealth of genetic information on the alcohol dehydrogenases of maize led to the cloning of their genes. From there we have been able to structurally characterize the genes and use this information to address other genetic phenomena such as the nature of allelic variation, transposable elements and some indication of the basis of mutations in plants. Although there is still much to be done to determine the structural change in many of the defined *Adh* mutants and variants, the next major area which must be developed is an understanding of the regulation of these genes. This may be approached in two ways. First, by analysis of the structure of natural and induced variants. Obviously this will provide clues only by relating structural alterations to phenotypic changes. Second, and more powerful, the suspected regulatory regions must be tested for activity *in vivo* by preparing constructs for transformation. The constructs are being prepared, the transformation system is required.

Acknowledgements

We wish to thank the many colleagues cited in the chapter who provided us with information on their unpublished data. During preparation of this manuscript W. L. G. was the recipient of a Harkness Fellowship at the University of California, Davis.

X. References

Bennetzen, J. L., Swanson, J., Taylor, W. C., Freeling, M., 1984: DNA insertion in the first intron of maize *Adh1* affects message levels: cloning of progenitor and mutant *Adh1* alleles. Proc. Nat. Acad. Sci., U.S.A. **13,** 4125—4129.

Benoist, C., Chambon, P., 1981: In vivo sequence requirements of the SV40 early promoter region. Nature **290,** 304—310.

Birchler, J. A., Schwartz, D., 1979: Mutational study of the alcohol dehydrogenase-1 FC^m duplication in maize. Biochem. Genet. **17,** 1173—1180.

Branden, C.-I. Eklund, H., Camillau, C., Pryor, A. J., 1984: Correlation of exons with structural domains in alcohol dehydrogenase. EMBO J. **3,** 1307—1310.

Breathnach, R., Chambon, P., 1981: Organisation and expression of eukaryotic split genes coding for proteins. Ann. Rev. Biochem. **50,** 349—383.

Brettell, R. I. S., Dennis, E. S., Scowcroft, W. R., Peacock, W. J., 1986: Molecular analysis of a somaclonal mutant of maize alcohol dehydrogenase. Mol. Gen. Genet. **202,** 235—239.

Chourey, P. S., Zurawski, D. B., 1981: Callus formation from protoplasts of a maize cell culture. Theor. Appl. Genet. **59,** 341—344.

Cooper, P., Ho, T. H. D., 1981: Heat shock proteins in maize. Plant Physiol. **71,** 215—222.

Corden, J., Wasylyk, B., Buchwalder, A., Sassone-Corsi, P., Kedinger, C., Chambon, P., 1980: Promoter sequences of eukaryotic protein coding genes. Science **209,** 1406—1414.

Dennis, E. S., Gerlach, W. L., Pryor, A. J., Bennetzen, J. L., Inglis, A., Llewellyn, D., Sachs, M. M., Ferl, R. J., Peacock, W. J., 1984: Molecular analysis of the alcohol dehydrogenase (*Adh1*) gene of maize. Nuc. Acid Res. **12,** 3983—4000.

Dennis, E. S., Sachs, M. M., Gerlach, W. L., Finnegan, E. J., Peacock, W. J., 1985: Molecular analysis of the alcohol dehydrogenase-2 (*Adh2*) gene of maize. Nuc. Acid Res. **13,** 727—743.

Dlouhy, S. R., 1980: Genetic, biochemical and physiological analysis involving the *Adh2* locus of *Zea mays*. Ph. D. Dissertation. Indiana University.

Doring, H.-P., Starlinger, P., 1984: McClintock's controlling elements now at the DNA level. Cell **39,** 253—259.

Doring, H. P., Freeling, M., Hake, S., Johns, M. A., Kunze, R., Merckelbach, A., Salamini, F., Starlinger, P., 1984: A *Ds*- mutation of the *Adh1* gene in *Zea mays*. Mol. Gen. Genet. **193,** 199—204.

Efstradiatis, A., Posakony, J. W., Maniatis, T., Lawn, R. M., O'Connell, C., Spritz, R. A., De Riel, J. K., Forget, B. G., Weissman, S. M., Slightom, J. L., Blechl, A. E., Smithies, O., Barelle, F. E., Shoulders, C. C., Proudfoot, N. J., 1980: The structure and evolution of the human B-globin gene family. Cell **21,** 653—668.

Ferl, R. J., Dlouhy, S. R., Schwartz, D., 1979: Analysis of maize alcohol dehydrogenase by native SDS two-dimensional electrophoresis and autoradiography. Mol. Gen. Genet. **169,** 7—12.

Ferl, R. J., Brennan, M. D., Schwartz, D., 1980: *In vitro* translation of maize alcohol dehydrogenase (EC 1.1.1.1.): Evidence for the anaerobic induction of messenger RNA. Biochem. Genet. **18,** 681—692.

Freeling, M., 1973: Simultaneous induction by anaerobiosis or 2,4-D of multiple enzymes specified by two unlinked genes: Differential *ADH1-ADH2* expression in maize. Mol. Gen. Genet. **127,** 215—227.

Freeling, M., 1974: Dimerization of multiple maize ADH's studied *in vivo* and *in vitro*. Biochem. Genet. **12,** 407—417.

Freeling, M., 1976: Intragenic recombination in maize: Pollen analysis methods and the effect of parental *Adh1+* isoalleles. Genetics **83,** 701—717.

Freeling, M., 1978: Allelic variation at the level of intragenic recombination. Genetics **89,** 211—224.

Freeling, M., Bennett, D. C., 1986: Maize *Adh1*. Annual Review of Genetics. In press.

Freeling, M., Birchler, J. A., 1981: Mutants and variants of the *alcohol dehydrogenase-1* gene in maize In: Setlow J. K., Hollaender A., (eds.), Genetic Engineering: Principles and Methods, pp. 223—264. New York: Plenum.

Freeling, M., Cheng, D. S.-K. 1978: Radiation-induced alcohol dehydrogenase mutants in maize following allyl alcohol selection of pollen. Genet. Res. **31,** 107—130.

Freeling, M., Schwartz, D., 1973: Genetic relationships between the multiple alcohol dehydrogenases of maize. Biochem. Genet. **8,** 27—36.

Gerlach, W. L., Pryor, A. J., Dennis, E. S., Ferl, R. J., Sachs, M. M., Peacock, W. J., 1982: cDNA cloning and induction of the alcohol dehydrogenase gene (Adh1) of maize. Proc. Nat. Acad. Sci., U.S.A. **79,** 2981—2985.

Grosveld, G. C., de Boer, E., Shewnamer, C. K., Flavell, R. A., 1982: DNA sequences necessary for transcription of the rabbit β-globin gene in vivo. Nature **295,** 120—126.

Hageman, R. H., Flesher, D., 1960: The effect of anaerobic environment on the activity of alcohol dehydrogenase and other enzymes of corn seedlings. Arch. Biochem. Biophys. **87,** 203—209.

Hake, S., Taylor, W. C., Freeling, M., 1984: Molecular analyses of genetically stable mutants of the *Adh1* gene of maize. Mol. Gen. Genet. **194,** 42—48.

Herrera-Estrella, L., Van den Broeck, G., Maenhaut, R., Van Montagu, M., Schell, J., 1984: Light-inducible and chloroplast-associated expression of a chimaeric gene introduced into *Nicotiana tabacum* using a Ti plasmid vector. Nature **310,** 115—120.

Johns, M. A., Strommer, J. N., Freeling, M., 1983: Exceptionally high levels of restriction site polymorphism in DNA near the maize *Adh1* gene. Genetics **105,** 733—743.

Kelley, P. M., Freeling, M., 1982: A preliminary comparison of maize anaerobic and heat-shock proteins. In: Schlesinger, M. J., Ashburner, M., Tissieres, A. (eds), Heat Shock From Bacteria to Man. pp. 315—329. Cold Spring Harbor, New York: Cold Spring Harbor Lab.

Kelley, P. M., Freeling, M., 1984a: Anaerobic expression of maize glucose phosphate Isomerase I. J. Biol. Chem. **259,** 673—677.

Kelley, P. M., Freeling, M., 1984b: Anaerobic expression of maize fructose-1,6-diphosphate aldolase. J. Biol. Chem. **259,** 14180—14183.

Kelly, J., Freeling, M., 1980: Purification of maize alcohol dehydrogenase-1 allozymes and comparison of their tryptic peptides. Biochim. Biophys. Acta **624,** 102—110.

Kozak, M., 1981: Possible role of flanking nucleotides in recognition of the AUG initiator codon by eukaryotic ribosomes. Nucleic Acids Res. **9,** 5233—5252.

Larkin, P. J., Scowcroft, W. R., 1981: Somaclonal variation — a novel source of variability from cell cultures for plant improvement. Theor. Appl. Genet. **60,** 197—214.

Laszlo, A., 1981: Maize pyruvate decarboxylase: An inducible enzyme. Ph. D. dissertation. University of California, Berkeley.

Lin, C. Y., Key, J. L., 1967: Dissociation and reassembly of polyribosomes in relation to protein synthesis in the soybean root. J. Mol. Biol. **26,** 237—247.

Llewellyn, D., Dennis, E. S., Peacock; W. J., 1985: The alcohol dehydrogenase genes of maize: expression studies in tobacco cells. In: Van Vloten-Doting, L., Groot G. S. P., and Hall, T. C. (eds.), Molecular form and function of the plant genome. pp. 593—607. New York: Plenum.

Lorz, H., Baker, B., Schell, J., 1985: Gene transfer to cereal cells mediated by protoplast transformation. Mol. Gen. Genet. **199,** 178—182.

Mayo, K. E., Warren, R., Palmiter, R. D., 1982: The mouse metallothionen-1 gene is transcriptionally regulated by cadmium following transfection into human or mouse cells. Cell **29,** 99—108.

Messing, J., Geraughty, D., Heidecker, G., Hu, N. T., Kridl, J., Rubenstein, I., 1983: Plant gene structure. In: Kosuge, T., Meredith, C. P., Hollaender, A. (eds.), Gentic Engineering of Plants, pp. 211—277. New York: Plenum.

Okimoto, R., Sachs, M. M., Porter, E. K., Freeling, M., 1980: Patterns of polypeptide synthesis in various maize organs under anaerobiosis. Planta **150,** 89—94.

Osterman, J. C., Schwartz, D., 1981: Analysis of a controlling element mutation at the *Adh* locus of maize. Genetics **99,** 267—273.

Peacock, W. J., Dennis, E. S., Gerlach, W. L., Llewellyn, D., Lorz, H., Pryor, A. J., Sachs, M. M., Schwartz, D., Sutton, W. D., 1983: Gene transfer in maize: Controlling elements and the alcohol dehydrogenase genes. In: Downey, K. (ed.), Proc. Miami Winter Symposium, pp. 311—325. New York: Academic Press.

Peacock, W. J., Dennis, E. S., Gerlach, W. L., Sachs, M. M., Schwartz, D., 1984: Insertion and excision of Ds controlling elements in maize. Cold Spr. Harb. Symp. Quant. Biol. **49,** 347—354.

Pelham, H. R. B., 1982: A regulatory upstream promoter element in the *Drosophila* Hsp70 heat shock gene. Cell **30,** 517—528.

Proudfoot, N. J., Brownlee, G. G., 1976: 3′ non-coding region sequences in eukaryotic messenger RNA. Nature **263,** 211—214.

Robertson, D. S., 1978: Characterization of a mutator system in maize. Mutat. Res. **51,** 21—28.

Sachs, M. M., Freeling, M., 1978: Selective synthesis of alcohol dehydrogenase during anaerobic treatment of maize. Mol. Gen. Genet. **161,** 111—115.

Sachs, M. M., Freeling, M., Okimoto, R., 1980: The anaerobic proteins of maize. Cell **20,** 761—768.

Sachs, M. M., Dennis, E. S., Gerlach, W. L., Peacock, W. J., 1986: Two alleles of maize *Alcohol dehydrogenase* 1 exhibit 3′ structural and Poly(A) addition Polymorphisms. Genetics, in press.

Saedler, H., Nevers, P., 1985: Transposition in plants: A molecular model. EMBO J. **4,** 585—590.

Scandalios, J. G., 1966: Genetic variation of *Adh* in maize. Genetics **54,** 359.

Schwartz, D., 1966: The genetic control of alcohol dehydrogenase in maize; gene duplication and repression. Proc. Nat. Acad. Sci., U.S.A. **56,** 1431—1436.

Schwartz, D., 1969a: An example of gene fixation resulting from selective advantage in suboptimal conditions. Amer. Nat. **103,** 479—481.

Schwartz, D., 1969b: Alcohol dehydrogenase in maize: Genetic basis for multiple isozymes. Science **164,** 585—586.

Schwartz, D., 1971: Genetic control of alcohol dehydrogenase — a competition model for regulation of gene action, Genetics **67,** 411—425.

Schwartz, D., 1973: The *Adh1-FCm* "operon". Maize Genet. Coop. News Lett. **47,** 53—54.

Schwartz, D., 1978: Allele-specific modification of the expression of the *Adh2* gene in maize by ethylene. Genetics **90,** 323—330.

Schwartz, D., Endo, T., 1966: Alcohol dehydrogenase polymorphisms in maize-simple and compound loci. Genetics **53,** 709—715.

Schwartz, D., Laughner, W. J., 1969: A molecular basis for heterosis. Science **166,** 626—627.

Schwartz, D., Osterman, J., 1976: A pollen selection system for alcohol dehydrogenase-negative mutants in plants. Genetics **83,** 63—65.

Strommer, J., Hake, S., Bennetzen, J., Taylor, W. C., Freeling M., 1982: Regulatory mutants of the maize Adh1 gene caused by DNA insertons..Nature **300,** 542—544.

Stuber, C. W., Goodman, M. M., 1983: Allozyme genotypes of popular and historically important inbred lines of corn, *Zea mays* L. ARS-ARR Southern Series **16,** 1—29.

Sutton, W. D., Gerlach, W. L., Schwartz, D., Peacock, W. J., 1984: Molecular analysis of *Ds* controlling element mutations at the *Adh1* locus of maize. Science **223,** 1265—1268.

Wieringa, B., Meyer, F., Reiser, J., Weissmann, C., 1983: Unusual splice sites revealed by mutagenic inactivation of an authentic splice site of the rabbit β-globin gene. Nature **301,** 38—43.

Wignarajah, K., Greenway, H., 1976: Effect of anaerobiosis on activities of alcohol dehydrogenase and pyruvate decarboxylase in roots in *Zea mays.* New Phytol **77,** 575.

Woodman, J. C., Freeling, M., 1981: Identification of a genetic element that controls the organ-specific expression of *Adh1* in maize. Genetics **98,** 354—378.

Chapter 5

The Molecular Genetics of Higher Plant Nitrate Assimilation

John L. Wray

Department of Biochemistry and Microbiology, University of St. Andrews, Irvine Building, North Street, St. Andrews, Fife, KY16 9AL, Scotland

With 2 Figures

Contents

I. Introduction

The nitrate assimilation pathway is the major point of entry of inorganic nitrogen into organic combination in most crop plants growing in well-aerated soils. This pathway converts nitrate to ammonium via a nitrate uptake mechanism and two enzymes, nitrate reductase (NR) and nitrite reductase (NiR), thereby making nitrogen available for a variety of biosyntheses, of which the most important quantitatively is protein synthesis.

Nitrate and hormone availability, light and end-products have all been shown to play a role in the regulation of this pathway and they presumably act together to match the *in vivo* rate of nitrate reduction to the physiological status of the plant (reviewed in Beevers and Hageman, 1980; Srivastava, 1980; Guerrero *et al.,* 1981; Huffaker, 1982; Wallace and Oaks, 1984). However, these physiological and biochemical studies have told us very little about the molecular mechanisms underlying the development and regulation of this important pathway. Such information is critical if we are to understand how the new techniques of molecular biology might be applied to improve the efficiency of the nitrate assimilation pathway.

Over the last eight years or so there has been considerable progress in the understanding of the genetics of nitrate assimilation, making it the best characterised metabolic pathway in higher plants. These genetic studies are an important first step in developing a molecular understanding of the pathway and are reviewed here in some detail against a background of the current biochemical information. I also consider some applications of these studies. In some places, where appropriate, I have drawn on the vast body of information available on the genetics of *Aspergillus* and *Neurospora* nitrate assimilation. The genetics of higher plant nitrate assimilation has recently been reviewed also by Dunn-Coleman *et al.* (1984) and Kleinhofs *et al.* (1985).

II. The Nitrate Assimilation Pathway

A. Nitrate Uptake

Little is known about the molecular basis of nitrate uptake. It is clear that uptake is an active process dependent on metabolic energy, since cyanide and anaerobic conditions, as well as uncouplers of oxidative phoshorylation, are inhibitory (Rao and Rains, 1976). Nitrate stimulation of nitrate uptake, after a lag in plants not previously exposed to nitrate (Minotti *et al.,* 1968; Jackson *et al.,* 1973; Rao and Rains, 1976), and the inhibitory effects of protein and RNA synthesis inhibitors, suggest that nitrate induces a specific nitrate uptake system. Nitrate uptake rates follow a dual-phase relationship with the concentration of nitrate available to the plant. Both phases appear to be hyperbolic in accordance with Michaelis-Menten kinetics and suggest the existence of at least two uptake mechanisms operating at high and low nitrate concentrations respectively (Rao and Rains, 1976; Doddema and Telkamp, 1979). Other studies emphasize the role of nitrate efflux in determining net uptake rate (Jackson *et al.,* 1976; Clement *et al.,* 1978; Deane-Drummond and Glass, 1983; Deane-Drummond, 1982).

Net nitrate uptake is dependent on an influx term determined primarily by external nitrate concentration and independent of internal nitrate concentration, and a variable efflux term directly proportional to internal nitrate concentration (Deane-Drummond and Glass, 1983). Glass *et al.*

(1985) have argued that the reduction in net nitrate uptake associated with nitrate loading of tissue is due to nitrate inhibition of tonoplast nitrate flux into the vacuole. This leads to increased cytoplasmic nitrate concentration and/or increased transport of nitrate to the shoot, together with increased efflux of nitrate. There is, however, evidence to suggest that nitrate efflux is not a futile "leak" of nitrate which is surplus to the requirements of the cell but that a specific carrier is involved (Deane-Drummond, 1985). These results have recently been interpreted in terms of a substrate cycle (Deane-Drummond, 1986). How they relate to the earlier data showing the existence of dual phase net-uptake systems is unclear at present.

B. Nitrate Reductase

Higher plant nitrate reductases are flavohaemomolybdoproteins which catalyse the two electron reduction of nitrate to nitrite (Hewitt and Notton, 1980). Almost all higher plants so far examined possess NADH-NR (EC 1.6.6.1) which has a pH optimum around 7.4 and a Michaelis constant for nitrate and NADH of ca. 200 μM and 2 μM respectively (Beevers and Hageman, 1980; Guerrero *et al.*, 1981). Some plants also possess a bispecific NAD(P)H-NR (EC 1.6.6.2) which uses either NADH or NADPH as electron donor.

NAD(P)H-NR has been found in rice seedlings (Shen *et al.*, 1976), maize scutella (Campbell, 1978), maize roots (Redinbaugh and Campbell, 1981), soybean cotyledons (Orihuel-Iranzo and Campbell, 1980) and soybean leaves (Evans and Nason, 1953; Jolly *et al.*, 1976; Campbell, 1976; Nelson *et al.*, 1984). Biochemical and physiological evidence suggests that the NAD(P)H-NRs are distinct species from the NADH-NRs with a higher Km for nitrate (4 mM) and a lower pH optimum (6.5) (Campbell, 1976). The two forms can be separated by chromatography on blue dextran Sepharose (Redinbaugh and Campbell, 1981) and show different developmental (Orihuel-Iranzo and Campbell, 1980) and induction (Shen *et al.*, 1976) patterns. The tropical legume, *Erythrina senegalensis*, is unique in prossessing only NAD(P)H-NR (Stewart and Orebamjo, 1979). The NAD(P)H-NRs may be smaller than the NADH-linked enzymes (Redinbaugh and Campbell, 1981; Streit *et al.*, 1985). Most work has been carried out on the NADH-nitrate reductases and this is discussed below.

i) Molecular Weight and Prosthetic Groups

Attempts to determine the molecular weight of the holoenzyme subunits have been complicated by their extreme sensitivity to proteolytic modification (Brown *et al.*, 1981; Wray and Kirk, 1981; Campbell and Wray, 1983; Wallace and Oaks, 1984). Their stability appears to differ between species and even between genotypes of the same species. Purification in the presence of proteinase inhibitors either reduces or eliminates molecular weight species of 20,000—75,000 seen after SDS polyacrylamide gel electrophoresis of several apparently homogeneous enzyme preparations (Notton and Hewitt, 1979; Mendel and Müller, 1980; Campbell and Wray,

1983; Redinbaugh and Campbell, 1983) and indicates subunit values of 100,000 in barley cv. Golden Promise (Campbell and Wray, 1983), 100,000 (Kuo *et al.*, 1980) or 110,000 (Kuo *et al.*, 1982) in barley cv. Steptoe, 105,000 and 114,000 (Nakagawa *et al.*, 1985) or 110,000 and 120,000 (Fido and Notton, 1984) in spinach and 115,000 in squash (Redinbaugh and Campbell, 1985). The variability seen is probably due at least in part to proteolytic modification but the data indicate a subunit molecular weight of around 110,000—120,000.

Estimated molecular weights of the holoenzyme vary from ca. 200,000 for spinach (Notton and Hewitt, 1979), tobacco (Mendel and Müller, 1980), barley cv. Golden Promise (Small and Wray, 1980) and wheat (Jones and Ní Mhuimhneacháin, 1985) through 220,000 or 230,000 for barley cv. Steptoe (Kuo *et al.*, 1980) or squash (Redinbaugh and Campbell, 1985) to 270,000 for spinach (Nakagawa *et al.*, 1985). This suggests that the native enzyme is a dimer.

Although the higher plant enzyme is usually assumed to be a homodimer there is no definitive biochemical evidence for this. Pan and Nason (1978) reported that the *Neurospora* nitrate reductase had subunits of unequal size (115,000 and 130,000 molecular weight), similar to the results from spinach. However, both protein species had the same N-terminal amino acid (glutamate) and similar peptide maps suggesting that they were produced by proteolytic interconversion of one to the other and that the enzyme was a homodimer. When attempts were made by Horner (1983) to reduce proteolysis by shortening the purification procedure in the presence of proteinase inhibitors, the subunit molecular weight of the *N. crassa* enzyme was 145,000. Thus the 115,000 and 130,000 molecular weight species seen by Pan and Nason (1978) are probably proteolytic derivatives of a larger subunit. Genetic evidence from *Neurospora crassa* (Tomsett and Garrett, 1980) and from *Aspergillus nidulans* (Cove, 1979) is consistent with the enzyme being a homodimer.

Direct and indirect evidence from a few NRs suggests that the flavin is flavin adenine dinucleotide (FAD) (Hewitt and Notton, 1980; Redinbaugh and Campbell, 1985). The activity of several NRs is stimulated by exogenous FAD (Maretski *et al.*, 1967; Schrader *et al.*, 1968), suggesting that in some cases it is readily dissociable and, in these instances at least, not covalently bound. Purified NR from spinach (Notton *et al.*, 1977), tobacco (Mendel and Müller, 1980), barley (Somers *et al.*, 1982) and squash (Redinbaugh and Campbell, 1985) have spectra indicative of the presence of a b-type cytochrome (cytochrome b_{557}). The first definitive evidence for the presence of molybdenum was obtained with the spinach enzyme (Notton and Hewitt, 1971 a) although it had previously been shown that in barley a non-functional form of the enzyme was synthesised in the presence of the molybdenum analogue, tungsten (Wray and Filner, 1970).

Prosthetic group stoichiometry has been determined so far for only one higher plant NR (from squash) and suggests the presence of one FAD, one haem and one Mo per 115,000 molecular weight subunit (Redinbaugh and Campbell, 1985). It is most likely that other higher plant NRs will fit this

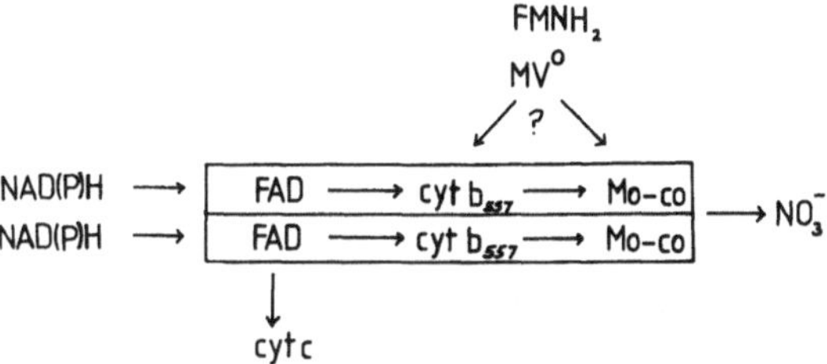

Fig. 1. Schematic representation of nitrate reductase showing probable sites of interaction of substrates and electron donors. FAD = Flavin adenine dinucleotide; $FMNH_2$ = Reduced flavin mononucleotide; MV = methyl viologen dye; Mo-co = molybdenum cofactor

pattern. Electron flow from NADH is generally accepted to be via flavin and haem to molybdenum, which acts as the terminal electron donor to nitrate (reviewed in Hewitt and Notton, 1980) (Fig. 1).

The redox centres of other electron transport proteins, such as flavocytochrome b_2 of bakers yeast (Naslin *et al.*, 1973; Jacq and Lederer, 1974) and the haemomolybdoprotein sulphite oxidase of rat liver (Johnson and Rajagopalan, 1977; Southerland and Rajagopalan, 1978) are organised in the proteins in domains (that is in independently folded, functionally intact regions of the polypeptide chain). The domains are considered to be held together by a flexible and loosely structured exposed region (hinge region) which is very sensitive to proteinase attack. We have suggested that the FAD and haem components of higher plant NR are contained in separate functional domains of the ca. 115,000 molecular weight subunit and that an additional domain may be involved in binding of molybdenum (Brown *et al.*, 1981). More recently Lé and Lederer (1983) have shown that the haem of the *N. crassa* NR is contained in a domain of 10,000—12,500 molecular weight which can be released from the native enzyme by chymotrypsin digestion.

ii) The Molybdenum Cofactor (Mo-co)

The molybdenum of probably all molybdoenzymes, except nitrogenase (Shah and Brill, 1977; Pienkos *et al.*, 1977), is carried on a dissociable, dialyzable and oxygen labile structure (Lee *et al.*, 1974; Amy and Rajagopalan, 1979), the molybdenum-cofactor (Johnson, 1980; Johnson *et al.*, 1980; 1984). This conclusion derives from the earlier observation that acid treatment of a variety of molybdoenzymes from diverse phylogenetic sources released a component which was able to reconstitute NADPH-NR activity *in vitro* from extracts of the nitrate reductase-minus mutant of *Neurospora crassa nit*-1 (Ketchum *et al.*, 1970; Nason *et al.*, 1971). The Mo-co has now been identified, on the basis of the characteristic fluorescence of its oxidation products, as a component of the NR of *Chlorella* (Solo-

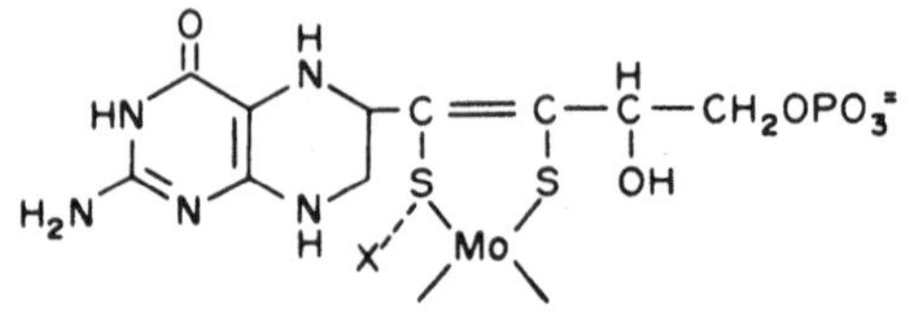

Fig. 2. Proposed structure of the molybdenum-cofactor (Johnson and Rajagopalan, 1982)

monson *et al.*, 1984) and squash (Redinbaugh and Campbell, 1985) as well as of *Neurospora crassa* (Kramer *et al.*, 1984).

a) Structure

Structural analysis of the functional Mo-co has been impossible due to its extreme lability when released from the protein environment. However, chemical, mass spectral and NMR studies of its human metabolic degradation product, urothione (Johnson and Rajagopalan, 1982), and of two stable fluorescent oxidation products derived from the molybdenum cofactor of chicken sulphite oxidase (Johnson *et al.*, 1980, 1984) have provided information on its probable structure (Fig. 2).

The Mo-co is considered to be a complex between molybdenum and a novel phosphorylated pterin, molybdopterin. The side chain contains at least four carbon atoms and most probably two sulphur atoms. The pterin presumably acts as a chelator of the metal, interfacing it to the protein and conferring on it biological activity (Wahl *et al.*, 1984). The nature of molybdenum ligation is unclear but the two side-chain sulphur atoms, and probably an oxygen atom of the phosphate group, perhaps supply at least part of the thiol and oxo ligands of molybdenum detected in several molybdoenzymes by EPR (Meriwether *et al.*, 1966; Bray, 1980) and EXAFS studies (Cramer *et al.*, 1981).

b) Role of Molybdenum Cofactor in Assembly of Nitrate Reductase

In addition to its role in catalysis, the Mo-co is responsible for dimerisation of the flavohaemoprotein subunits of NR. Reconstitution of NADPH-NR from *N. crassa nit*-1 extract by molybdoenzyme-derived component is in fact due to dimerisation of the monomer flavohaemoprotein subunits present in mutant extracts (Ketchum *et al.*, 1970). This reconstitution process has recently been examined in detail by Kramer *et al.* (1984) who showed that the molybdopterin moeity of the *nit*-1 Mo-co is defective. Other evidence that the molybdopterin moeity of the Mo-co is responsible for dimerisation ability of the Mo-co comes from the observations that demolybdo-molybdenum cofactor, produced either by molybdenum starvation (Hewitt *et al.*, 1977; Gewitz *et al.*, 1981) or mutation (Mendel *et al.*, 1981) is able to effect dimerisation, as is the tungsten analogue of the Mo-co (Wray and Filner, 1970). Dimerisation appears to involve interaction of the Mo-co with a protein thiol on the flavohaemoprotein subunit (which may also function as a molybdenum ligand) since *nit*-1 flavohaemoprotein subunits treated with the thiol-blocker, N-ethylmaleimide, are not dimerised by exogenous Mo-co (Wahl *et al.*, 1984).

Recent refinements have allowed this *in vitro* reconstitution phenomenon to function as a quantitative assay for higher plant Mo-co (Mendel, 1983; Mendel *et al.*, 1985).

iii) *The Catalytic Activities Associated with the Nitrate Reductase Holoenzyme*

In addition to the overall physiological reaction (NADH dependent nitrate reduction) the higher plant enzymes carry a dehydrogenase (diaphorase) function (which allows the *in vitro* transfer of electrons from NADH to a variety of electron acceptors such as nitroblue tetrazolium, dichlorophenol-indophenol and cytochrome c) as well as reduced flavin mononucleotide (FMNH) and reduced viologen dye nitrate reductase activity. These so-called partial activities are considered to be catalysed by specific regions of the NR molecule (Fig. 1).

The dehydrogenase function, usually measured as cytochrome c reductase (CR) activity, is FAD-dependent in *N. crassa* (Garrett and Nason, 1969). It is not clear whether haem is obligatorily involved (Fido *et al.*, 1979). Removal of molybdenum (Gewitz *et al.*, 1981) or substitution by tungsten (Wray and Filner, 1970; Notton and Hewitt, 1971 b) does not affect the CR activity of NR. These results suggest that the dehydrogenase function is catalyzed by the proximal part of the electron transport chain and that it is an activity of the flavohaemoprotein subunit, independent of the Mo-co. We (Brown *et al.*, 1981; Wray and Kirk, 1981) and others (Hamano *et al.*, 1984 a, b) have described a ca. 40,000 molecular weight CR species which is released from barley NR after proteolytic attack by an endogenous leupeptin-inhibited cysteine endoproteinase (Miller and Huffaker, 1981). We have suggested that this species probably carries at least the FAD domain together with the NADH binding site (dinucleotide fold) (Thompson *et al.*, 1975) of the native NR flavohaemoprotein subunit (Brown *et al.*, 1981). Since this species lacks haem (Campbell *et al.*, 1984) it may be released by proteolysis of a hinge region located between the flavin and haem domains of the subunit.

Heat treatment at 45° for 10 min inactivates both NADH-NR and CR activities of the barley enzyme but has little effect on FMNH-NR activity (Wray and Filner, 1970). Both NADH-NR and CR activity of the spinach enzyme are protected against heat inactivation by FAD (Relimpio *et al.*, 1971). These results suggest that the FAD-protected, heat labile part of the subunit (the flavin domain) is not required for the expression of FMNH-NR activity and that electrons are donated from FMNH (and probably from reduced viologen dye) to a later, but still undefined, site in the electron transport chain. Since NR partial activities are molybdenum dependent (Wray and Filner, 1970), FMNH-NR and reduced viologen dye NR activity are functions of the distal part of the electron transport chain.

Nothing is known about how the product of the NR apoprotein gene is processed to generate the flavohaemoprotein subunit. Insertion of FAD and haem occurs in some as yet undefined way, either before or after interaction with the Mo-co causes dimerisation and leads to the formation of the holoenzyme. The coding sequence of the NR apoprotein gene is at least 3 kb in length and represents a large target for mutagenic attack. Base substitution or deletion may be expected to affect the expression of the activities of the NR molecule by abolishing or altering domain function or by interfering with domain-domain interactions. In some of the studies discussed below attempts have been made to locate the approximate site of mutations within the apoprotein gene by measuring the partial activities of NR.

C. Nitrite Reductase

The assimilatory nitrite reductases (ferredoxin nitrite oxidoreductase, EC 1.7.7.1) of higher plants catalyse the six electron reduction of nitrite to ammonium within the chloroplast (Guerrero *et al.*, 1981). The enzyme has been purified from squash (Hucklesby *et al.*, 1967), spinach (Vega and Kamin, 1977; Ida, 1977), barley (Serra *et al.*, 1982) and wheat (Small and Gray, 1984) and is a monomer with a molecular weight of 60,000—63,000. The spinach enzyme is best characterised and contains one tetranuclear iron-sulphur cluster (Fe_4S_4) and one sirohaem per 60,000 molecular weight peptide chain (Vega and Kamin, 1977; Lancaster *et al.*, 1979). Sirohaem, dimethylurotetrahydroporphyrin, is an iron tetrahydroporphyrin of the isobacteriochlorin type. Mossbauer spectroscopy of the oxidised NiR reveals the presence of exchange interactions between the ferrihaem and the $S = OFe_4S_4$ cluster indicating that the two centres are chemically linked (Wilkerson *et al.*, 1983). This suggests that the nitrate reducing centre of the spinach NiR is the coupled Fe_4S_4/sirohaem pair.

Nitrite reductase of oat has been shown to be nuclear-encoded (Heath-Pagliuso *et al.*, 1984). The 60,500 molecular weight NiR of wheat is synthesised on cytoplasmic ribosomes as a 64,000 molecular weight precursor (Small and Gray, 1984). Several other chloroplast-located proteins have been shown to be synthesised as higher molecular weight forms in the cytoplasm before transfer into the chloroplast where they are processed to the mature form (Highfield and Ellis, 1978; Grossman *et al.*, 1982).

III. Genetics of Nitrate Assimilation

A. Introduction

Mutants altered in the nitrate assimilation pathway have been isolated at the level of both the cell/protoplast and the intact plant. The latter procedure directly gives whole mutant plants which are usually fertile, and overcomes the problems which sometimes exist in the regeneration of plants from mutant cell lines.

Two approaches have been used to identify NR-defective individuals within populations of mutagenised cells/protoplasts or whole plants. The first involves a non-selective total isolation procedure and has been used to isolate NR-defective mutants at the protoplast level in *Hyoscyamus muticus* (Strauss *et al.*, 1981) and at the whole plant level in barley (Warner *et al.*, 1977). The other approach uses chlorate as a selective agent, based on the suggestion of Åberg (1947) that chlorate toxicity in wheat is due to the NR-catalysed reduction of chlorate to toxic chlorite; individuals which, for whatever reason, lack a catalytically functional NR would be expected to be chlorate resistant and could be selected as such.

There is abundant evidence to support Åberg's suggestion that chlorate toxicity is mediated through the catalytic action of NR. Thus, urea grown wheat plants (expected to have little or no NR activity) are less sensitive to chlorate than are nitrate-grown plants. Light, which increases NR activity, also increases chlorate sensitivity (Liljestrom and Åberg, 1966). In soybean, chlorate sensitivity of leaf parts correlates with their NR activity (Weaver, 1942; Harper, 1981 a). More direct evidence comes from the demonstration that *Chlorella* (Solomonson and Vennesland, 1972; Vega, 1972), tomato (Hofstra, 1977) and tobacco (Zabala and Filner, 1980) NRs can use chlorate as a substrate. However Åberg's suggestion is not consistent with the wide-spread occurrence of chlorate-resistant, higher plant cell-lines which still possess NR activity (see for example Müller and Grafe, 1978; Murphy and Imbrie, 1981; Buchanan and Wray, 1982).

Further evidence that the basis of chlorate resistance may not be as straightforward as originally thought comes from *Aspergillus* where chlorate-sensitive mutants lacking NR activity have been reported (Cove, 1976 a). In this organism both NR and the product of the regulatory gene, *nirA*, are involved in the catabolism of some nitrogen compounds and also in the mediation of chlorate toxicity. Cove has suggested that chlorate is toxic because it mimics nitrate in mediating, via NR and the *nirA* gene product, a shut-down in nitrogen catabolism. Since chlorate cannot act as a nitrogen source, nitrogen starvation ensues. Thus in this organism at least chlorate toxicity may not be mediated by the catalytic function of NR.

Whatever the mechanism of chlorate resistance may be, chlorate has proved useful as a selective agent in the isolation of NR-defective mutants. Its first reported use in higher plants was in the early 1970s (Oostindier-Braaksma, 1970; Laan *et al.*, 1971; Oostindier-Braaksma and Feenstra, 1972).

Chlorate also acts as a nitrate analogue at the level of uptake. Nitrate competes with chlorate for uptake into *Arabidopsis thaliana* (Doddema and Telkamp, 1979). In barley, depletion of chlorate or nitrate from uptake media over 2—6 h by seedlings was found to be dependent on combined nitrate plus chlorate concentration, and total anion uptake was equivalent at different nitrate to chlorate ratios. Lineweaver-Burk plots of the interaction between nitrate and chlorate were characteristic of competitive inhibition (Deane-Drummond and Glass,

1983). One might therefore expect that some mutations within nitrate uptake could lead to chlorate resistance.

Higher plant nitrate assimilation mutants have been analysed using both genetic and biochemical approaches. Conventional genetic analysis by crossing fertile mutant plants has allowed the testing of allelism, as well as determination of the recessive or dominant character of the mutation and the type of inheritance. In those cases where only mutant cell lines are available more limited genetic analysis has been carried out by complementation in somatic hybrids. It should be emphasized however that definitive allocation of a function to a locus requires analysis of more than one or a few mutant alleles at that locus.

B. Nitrate Uptake Mutations

In *Aspergillus,* mutations at the *crnA* locus lead to reduced net uptake of nitrate into conidiophores and young mycelia. Loss of function confers resistance to chlorate and bromate without any obvious nutritional impairment (Tomsett and Cove, 1979; Brownlee and Arst, 1983). Only two putative nitrate uptake mutants, isolated on the basis of chlorate resistance, have been described from higher plants. These are the whole plant allelic mutants B1 and B3 of *Arabidopsis thaliana* (Oostindier-Braaksma and Feenstra, 1973). The mutation is monogenic and recessive and the gene locus has been designated *chl*1.

Net uptake of nitrate in wild-type *Arabidopsis* is biphasic, each phase showing Michaelis-Menten kinetics. Phase I has a Km of 0.04 mM and phase II a Km of 25 mM for nitrate. In the B1 mutant, phase II does not follow Michaelis-Menten kinetics and the net uptake of nitrate in the phase II concentration range is considerably lower than the wild type. It is interesting that this defect in phase II confers sufficient resistance to 13 mM chlorate to allow B1 to be distinguished from wild-type plants (Doddema and Telkamp, 1979). B1 is still able to take up chlorate (Oostindier-Braaksma and Feenstra, 1973) as well as nitrate. It would be useful to know whether the ability to select nitrate uptake mutants is dependent on the concentration of chlorate used in the screen. It might be that at higher chlorate levels mutation in one of the uptake systems would be insufficient to allow selection at the whole plant level.

Since the phase I uptake mechanism is unaffected in B1, Doddema and Telkamp (1979) have concluded that uptake of nitrate in *Arabidopsis* is mediated by (at least) two independent uptake mechanisms. If this biphasic uptake mechanism is common in higher plants then it is unlikely that mutants completely defective in nitrate uptake could be selected at the whole plant level in a one-step chlorate screen. Individuals within M_2 populations simultaneously defective in both (or more than two) uptake genes are likely to be extremely rare.

C. Nitrate Reductase Mutations

i) Apoprotein Gene Mutants

Extensive genetic analysis in lower eukaryotes, particularly in *Aspergillus nidulans* (reviewed by Cove, 1979) but also in *Neurospora crassa* (Sorger and Giles, 1965; Tomsett and Garrett, 1980) and *Penicillium chrysogenum* (Birkett and Rowlands, 1981), indicates the presence of a single gene locus coding for the apoprotein of NADPH-NR (EC 1.6.6.3) and suggests the enzyme is a homodimer. This latter conclusion is supported by studies on the enzyme purified from *Neurospora* (Pan and Nason, 1978). The situation is rather more complicated in higher plants due to the presence of both constitutive and nitrate-inducible forms of NR and of species which are either NADH-specific (EC 1.6.6.1) or bispecific for NADH and NADPH (EC 1.6.6.2) as electron donor. The isolation and characterisation of apoprotein gene mutants in higher plants has contributed greatly to our understanding of the relationships between these different types and has also stimulated studies into the characterisation of the forms of NR present in wild-type plants.

A wide range of NR mutants having properties consistent with defects in an apoprotein gene locus have been isolated in higher plants (Table 1). The evidence that these are indeed apoprotein gene mutants is stronger in some cases, for example in *Nicotiana* species and *Hordeum vulgare,* than in others where interpretation rests solely on the demonstration of the retention of xanthine dehydrogenase activity. In *N. tabacum* and *Hordeum vulgare* the evidence is particularly strong that the apoprotein of NADH-NR is coded for by a single gene, indicating that the enzyme is a homodimer. Similar conclusions for other higher plant NRs are weakened by the relatively small number of mutants which have been isolated and characterised.

a) *Nicotiana tabacum*

Detailed biochemical and genetic analysis has been carried out in *Nicotiana* spp., particularly with the allotetraploid *N. tabacum* L. var Gatersleben (Müller and Grafe, 1978). NR-minus mutants were isolated by screening mutagenised amphihaploid (n = 24) cell suspensions for resistance to 20 mM potassium chlorate. Thirty six lines retaining xanthine dehydrogenase activity (Mendel and Müller, 1976; Müller and Grafe, 1978; Mendel and Müller, 1979; Müller, 1983; pers. comm.) were designated *nia* in accordance with the gene symbol used in *Aspergillus* for those mutants which were NR-minus but had xanthine dehydrogenase activity (Pateman *et al.,* 1964; Cove and Pateman, 1969). All these mutants had lost the CR activity catalysed by the flavohaemoprotein subunit of NR and all except three had lost all three (NADH-, FMNH-, BVH-) NR activities. The exceptions are mutants Nia95, Nia26 and Nia28 which, although lacking CR activity, still show inducible FMNH and BVH-NR activities. These biochemical data have been interpreted to show that the *nia* mutants are altered in the gene coding for the apoprotein of NR. The differing enzy-

Table 1. Higher plant nitrate assimilation mutants

Species	Gene	Mutant Number	Function	Phenotype	Sexual Transmission to Seed Progeny	Isolation Procedure[a]	Reference
Arabidopsis thaliana	*chl*2 (chromosome 2)	B 2-1	?	Deficient in NR and XDH; poor growth on nitrate; chlorate resistant	yes	1	Braaksma and Feenstra, 1982 a
	*chl*3 (chromosome 1)	B 29 B 31-2, B 33, B 35 (four alleles)	?	Deficient in NR and XDH; strong growth on nitrate; chlorate resistant	yes	1	as above
	rgn (chromosome 1)	B 25	Molybdenum cofactor	Deficient in NR and XDH; poor growth on nitrate; chlorate resistant	yes	1	as above
	cnx (chromosome 5)	B 73	Molybdenum cofactor	Deficient in NR and XDH; poor growth on nitrate; chlorate resistant	yes	1	as above
	nd	B 31-1	?	Deficient in NR; strong growth on nitrate; chlorate resistant	yes	1	as above
	*chl*1	B 1, B 3 (two alleles)	Nitrate uptake gene	Decreased uptake of nitrate; retains NR and XDH; chlorate resistant	yes	1	as above
	nd	B 36	?	Deficient in NR; strong growth on nitrate; chlorate resistant	yes	1	as above
	nd	B 40	?	Deficient in NR; strong growth on nitrate; chlorate resistant	yes	1	as above

Species	Gene	Mutant Number	Function	Phenotype	Sexual Transmission to Seed Progeny	Isolation Procedure[a]	Reference
Glycine max L. Merr cv. Williams	*nr*$_1$	LNR-2 LNR-3 and LNR-4 (three alleles)	Possible apoprotein or regulatory gene	Constitutive-NR minus: (acetaldehyde oxime-evolution minus) retains XDH and inducible NR; growth on nitrate; chlorate resistant	yes	2	Nelson *et al.*, 1983, 1984, and Ryan *et al.*, 1983
Hordeum vulgare cv. Steptoe	*nar*1	9 Alleles initially coded Az 1, 2 etc.	Apoprotein gene	Deficient in NADH-NR; retains XDH; growth on nitrate	yes	3	Warner *et al.*, 1977; Kleinhofs *et al.*, 1980
	*nar*2	Az 34	Molybdenum cofactor	Deficient in NR and XDH; reduced growth on nitrate	yes	3	Kleinhofs *et al.*, 1980
	*nar*4	Az 72	Molybdenum cofactor	Deficient in NR and XDH; reduced growth on nitrate	yes	3	Kleinhofs *et al.*, 1985
cv. Winer	*nar*1	Xno 24	Apoprotein gene	Deficient in NR; chlorate resistant, retains XDH; growth on nitrate	yes	4	Tokarev and Shumny, 1977; Shumny and Tokarev, 1982; Kleinhofs *et al.*, 1983; Somers *et al.*, 1983
	*nar*3	Xno 18, Xno 19 (two alleles)	Molybdenum cofactor	Deficient in NR and XDH; chlorate resistant, growth on nitrate	yes	4	as above
cv. Maris Mink	*nar*2	R 9201, R 9401 (two alleles)	Molybdenum cofactor	NR and XDH minus; chlorate resistant; no growth on nitrate; conditional lethal	yes	5	Bright *et al.*, 1983; Steven, unpublished; Wray *et al.*, 1985

Species	Gene	Mutant Number	Function	Phenotype	Sexual Transmission to Seed Progeny	Isolation Procedure[a]	Reference
	nd	11301	Molybdenum cofactor	as above	yes	5	as above
		12202	Molybdenum cofactor	as above	yes	5	as above
Hyoscyamus muticus	*cnx*A	MA-2	Molybdenum cofactor	NR and XDH minus; no growth on nitrate, chlorate resistant; Mo repairable	no	6	Strauss *et al.*, 1981; Lazar *et al.*, 1983
		O, Q	Molybdenum cofactor	as above	no	7	H. Fankhauser, pers. comm.
	nd	I (complementation group B)	Molybdenum cofactor	as above	no	7	as above
	nd	I_2D 12 (complementation group C)	Molybdenum cofactor	NR and XDH minus; no growth on nitrate; chlorate resistant; Mo repairable	no	6	Strauss *et al.*, 1981; Fankhauser *et al.*, 1984
	nd	X 1VE 9	Molybdenum cofactor	NR and XDH minus; no growth on nitrate; chlorate resistant	no	6	Fankhauser *et al.*, 1984; Strauss *et al.*, 1981
		T, C (complementation group D)		as above		7	H. Fankhauser, pers. comm.
	nd	V 1C 2 (complementation group E)	Molybdenum cofactor	as above	no	6	Strauss *et al.*, 1981; Fankhauser *et al.*, 1984

Species	Gene	Mutant Number	Function	Phenotype	Sexual Transmission to Seed Progeny	Isolation Procedure[a]	Reference
Nicotiana plumbaginifolia	*nia*	NA (7 lines)	Apoprotein gene	NR minus; retains XDH; no growth on nitrate; chlorate resistant	no	8	Marton *et al.*, 1982a, b
		26 lines	Apoprotein gene	NR minus; retains XDH; no growth on nitrate; chlorate resistant	yes	9	Negrutiu *et al.*, 1983
	*cnx*A	NX 1, NX 9 (two alleles)	Molybdenum cofactor	NR and XDH minus; no growth on nitrate; chlorate resistant; Mo repairable	no	8	Marton *et al.*, 1982a, b; Xuan *et al.*, 1983
		CNX 20 CNX 82 (two alleles)	Molybdenum cofactor	NR and XDH minus; no growth on nitrate; chlorate resistant; Mo repairable	yes	9	Negrutiu *et al.*, 1983; Dirks *et al.*, 1985
	*cnx*B	NX 24	Molybdenum cofactor	NR and XDH minus; no growth on nitrate; chlorate resistant	no	8	Marton *et al.*, 1982a, b
		CNX 27	Molybdenum cofactor	NR and XDH minus; no growth on nitrate; chlorate resistant	no	9	Negrutiu *et al.*, 1983; Dirks *et al.*, 1985
	cnxC	NX 21	Molybdenum cofactor	NR and XDH minus; no growth on nitrate; chlorate resistant	no	8	Marton *et al.*, 1982a, b
	*cnx*D	CNX 103	Molybdenum cofactor	NR and XDH minus; no growth on nitrate; chlorate resistant	yes	9	Negrutiu *et al.*, 1983; Dirks *et al.*, 1985

Species	Gene	Mutant Number	Function	Phenotype	Sexual Transmission to Seed Progeny	Isolation Procedure[a]	Reference
Nicotiana tabacum cv. Gatersleben	*nia* 1 *nia* 2	36 alleles	Duplicate apoprotein genes	NR-minus; no growth on nitrate; chlorate resistant, retains XDH	yes-fertile plants regenerated from 15 cell lines	10	Müller and Grafe, 1978; Müller, 1983
	*cnx*A 1 *cnx*A 2	Cnx 68 Cnx 101 Cnx 109 Cnx 135 (four alleles)	Duplicate molybdenum cofactor genes	NR and XDH minus; no growth on nitrate; chlorate resistant; Mo repairable	yes-fertile plants regenerated from Cnx 135	10	Müller and Grafe, 1978; Grafe and Müller, 1983; A. Müller, pers. comm.
cv. Xanthi	*cnx*B	O 42, P 12 P 31, P 47 (four alleles)	Molybdenum cofactor	NR and XDH minus, no growth on nitrate; chlorate resistant	no	11	Buchanan and Wray, 1982; Xuan *et al.*, 1983; Mendel *et al.*, 1984
cv. Xanthi	nd	clr 19	Apoprotein gene?	NR minus, retains XDH; no growth on nitrate; chlorate resistant	no	12	Evola, 1983 a, b
		clr 0	Molybdenum cofactor	NR and XDH minus; no growth on nitrate; chlorate resistant	no	13	as above
Petunia hybrida var. Mitchell	nd	line 1	Possible apoprotein gene	NR minus; no growth on nitrate; chlorate resistant; retains XDH	no	14	Steffen and Schieder, 1984
	nd	line 2	Molybdenum cofactor	NR and XDH minus; no growth on nitrate; chlorate resistant	no	14	as above
	nd	lines 3 and 4	Molybdenum cofactor	NR and XDH minus; no growth on nitrate; chlorate resistant; Mo repairable	no	14	as above

Species	Gene	Mutant Number	Function	Phenotype	Sexual Transmission to Seed Progeny	Isolation Procedure[a]	Reference
Pisum sativum cv. Juneau	*nar*1	A317, A334 (two alleles)	Possible apoprotein gene	Deficient in NR	yes	15	Kleinhofs *et al.*, 1978; Warner *et al.*, 1982
	*nar*2	A300	Molybdenum cofactor	Deficient in NR and XDH	yes	15	as above
cv. Rondo	nd	E1	Molybdenum cofactor	Deficient in NR and XDH; chlorate resistant	yes	16	Feenstra and Jacobsen, 1980; Jacobsen *et al.*, 1984

NR-deficient mutants have also been reported in *Datura innoxia* (King and Khanna, 1980) and *Rosa damascena* (Murphy and Imbrie, 1981) cell cultures. ENU, N-ethyl-N-nitrosourea; EMS, ethylmethane sulphonate; MNNG, N-methyl-N′-nitro-N-nitrosoguanidine, nd — not designated.

[a] *Isolation Procedures*

1 6 day old M_2 plants (40 mM EMS in M_1) resistant to 13 mM $NaClO_3$.
2. 10—12 day old plants derived from seed mutagenised for four successive generations with a variety of chemical and physical mutagens and screened for decreased visual damage in the presence of 0.1 mM $KClO_3$. Rescreened for lowered *in vivo* NR.
3 7 day old M_2 plants (1 mM azide pH 3 for 2 h in M_1) screened for lowered *in vivo* NR activity.
4 7 day old M_2 plants (0.25 % EMS for 10—12 h in M_1) screened for resistance to 8 mM $KClO_3$ for 2 days.
5 6 day old M_2 plants (1 mM azide pH 3 for 2 h in M_1) screened for resistance to 10 mM $KClO_3$ for 9 days.
6 Haploid mesophyll protoplasts mutagenised with 20 mg. l^{-1} MNNG and screened for amino acid auxotrophy followed by nitrate non-utilization.
7 Haploid mesophyll protoplasts mutagenised and screened for resistance to chlorate.
8 Haploid protoplasts mutagenised with either a ^{60}Co source (0.04 Gy. sec^{-1}) or ENU and screened for resistance to 40 mM $KClO_3$.
9 Haploid protoplasts screened for spontaneous resistance to 125 mM $KClO_3$.
10 Amphihaploid cell suspension mutagenised with 0.25 mM ENU and screened for resistance to 20 mM $KClO_3$.
11 Amphihaploid cell suspension mutagenised with 0.4 % EMS for 1 h and screened for resistance to 20 mM $KClO_3$.
12 Amphihaploid cell suspension mutagenised with 0.5 mM ENU and screened for resistance to 20 mM $KClO_3$.
13 Amphihaploid cell suspension screened for spontaneous resistance to 20 mM $KClO_3$.
14 Cell colonies derived from haploid mesophyll protoplasts mutagenised with X-rays (1000 R) screened for resistance to 100 mM $KClO_3$.
15 M_2 plants (1 mM azide pH 3 for 2 h in M_1) screened for lowered *in vivo* NR activity.
16 12 day old M_2 plants (0.3 % EMS for 4 h in M_1) screened for resistance to 20 mM $KClO_3$.

matic activities possessed by mutants may be due to point mutations at different places within the *nia* gene locus.

Fifteen of the thirtysix *nia* mutants have been regenerated to fertile amphidiploid plants and have been analysed genetically through sexual crosses. As anticipated from the biochemical evidence, all fifteen mutants proved allelic (Müller, 1983). Segregation among F_2 progeny from crosses between mutant and wild-type plants and among test-cross progeny showed the regenerated plants to be homozygous, double mutants. The NR-minus phenotype is conferred by two unlinked recessive nuclear mutations which define a pair of duplicate loci *(nia*1 *nia*2).

This interesting finding may be explained by the fact that *N. tabacum* (2n = 48) is believed to have arisen by hybridisation between two diploid progenitor species *N. sylvestris* (2n = 24) and *N. tomentosiformis* (2n = 24) (Gray *et al.*, 1974). One of the NR apoprotein gene loci is presumably derived from each parent. Genetic analysis showed that both loci (*nia*1 and *nia*2) mutated after initiation of the parental cell culture, most probably after treatment with mutagen, since mutants were recovered only after mutagen treatment. The low frequency of recovered NR-minus mutants (about 10^{-7}) is consistent with simultaneous induction of two independent mutations and is much higher (10^{-3} to 10^{-4}) in similar work on monoploid cells of the true haploid *Nicotiana* species, *N. plumbaginifolia* (Marton *et al.*, 1982a) discussed below. This shows that both homeologous gene loci are functional within *N. tabacum* and raises the possibility that two types of apoprotein are synthesised. Random assembly of flavohaemoprotein subunits might generate three types of NADH-NR. Since constitutive NR activity as well as nitrate inducible NR activity is missing from *nia* mutants (Müller and Grafe, 1978; Müller, 1983) it is likely that their apoproteins are the products of the same gene.

Construction of mutant plants carrying different numbers (up to four) of *nia*+ genes, single mutants and even plants possessing only one wild-type allele at either of the *nia* loci were all normal with respect to growth on nitrate, growth on soil and NR level at most developmental stages. Only at the early seedling stage did the NR activity respond to the number of *nia* genes, as shown by the *in vivo* NR activity and chlorate sensitivity. Thus the initially constitutive expression of the *nia*+ genes changes during seedling development to a strictly regulatory type of gene expression (Müller, 1983).

A further NR-minus chlorate resistant mutant has been isolated from mutagenised protoplasts of a dihaploid plant of *N. tabacum* L. var Xanthi (Evola, 1983a). This mutant, *clr*19, which retains xanthine dehydrogenase activity, lacks nitrate-inducible CR activity but possesses a very low level of nitrate inducible BVH-NR activity, suggesting that it might be an apoprotein gene mutant altered in the proximal part of the electron transport chain. There is no further biochemical information on this mutant and, although *clr*19 scions flowered when grafted on wild-type stocks it is not known whether they are allelic to Gatersleben mutants.

b) *Nicotiana plumbaginifolia*

Putative NR-minus apoprotein gene mutants have been isolated from the true haploid *Nicotiana* species *N. plumbaginifolia* (n = 10). Marton *et al.* (1982a) isolated 36 chlorate-resistant clones of which 29 were fully deficient in NR. Of nine clones examined, five (designated NA) possessed xanthine dehydrogenase activity and were allelic as shown by non-complementation in somatic hybrids (Marton *et al.*, 1982b). Attempts to reconstitute NR activity by cohomogenisation of cells of each of the five lines with the apoprotein gene mutant, Nia63, of *N. tabacum* var. Gatersleben were unsuccessful, suggesting that they may also be apoprotein gene mutants (Marton *et al.*, 1982a). In *N. plumbaginifolia* selection for resistance of haploid protoplasts to potassium chlorate (40mM) produced spontaneous mutants at a frequency of 10^{-5} to 10^{-6} (Negrutiu *et al.*, 1983). In *Aspergillus, nia* mutants selected via chlorate resistance without prior mutagenesis are more likely to be deletion mutants (Cove, 1976b). Twenty six lines which were NR-minus but retained xanthine dehydrogenase activity were classified as apoprotein gene mutants (Negrutiu *et al.*, 1983). Plants regenerated from twelve of these lines were shown to express and transmit the NR-minus phenotype to the progeny in a Mendelian fashion. The plants were allelic and diploid and the mutation was inherited as a single recessive nuclear gene (Dirks *et al.*, 1985).

c) *Petunia hybrida*

Steffen and Schieder (1984) have recently reported the isolation of an NR-minus chlorate-resistant (100mM) line of *Petunia hybrida* var. Mitchell. The line retained xanthine dehydrogenase activity and was tentatively classed as an apoprotein gene mutant. Plants could not be regenerated.

d) *Hordeum vulgare*

Whole plant mutants have been directly isolated in barley, *Hordeum vulgare* L. cv. Steptoe (Kleinhofs *et al.*, 1978). Mutagenesis in the M_1 was carried out with sodium azide at low pH. Kleinhofs and coworkers initially attempted to screen for loss of NR activity in M_2 seedlings through the use of chlorate. Although chlorate resistant seedlings were identified at a frequency of 6 per 10,000, all appeared lethal and attempts to transplant and propagate these seedlings failed.

Subsequently NR mutants were identified by a rapid semiquantitative *in vivo* NR assay of individual seedling leaves. Seedlings showing 10 % or less of control NR activity were classified as being NR defective (Warner *et al.*, 1977). Nine NR-deficient mutants were shown to be allelic when tested by reciprocal crossing in all possible combinations and the locus represented by these mutants (Az12, 13, 23, 28, 29, 30, 31, 32 and 33, subsequently renumbered a, ... i) was designated *nar*1. The *nar*1 gene is a nuclear gene (Kleinhofs *et al.*, 1980). These mutants have NR activities ranging between 2 and 7 % of the wild type when assayed *in vitro* with NADH as electron donor (Somers *et al.*, 1983).

Several lines of evidence together show unequivocably that the *nar*l locus is the apoprotein gene locus. Some *nar*l alleles possess no nitrate inducible CR activity (carried by the proximal part of the electron transport chain), some have levels similar to wild type, whilst one (*nar*ld) has an extremely high level. All mutants except one (*nar*lh) lack nitrate-inducible FMNH-NR activity, catalysed by the distal end of the electron transport chain (Kleinhofs *et al.*, 1980). NR CRM levels correlated with the level of NR-associated enzyme activities (Kuo *et al.*, 1981; Somers *et al.*, 1983). These results are consistent with the suggestion that the alleles are due to mutations at different sites within the *nar*l locus. These mutations produce mutant flavohaemoprotein subunits which are responsible for the different levels and types of subunit-associated enzyme activity (Kleinhofs *et al.*, 1980). Cleveland mapping of the *nar*ld CR species shows that it contains a glutamic acid substitution when compared to the wild-type NR (Kuo *et al.*, 1984). Further evidence that the flavohaemoprotein subunit is defective in *nar*l alleles comes from the observation that functional Mo-co, released from bovine xanthine oxidase by heat treatment, cannot reconstitute NR activity *in vitro* from extracts of any of the *nar*l alleles (Narayanan *et al.*, 1984). A *nar*l allele Xno24, designated *nar*lj (Kleinhofs *et al.*, 1983), has also been isolated from the barley cultivar Winer (Shumny and Tokarev, 1982).

When grown to maturity with nitrate as sole nitrogen source, the *nar*l alleles *nar*la and *nar*lb possessed as much dry weight and reduced nitrogen per plant as the wild-type control. Similar results were obtained with aseptically grown excised embryos (Warner and Kleinhofs, 1981). When grown in the field the total vegetative dry weight, total reduced nitrogen and percent grain protein of the mutants was the same as the wild type, whereas grain yield was lower (Oh *et al.*, 1980). These results indicate that *nar*l alleles with NR levels less than 10% of wild type can utilise nitrate as sole nitrogen source and perform almost as well as wild-type plants.

It might be thought that the ability of these *nar*l alleles to utilise nitrate was due to residual NADH-NR activity present as a result of leaky mutations within the *nar*l locus. However, very surprisingly, the activity present in these alleles appears to be that of a bispecific NAD(P)H-linked enzyme which cannot be detected in wild-type plants (Dailey *et al.*, 1982a, b). This bispecific NAD(P)H-NR cross-reacted with antiserum raised against the wild-type NADH-NR but with a much lower specificity, perhaps due in part to its slightly smaller subunit molecular weight (A. Kleinhofs, pers. comm.). Although the NAD(P)H-NR plays a significant role in nitrate reduction in *nar*l mutants, the mechanism which allows it to be expressed in these mutants is not clear.

e) *Pisum sativum*

Two whole plant mutants have been isolated in pea, *Pisum sativum*, by screening in the M_2 for low *in vivo* NR activity (Kleinhofs *et al.*, 1978;

Warner *et al.*, 1982). The mutant plants (A317 and A334) had less than 6 % of the wild-type *in vitro* NADH-NR and FMNH-NR activity but still retained inducible NADH-CR activity. The mutations of A317 and A334 are monogenic, allelic, express incomplete dominance and have been designated *nar*1. The mutants possess xanthine dehydrogenase activity. The most likely interpretation of these results is that the *nar*1 locus is the apoprotein gene locus. Mutations in both alleles appear to be associated with the distal end of the electron transport chain since FMNH-NR activity is impaired whilst the NADH-CR activity associated with the proximal end of the electron transport chain is not. Nitrite reductase activity was inducible, as in the wild type, but activities were higher, perhaps reflecting the much higher nitrate levels which accumulated in the mutants (Warner *et al.*, 1982).

f) *Glycine max*

Mutant lines of soybean have been invaluable in understanding the relationship between the different forms of NR which exist in this species (Evans and Nason, 1953; Jolly *et al.*, 1976; Campbell, 1976; Lahav *et al.*, 1976; Kakefuda *et al.*, 1983). Whole plant mutants were obtained by screening M_2 seed, derived from soybean (*Glycine max* L. Merr. cv. Williams) seed which had been treated for four successive generations with various chemical and physical mutagens (designated the M_1 seed) for resistance to 0.1 mM potassium chlorate. Chlorate toxicity was assessed by necrosis of the cotyledon margins and/or chlorosis and necrosis of the unifoliate leaves with subsequent stunted leaf expansion. Forty-nine potential NR-deficient plants were selected from 12,000 M_2 seedlings. Thirty-eight of these produced seed. The M_3 plants were subjected to a second chlorate screen and leaf *in vivo* NR activity was determined in resistant plants. Selected M_3 plants were harvested and three lines with decreased NR activity (LNR-2, LNR-3 and LNR-4) were identified in the M_4 (Nelson *et al.*, 1983). The three lines are allelic (Ryan *et al.*, 1983).

Wild-type soybean possesses two main types of NR, a form present in urea-grown plants which has been called constitutive since it does not require nitrate for expression, and a form which is nitrate inducible. In young soybean leaves grown on nitrate the constitutive NR makes up approximately 50 % of the total activity. The NR activities of the mutant lines are approximately 50 % of the wild-type activity (assayed *in vivo* and *in vitro* with NADH) when grown on nitrate as sole nitrogen source, but NR activity was absent from leaves of the urea-grown mutant. Thus the mutants lack the constitutive NR activity but still retain the nitrate inducible NR activity (Nelson *et al.*, 1983). The nitrate inducible NR was purified from nitrate-grown mutant plants and shown to be a NADH-NR (EC 1.6.6.1) with a pH optimum of 7.5 (Streit *et al.*, 1985). This enzyme, which is present in all other plant species examined, had not previously been identified in soybean due to the presence of the constitutive NR. The nitrate grown mutant is a source of pure soybean NADH-NR.

The urea-grown soybean mutant LNR-2 (designated nr_1) had also lost *in vitro* FMNH-NR, and CR activity was much reduced. Since the xanthine dehydrogenase activity of nr_1 was unaffected, Nelson *et al.*, (1984) suggested that the mutation in nr_1 was in the apoprotein gene of the constitutive NR. Evolution of acetaldehyde oxime which occurs in wild-type soybean plants under anaerobic conditions (Harper, 1981 b; Mulvaney and Hageman, 1984) did not occur in the nr_1 mutant, indicating a role for the constitutive NR in gas evolution. Absence of constitutive NR and evolution of acetaldehyde oxime are controlled by a single recessive nuclear gene (Ryan *et al.*, 1983). The absence of constitutive NR activity in the mutants did not increase nitrate levels or decrease reduced-nitrogen concentrations in the plants. The ability of these soybean mutants to maintain apparently normal nitrogen metabolism, despite lowered NR activity, is similar to the *nar*1 barley mutants (Warner and Kleinhofs, 1981).

Immunological and blue-dextran Sepharose chromatographic studies have subsequently shown the constitutive NR to consist of two species (Robin *et al.*, 1985; Streit *et al.*, 1985). One form, eluted with NADPH from blue-Dextran Sepharose, is more active with NADPH than NADH as electron donor, has a relatively high Km for nitrate and a pH optimum of 6.5 and is immunologically identical to the NAD(P)H-NR described by Jolly *et al.* (1976) and Campbell (1976). The other form, eluted with NADH, is more active with NADH, has a pH optimum of 6.5 and is the NADH-NR described by Jolly *et al.* (1976). The nitrate-inducible NR has a sedimentation coefficient of 7.6 like most other NADH-NRs examined. The constitutive NRs had sedimentation coefficients of 5.6 (NADPH-eluted) and 6.0 (NADH-eluted) and had higher mobilities on polyacrylamide gel than the nitrate-inducible NADH-NR.

Since loss of constitutive NR activity is due to a mutation in a single recessive nuclear gene, the expression of both forms of constitutive NR must be controlled by the same gene. This would imply that they share common flavohaemoprotein subunits or that the nr_1 mutation is in a regulatory, rather than a structural, gene locus. The functions of these constitutive NRs are unknown but they are active *in vivo* (Ryan *et al.*, 1983).

g) *Arabidopsis thaliana*

Whole plant mutants have been isolated by screening M_2 plants for resistance to 13 mM chlorate (Braaksma and Feenstra, 1982 a). Whilst the genetic analysis of these mutants is the most sophisticated so far carried out in higher plants, the nature of some of the mutations is unclear.

Braaksma and Feenstra (1982 a) have argued that the mutations in *chl2* and *chl*3 probably do not affect the Mo-co and that both are apoprotein gene mutations. Since *Arabidopsis* apparently possesses only NADH-NR this would suggest that the situation in *Arabidopsis* is different from all other eukaryotic NR systems where biochemical and genetic analysis indicates the presence of only one type of apoprotein subunit in individual NR species.

The argument that *chl2* is an apoprotein gene locus is based solely on the observation that the 8S NR/CR peak seen after sucrose gradient analysis of *chl2* extracts has a lower ratio of NR to CR activity than is seen in the wild type, indicating perhaps some lesion in the apoprotein which affects CR activity. However, xanthine dehydrogenase activity in *chl2* is also reduced to less than half that of the wild type, suggesting a defect in the Mo-co.

Their argument that *chl3* is also altered in an apoprotein gene is based on the observation that there are irregularities in the aggregation of the enzyme complex and that this might be caused by an altered flavohaemoprotein subunit. Whilst this is a possibility, it is known that aggregation is also a function of Mo-co structure and does not explain why xanthine dehydrogenase levels in this mutant are also lowered compared to the wild type. Thus it is not clear whether either *chl2* or *chl3* represent apoprotein gene mutations. At least part of the problem may be attributed to the leaky nature of the mutants. *chl2* and *chl3* possess approximately 10 % and 20 % of the wild-type NR activity respectively.

ii) Molybdenum Cofactor Mutants

Mutations within Mo-co genes were first described in the lower eukaryote, *Aspergillus nidulans* (reviewed in Cove, 1979). Some mutants selected for their inabililty to utilise nitrate as a nitrogen source were also unable to utilise hypoxanthine (Pateman *et al.*, 1964). Both nitrate reductase and xanthine dehydrogenase were undetectable or present at very low levels in these mutants and, since both were molybdoenzymes, the defect was suggested to be within a molybdenum cofactor shared by these two proteins (Cove, 1963; Pateman *et al.*, 1964). These mutants were thus designated *cnx* (*c*ofactor for *n*itrate reductase and *x*anthine dehydrogenase).

Heterokaryon complementation tests for nitrate utilization have been carried out for over four hundred independently-isolated *cnx* mutants. The *cnx* mutants fall into seven complementation groups, designated A, B, C, E, F, G and H (Cove, 1963; Rever, 1966; Hartley, 1969, 1970). *cnx* A, B and C mutants show an overlapping complementation pattern and are closely linked. *cnx*A and *cnx*C mutants may involve two distinct genes, with *cnxB* mutants lacking both functions. *cnx*A, B and C mutations are unlinked to *cnx*E, *cnx*F, *cnx*G and *cnx*H which are in turn unlinked to one another (Cove, 1979). More recently mutants at a further *cnx* locus, *cnx*J, have been described. Arst *et al.*, (1982) suggest that the *cnx*J gene product plays a regulatory role in Mo-co synthesis. Thus at least six and possibly seven genes can mutate to give the *cnx* phenotype. Mutations within the Mo-co have been described also in *N. crassa* (Sorger and Giles, 1965; Coddington, 1976; Tomsett and Garrett, 1980), *Penicillium chrysogenum* (Birkett and Rowlands, 1981) and in prokaryotes, for example, *Escherichia coli* (Ruiz-Herrera *et al.*, 1969; Stewart and MacGregor, 1982).

A consideration of the structure (Fig. 2) and function of the Mo-co provides an explanation of why so many *cnx* loci are involved in its biosynthesis and suggests that two metabolic pathways are probably required. One pathway, leading from mainstream pterin metabolism, involves modification of the pterin nucleus to generate the phosphorylated pterin derivative, molybdopterin. The other pathway, some steps of which are speculative, is envisaged to include molybdate uptake into the cell, intracellular transport and storage of molybdate, generation of a form of molybdenum which is in the correct redox and ligandable state, and finally liganding of molybdenum to molybdopterin to generate the functional Mo-co. Whether this last step occurs before or after interaction of molydopterin with flavohaemoprotein subunits is not known. Some of the *cnx* loci will be involved in synthesis of the phosphorylated carbon side chain and in attachment of the sulphur atoms of molybdopterin (it is unlikely that they will be involved in synthesis of the pterin nucleus itself since mutations in such loci would be expected to be lethal) and some will be involved in what may be loosely described as molybdate metabolism/processing and in liganding of molybdenum to molybdopterin.

Arst *et al.* (1970) have shown that high concentrations of molybdate supplied to growing cultures of *cnx*E mutants partially restored NR and xanthine dehydrogenase activity (*in vivo* repair) and they suggested that the *cnx*E gene product might be involved in the insertion of molybdenum into the cofactor. Since some *cnx*H mutations are temperature-sensitive, this locus may specify a structural component of the Mo-co (MacDonald and Cove, 1974) but what this might be is unclear. Evidence for a molybdate reduction step has recently come from work with *E. coli* (Campbell *et al.*, 1985). Mutations in this step have been isolated but not yet characterised.

Mo-co mutants have been described in a wide variety of higher plants (Table 1). Identification of mutants as *cnx* has depended largely on the demonstration of the pleiotropic loss of NR and xanthine dehydrogenase activity with the retention of a functional flavohaemoprotein subunit.

a) *Nicotiana*

Molybdenum cofactor mutants have been isolated in *N. tabacum* L. var Gatersleben (Müller and Grafe, 1978) and var Xanthi (Buchanan and Wray, 1982; Evola, 1983 a) and in *N. plumbaginifolia* (Marton *et al.*, 1982 a; Negrutiu *et al.*, 1983).

Four NR-minus chlorate resistant lines of *N. tabacum* L. var Gatersleben were isolated in three independent experiments (Müller and Grafe, 1978). The lines lacked xanthine dehydrogenase activity but still possessed the CR activity catalysed by the proximal part of the electron transport chain of the flavohaemoprotein subunit (Mendel and Müller, 1976, 1979; pers. comm.). The lines were designated *cnx* in accordance with the terminology used with *A. nidulans* mutants of this type (Pateman *et al.*, 1964).

Since it proved impossible to regenerate fertile plants from these four *cnx* lines to allow conventional genetic analysis, complementation analysis of the *cnx* lines and of the Nia 115 line was carried out by somatic hybridization (Grafe and Müller, 1983). All *nia* + *cnx* combinations resulted in the formation of cell colonies capable of growing with nitrate as sole nitrogen source and possessing NR activity. However no nitrate-utilising colonies were formed from any of the *cnx* + *cnx* combinations.

The finding that all *nia* + *cnx* combinations studied are complementary shows that all the mutants involved are recessive. Grafe and Müller (1983) therefore concluded that the failure of the 4 *cnx* mutants to complement each other is not due to dominance but due to allelism. Thus the 4 *cnx* mutants tested are allelic to each other, not allelic to Nia 115 (nor to other *nia* mutants), and represent four alleles at a gene locus designated *cnx*A (Grafe and Müller, 1983). Since these *cnx* mutants occurred in amphihaploid cells and the *nia* mutants have been shown to carry two unlinked mutations which affect the duplicate structural genes for the NR apoprotein, Grafe and Müller (1983) have argued that these *cnx* mutants are also double mutants in duplicate genes *(cnx*A1, *cnx*A2).

Cnx 135 has now been regenerated to fertile plants. Results of crosses between Cnx 135 and wild type confirm that Cnx 135 is a double mutant in a pair of duplicate loci *(cnx*A1, *cnx*A2) (Müller, pers. comm.).

Four further NR-minus Mo-co mutants have been isolated from amphihaploid cells of *N. tabacum* L. var Xanthi by Buchanan and Wray (1982). Thirty-nine chlorate-resistant lines were isolated but only four lines O 42, P 12, P 31 and P 47 were unable to grow on nitrate as sole nitrogen source. These four lines lacked both NR and xanthine dehydrogenase activity. *In vitro* complementation between a *nia* mutant (Nia 63) and each of the four *cnx* lines (that is formation of NR activity due to assembly of NR molecules when cells of the two NR-minus partners are cohomogenized) and retention of nitrate inducible CR activity indicates that they possess functional flavohaemoprotein subunits (Buchanan and Wray, 1982; Mendel *et al.*, 1984). Genetic analysis has depended on somatic hybrids since it proved impossible to regenerate plants. All four lines fail to complement each other but complement an *N. tabacum cnx*A mutant (Cnx 68) as well as a *nia* mutant suggesting that they are allelic and recessive (Xuan *et al.*, 1983). The gene locus has been designated *cnx*B.

A total of eight chlorate-resistant NR-minus lines which lack xanthine dehydrogenase activity have been isolated in *N. plumbaginifolia*. Plants could not be regenerated from four lines designated NX (Marton *et al.*, 1982 a), and complementation in somatic hybrids showed they belonged to three complementation groups (NX 1 and NX 9; NX 21; NX 24) (Marton *et al.*, 1982 b). The other four lines, which were isolated without a mutagenic step, also belonged to three complementation groups (CNX 20 and CNX 82; CNX 27; CNX 103) (Negrutiu *et al.*, 1983; Dirks *et al.*, 1985) as shown by complementation in somatic hybrids. The mutations are recessive.

b) Complementation Analysis between *Nicotiana* Molybdenum Cofactor Mutants

The eight *N. tabacum* and eight *N. plumbaginifolia* Mo-co mutants described above represent a total of four complementation groups, *cnx*A, *cnx*B, *cnx*C and *cnx*D, as determined by complementation analysis in somatic hybrids (Xuan *et al.*, 1983; Dirks *et al.*, 1985). The relationship between the individual alleles is shown in Table 1. Fertile homozygous diploid plants have been regenerated from all *N. plumbaginifolia* CNX mutants, except CNX 27, and genetic analysis by sexual crosses confirms their assignment to the complementation groups (Dirks *et al.*, 1985). Individual segregation in the F_2 generation of the cross between CNX103 and CNX 20 suggests a linkage between these *cnx* loci. The *nia* gene and the *cnx* gene responsible for the mutant character of CNX 20 are either located on different chromosomes or the loci are far enough apart that they behave as though they are unlinked (Dirks *et al.*, 1985).

c) The Function of *Nicotiana cnx* Genes

Biochemical analysis of the *Nicotiana* Mo-co mutants has been carried out in an attempt to pinpoint the defect in Mo-co synthesis and thus determine the likely function of these genes.

*cnx*A alleles of *N. tabacum* produce a CR species the same size as wild-type NR, suggesting that these mutants contain a Mo-co which is defective in catalysis but which is still able to mediate dimerisation of flavohaemoprotein subunits (Mendel and Müller, 1979). Reconstitution of NR activity from *N. crassa nit*-1 extract by Mo-co from *N. tabacum* and *N. plumbaginifolia cnx*A alleles in the presence, but not the absence, of exogenous molybdate (20 mM) (Mendel *et al.*, 1981; Mendel *et al.*, 1986) and restoration of NR activity to all *cnx*A alleles tested by growth on medium supplemented with between 0.3 mM and 5 mM molybdate (Mendel *et al.*, 1981; Marton *et al.*, 1982a; Negrutiu *et al.*, 1983) indicates that the inactive Mo-co can be reactivated both *in vivo* and *in vitro* by unphysiologically high concentrations of molybdate. This suggests that the defect in the Mo-co of *cnx*A alleles is due to the absence of catalytically active molybdenum within the cofactor and that the molybdopterin moiety of the Mo-co is probably not impaired by the mutation. The defect is unlikely to be in molybdenum uptake since molybdenum levels of *cnx*A alleles of *N. tabacum* and *N. plumbaginifolia* are similar to wild type (Mendel *et al.*, 1984; Mendel *et al.*, 1986).

Extraction of cells of *N. tabacum* and *N. plumbaginifolia* with *cnx*A alleles in the presence of reduced glutathione, EDTA and 20 mM molybdate results in the reconstitution of NR activity (Mendel and Müller, 1985; Mendel *et al.*, 1986). The simplest interpretation of these data is that the *cnxA* gene product catalyses the liganding of molybdenum to the molybdopterin moiety of the cofactor. Mendel and Müller (1985) speculate that the reduced glutathione may supply thiol ligands required for molybdenum liganding, that EDTA may be involved in molybdenum chelation and that, in the presence of these compounds, the enzymatic insertion step can be bypassed in the mutant. The *cnx E* mutant of *A. nidulans* is phenotypically similar to *cnx*A (Arst *et al.*, 1970).

*cnx*B alleles of both *N. tabacum* and *N. plumbaginifolia*, unlike *cnxA* alleles, cannot be repaired *in vivo* by growth in the presence of unphysiologically high levels of molybdate (Buchanan and Wray, 1982; Marton *et al.*, 1982a; Negrutiu *et al.*, 1983). Since the Mo-co of *cnx*B alleles of *N.*

tabacum is capable of *in vivo* dimerisation of *cnx*B flavohaemoprotein subunits (Buchanan and Wray, 1982) and, in the presence of exogenous molybdate, is able to reconstitute NR activity from *N. crassa nit*-1 extract, we originally concluded that the defect in *cnx*B alleles was not in molybdopterin biosynthesis but in some undefined aspect of molybdenum processing (Mendel *et al.,* 1984). However, subsequent studies suggest that these properties are allele-specific rather than locus-specific since, in contrast, the Mo-co of the *N. plumbaginifolia cnx*B allele, NX24, can neither dimerise its flavohaemoprotein subunits nor reconstitute NR activity from *N. crassa nit*-1 extract (Mendel *et al.,* 1986). This suggests that the *cnx*B mutation lies within the molybdopterin biosynthetic pathway and that NX24 represents a null allele at this locus. This is supported by the observation that fluorescent oxidation products of molybdopterin (Kramer *et al.,* 1984) cannot be detected in NX24 extracts but are present in extracts of both *N. tabacum cnx*B alleles and wild type (Mendel *et al.,* 1986; R. Mendel, pers. comm.). Two types of allele have also been identified at the *A. nidulans cnx*B, *cnx*F, *cnx*G and *cnx*H loci. Dimerisation of flavohaemoprotein subunits occurs in only one of the two types (MacDonald *et al.,* 1974).

The *cnx*C allele, NX21, of *N. plumbaginifolia* possesses a Mo-co which is not only defective in catalysis (Marton *et al.,* 1982a) but also in dimerisation ability, since it cannot dimerise the flavohaemoprotein subunits present in NX21 nor reconstitute NR activity from *N. crassa nit*-1 extract. The amounts of fluorescent oxidation products of Mo-co are low, suggesting the defect is in molybdopterin synthesis (Mendel *et al.,* 1986).

The single *cnx*D allele, CNX103, of *N. plumbaginifolia* cannot be repaired *in vivo* by unphysiologically high levels of molybdate (Negrutiu *et al.,* 1983) but is otherwise uncharacterised.

d) *Petunia hybrida*

Three Mo-co defective lines representing two putative *cnx* loci have been identified by screening mutagenised *Petunia* cell suspensions for resistance to chlorate. Two of the lines (3 and 4) are allelic as determined by complementation in somatic hybrids and are molybdenum-repairable. The other line (line 2) is non-allelic to lines 3 and 4, and NR activity cannot be restored by growth on unphysiologically high levels of molybdate. All three lines lack NR and xanthine dehydrogenase activity (Steffen and Schieder, 1984).

e) *Hyoscyamus muticus*

Unlike all other selection procedures in tissue culture, which have relied on chlorate screening, mutants in *H. muticus* have been isolated using a non-selective total isolation procedure pioneered by Beadle and Tatum (1945) and subsequently applied successfully to the isolation of mutants in *Datura innoxia* (Savage *et al.,* 1979) and the moss *Physcomitrella* (Ashton and Cove, 1977).

Four clones, MA 2, I_2 D 12, VIC 2 and XIVE 9 were isolated on the basis of a growth requirement for casein hydrolysate and were subsequently shown to be both nitrate non-utilizers and resistant to 50 mM chlorate (Strauss *et al.*, 1981; Gebhardt *et al.*, 1981; Fankhauser *et al.*, 1984). All four lines are unable to grow on hypoxanthine as sole nitrogen source and lack xanthine dehydrogenase activity (Fankhauser *et al.*, 1984). Both MA 2 and I_2 D 12 are molybdate repairable *in vivo* (Fankhauser *et al.*, 1984).

Recently further Mo-co defective lines (O, Q, I, T and C) have been isolated on the basis of chlorate resistance and they, together with the lines discussed above, have been analysed in some detail by complementation in somatic hybrids and by *in vitro* complementation by co-homogenisation of mutant lines (H. Fankhauser, pers. comm.). These studies point to the existence of a total of four complementation groups.

The molybdenum-repairable lines fall into three complementation groups which have been designated A (MA-2, O and Q), B (I) and C (I_2 D 12). Mutant I of group B complements neither A nor C. This overlapping complementation pattern may signify the existence of two mutable loci (A and C) determining molybdenum ligand binding, with B being a double mutant. Alternatively all three mutants could be present within one locus with A and C complementing intragenically. This type of organisation has not been described in other plants but is similar to the *nit*-9 ABC locus of *Neurospora* (Dunn-Coleman, 1984a). Lines XIVE 9 plus T and C plus VIC 2 have been assigned to groups D and E respectively and are presently uncharacterised. Lack of complementation between MA 2 and a *N. tabacum cnx*A allele, Cnx 68, in interspecific somatic hybrids suggests that the complementation group A mutation is equivalent to the *cnx*A mutation of *N. tabacum* (Lazar *et al.*, 1983).

f) *Hordeum vulgare*

Molybdenum-cofactor defective whole plant barley mutants representing at least three gene loci have been identified. One of these, the *nar*2 locus, is represented by the allele *nar*2a (previously Az 34) which was isolated by screening individual seedling leaves for lowered *in vivo* NR activity (Warner *et al.*, 1977; Kleinhofs *et al.*, 1978). It is a poor source of functional Mo-co in the *in vitro* reconstitution of NR from a barley apoprotein source (Narayanan *et al.*, 1984) and since it still possesses low levels of xanthine dehydrogenase activity (Somers *et al.*, 1983) and NR activity (8 % of the wild-type level), it appears to be a leaky Mo-co mutant.

Unlike *nar*2 a, other Mo-co mutants have been isolated on the basis of chlorate resistance. Selection for resistance to 8 mM chlorate within an M_2 population of barley cv. Winer led to the isolation of two mutants (*Xno*18 and *Xno*19) which had low levels of both NR (Shumny and Tokarev, 1982) and xanthine dehydrogenase activity (Somers *et al.*, 1983). Like *nar*2 a, both *Xno* 18 and *Xno* 19 were poor sources of functional Mo-co (Narayanan *et al.*, 1984), and genetic analysis by sexual crossing showed them to be allelic to each other but not to either *nar*1 (Shumny and Tokarev, 1982) or *nar*2a (Kleinhofs *et al.*, 1983). *Xno*18 and *Xno*19 thus represent alleles of a further Mo-co gene locus which has been designated *nar*3.

A much more rigorous selection procedure than that used by either Kleinhofs and co-workers (Warner *et al.*, 1977; Kleinhofs *et al.*, 1978) or Shumny and Tokarev (1982) has led to the recovery of four further Mo-co mutants, R (= Rothamsted) 9401, 9201, 11301 and 12202 (Bright *et al.*, 1983; Wray *et al.*, 1985; B. Steven, pers. comm.). M_2 seedlings (mutagenised with azide in the M_1) were screened for 8 days with 10 mM chlorate and the selected plants were grown hydroponically with ammonium ions as sole nitrogen source. Maintenance of the plants to flowering was extremely difficult and several of the selections were self-infertile. However, three lines were recovered as heterozygotes after crossing with the cv. Golden Promise, whilst R12202 was fertile and was recovered in the homozygous state.

All mutants lacked xanthine dehydrogenase activity and NR activity (Bright *et al.*, 1983; B. Steven, unpublished). R9201 and R9401 are allelic whilst R11301 represents a separate locus. R12202 has not yet been tested (B. Steven, unpublished).

R9201/R9401 are allelic to *nar*2a (previously Az34) (A. Kleinhofs, pers. comm.) and thus represent two further alleles at this locus, *nar*2b and *nar*2c. R9401 lacks dimerised flavohaemoprotein subunits and its Mo-co is unable to reconstitute NR activity from *N. crassa nit*-1 extract (Wray *et al.*, 1985). The results obtained from R9201 are similar (B. Steven, unpublished). One may draw the tentative conclusion from this that these *nar*2 alleles possess a defective Mo-co which is unable to efficiently dimerise flavohaemoprotein subunits, and that the *nar*2 locus probably specifies a step in molybdopterin biosynthesis. Evidence from the *nar*2a allele, Az34, is consistent with this suggestion (Narayanan *et al.*, 1983; 1984).

The Mo-co of R11301 is capable of reconstituting NR activity from *nit*-1 extract in the presence of exogenous molybdate and thus this mutant is unlikely to be altered in molybdopterin synthesis (B. Steven, unpublished). The biochemical evidence supports the conclusion drawn from genetic analysis that R11301 is defective in a different step in Mo-co synthesis from R9401. This step is unlikely to be equivalent to that in *N. tabacum cnx*A mutants since neither NR nor xanthine dehydrogenase activity is restored to R11301 (nor to R9401 or R9201) by growth on unphysiologically high levels of molybdate (B. Steven, unpublished).

The R series of mutants discussed above lack NR activity and are conditional lethal mutants. They are able to grow on reduced forms of nitrogen but show no growth on nitrate as sole nitrogen source and eventually die (Bright *et al.*, 1983; B. Steven, unpublished). In contrast, all other barley mutants discussed above have residual NR activity and grow to different extents with nitrate as sole nitrogen source.

g) *Arabidopsis thaliana*

Arabidopsis thaliana was the first plant species in which mutations in nitrate assimilation were described (Oostindier-Braaksma, 1970; Oostindier-Braaksma and Feenstra, 1973). These whole plant mutants have NADH-NR activities ranging from 1—50 % of the wild-type level (Braaksma and Feenstra, 1982a). Two Mo-co mutants, B-25 (designated *rgn*) and B-73 (designated *cnx*), have been described which have the lowest NR activities of these mutants and also show the poorest growth on nitrate as sole nitrogen source. Most of the other mutants grow as well on nitrate as sole nitrogen source as the wild type.

The xanthine dehydrogenase activity of *rgn* is about 10—30 % of the wild type and the CR activity of the flavohaemoprotein subunits sediments at 4S, suggesting that the subunits are not dimerised (Braaksma and Feenstra, 1982b). This perhaps implies a defect in the molybdopterin moiety of the Mo-co, although the mutant still retains appreciable levels of xanthine dehydrogenase activity. *cnx* has only 10 % of the wild-type xanthine dehydrogenase activity but has dimerised flavohaemoprotein subunits. Since NR activity and ability to grow on nitrate can be increased by growth on unphysiologically high levels of molybdate, it is probable that *cnx* has a mutation equivalent to that of *N. tabacum cnx*A and that it is defective in the insertion of molybdenum into the Mo-co.

Revertants of *rgn* have been isolated which carry mutations in the same suppresser gene *(su)* which is unlinked to *rgn*. These reverse mutants can grow on nitrate as sole nitrogen source and are susceptible to chlorate (Braaksma and Feenstra, 1982c). The revertants have 0.4 to 1.5 times the wild-type NR activity, and the ability to assemble NR from its subunits, missing from *rgn*, has been restored. This appears to be the only case of suppressor genes so far reported in higher plants, although the nature of the suppressor mutation is unclear.

h) *Pisum sativum*

A leaky Mo-co mutant lacking xanthine dehydrogenase activity but retaining 20 % of the wild-type level of NADH-NR activity has been isolated by screening an M_2 population of pea seedlings for lowered *in vivo* NR activity (Warner *et al.*, 1982). The nitrate inducible CR activity and the nitrite reductase activity of the mutant, A300, were double the wild-type induced value probably reflecting the increased level of nitrate present. The mutation in A300 designated *nar*2, is monogenic, exhibits incomplete dominance and is unlinked to the *nar*1 locus. The E1 mutant has lowered NR and xanthine dehydrogenase levels and is probably a Mo-co mutant (Feenstra and Jacobsen, 1980; Jacobsen *et al.*, 1984).

D. Nitrite Reductase Mutations

No NiR apoprotein gene mutants have been identified so far from higher plants and it is, of course, unlikely that they could be identified by chlorate screening since, like wild-type plants, NiR-minus plants possess NR and would be expected to be sensitive to chlorate. The use of replica plating techniques has enabled NiR structural gene mutants to be isolated from *A. nidulans* (Pateman *et al.*, 1967) and *N. crassa* (Coddington, 1976) on the basis of their growth requirements, since such mutants are unable to utilise either nitrate or nitrite as sole nitrogen source but grow like wild type on ammonium. Replica plating of this type is difficult with higher plant cells but other approaches to the isolation of these mutants do exist. At the whole plant level one could screen for individuals within M_2 populations which accumulate nitrite after the application of nitrate, as has also been suggested by Dunn-Coleman *et al.* (1984). The major difficulty here is that nitrite is toxic, but it might be possible to devise screens based on short-term exposure of plants to low levels of nitrate which allow NiR-minus mutants to be recovered. At the callus level a total isolation procedure might be used. Such an approach has already yielded pantothenate auxotrophs in *Datura innoxia* (Savage *et al.*, 1979), and lines auxotrophic for histidine, tryptophan, leucine or nicotinamide, as well as NR-minus nitrate non-utilizers, in *Hyoscyamus muticus* (Gebhardt *et al.*, 1981; see also Chapter 10, this Volume).

E. Regulatory Alterations in Nitrate Assimilation

Induction of NR and NiR activity by nitrate in *A. nidulans* is mediated by the product of the *nir*A locus which acts in a positive regulatory fashion such that it is required for the expression of the *nia*D and *nii*A apoprotein genes and is active only in the presence of nitrate (Pateman *et al.*, 1967; Cove, 1979). *nir*A$^-$ strains grown in either the presence or absence of nitrate produce low levels of both NR and NiR equivalent to those of the wild-type strain grown in the absence of nitrate. *Nir*A^c strains have constitutive synthesis of both NR and NiR. Repression by ammonium or glutamine (nitrogen metabolite repression) is mediated by the *are*A gene product (Arst and Cove, 1973). A locus performing the same or similar function has been identified also in *N. crassa* (Tomsett and Garrett, 1980; Tomsett *et al.*, 1981).

Nitrate reductase regulatory gene mutants have not been isolated, or at least have not been recognised, in higher plants so far. The only possible exception to this generalisation may be the nr_1 mutants of soybean (Nelson *et al.*, 1984). There is, however, some evidence that mutations within the apoprotein gene, or within Mo-co genes, lead to regulatory perturbations which may have their parallels in both *A. nidulans* and *N. crassa.*

In two *cnx*A alleles of *N. tabacum* Cnx 68/2 (a subline of Cnx 68), and Cnx 101 (Mendel and Müller, 1979), NR apoprotein is produced constitutively, as measured by the presence and amount of a 7.7 S CR species in nitrate-less cell culture and, in the case of Cnx 68/2 and Cnx 109, NiR activity is also produced constitutively. NiR levels were still inducible in Cnx 101 (and also in Nia 102) but the activity present, both under uninduced and induced conditions, was double that of the wild-type values. NR-associated CR activity was also produced constitutively in *cnx*B alleles (Mendel *et al.*, 1984). Whether the constitutive levels of NR-associated CR and of NiR activity are a consequence of the Mo-co mutation (or the *nia* mutation in the case of Nia 102) is not clear. However in *A. nidulans* some mutations causing loss of a functional NR molecule lead to constitutive synthesis of both NR and NiR and suggest an autoregulatory role for functional NR molecules in the regulation of synthesis of both NR and NiR (Pateman *et al.*, 1967; Cove and Pateman, 1969).

Cove (1979) has proposed a model to accommodate these findings. According to this model, in the absence of nitrate NR interacts with the *nir*A gene product converting it into a form which does not allow the *nia*D and *nii*A genes to be expressed. However, in the presence of nitrate NR binds to its substrate and can no longer interact with the *nir*A gene product, thus allowing it to mediate expression of the *nia*D and *nii*A genes. Many mutants possessing defective NR molecules, due to mutation either in the apoprotein gene or a Mo-co gene, would no longer be able to inactivate the *nir*A gene product and would show constitutive synthesis of both NR and NiR. Those mutants which possess wild-type regulation have catalytically defective NR molecules which can still interact with the *nir*A gene product in the wild-type manner.

Whilst the data from *N. tabacum* provides some evidence for an autoregulatory role for NR, there is no *a priori* reason to suppose that the regulatory networks involved in controlling NR and NiR gene expression in higher plants are the same as those operating in lower eukaryotes and there is in fact little corroborative evidence from other sources. Thus NR-associated CR and/or NiR levels are not constitutive in *nar*1 and *nar*2 alleles of barley cv. Steptoe (Kleinhofs *et al.*, 1980), in the Mo-co mutant R 9401 of barley cv. Golden Promise (Bright *et al.*, 1983), in the nr_l mutant of soybean (Nelson *et al.*, 1984) nor in the *nar*1 and *nar*2 mutants of pea (Warner *et al.*, 1982). However NR-associated CR and/or NiR levels have not been assayed from uninduced tissue of the large number of other mutants so far isolated.

There is evidence to suggest a link between the synthesis of functional NR molecules and of the Mo-co. The Mo-co of *N. tabacum* (Mendel *et al.*, 1982) and of barley (Narayanan *et al.*, 1984; Mendel *et al.*, 1985) is nitrate inducible. Some *N. tabacum nia* alleles, however, have a Mo-co which is no longer nitrate-inducible whilst others, which retain residual benzyl viologen-NR activity, still possess a nitrate-inducible Mo-co (Mendel *et al.*, 1982). The authors have speculated that regulation is normal in the latter group due to the formation of an assembled, but catalytically-defective, NR holoenzyme which is still able to bring about wild-type regulation, whilst the defect in the apoprotein of the former class does not allow assembly and wild-type regulation is lost. This hypothesis suggests that the NR holoenzyme acts in a positive way to regulate Mo-co synthesis. This is supported by the observation that Mo-co synthesis in *cnx*A (Mendel *et al.*, 1982) and *cnx*B (Mendel *et al.*, 1984) alleles of *N. tabacum* which produce an inactive NR holoenzyme constitutively is also constitutive. The possibility that this phenomenon exists in other species appears not to have been examined yet.

F. Conclusions

A great number of mutants altered in nitrate reduction (due either to *nia* or *cnx* mutants) have been isolated usually on the basis of chlorate resistance (Table 1). Selection pressure appears to play a role in the type of mutant obtained at the whole plant level. Rigorous screening of M_2 barley plants by Bright *et al.* (1983) (10 mM chlorate for nine days) led to the recovery of only Mo-co mutants (B. Steven, unpublished), whilst the least rigorous selection procedure (direct selection of barley plants showing decreased levels of NR (Kleinhofs *et al.*, 1980) recovered many more apoprotein gene mutants than Mo-co mutants.

At least part of the reason may be due to the fact that barley (or at least the cultivar Steptoe) possesses a second NR apoprotein gene (encoding the NADPH-linked enzyme) which is expressed only in the presence of a defect in the NADH-NR apoprotein gene (Dailey *et al.*, 1982a, b). NADH-NR apoprotein gene mutants possessing a functional NADPH-NR apoprotein gene are still chlorate sensitive (Warner and Kleinhofs, 1981). Whilst double apoprotein gene mutants would be expected to be chlorate resistant, their presence in M_2 populations is likely to be extremely low. Such mutants isolated at the cell culture level in *N. tabacum* occur at a frequency of 10^{-7} (Müller, 1983). This argument suggests that the Mo-co mutants isolated by Bright *et al.* (1983) are the result of single gene mutations.

This does not explain, however, why so few Mo-co mutants were recovered by Kleinhofs *et al.* (1980) nor why they are often poorly represented when selection is done at the cell/protoplast level (Müller and Grafe, 1978; Marton *et al.*, 1982a; Negrutiu *et al.*, 1983) since there are likely to be at least six Mo-co gene loci compared to only one or two structural gene loci. One possible explanation is that *nia* loci are more prone to mutagenesis than are *cnx* loci. Alternatively it may be related to the underlying mechanism involved in chlorate resistance. Cove (1976b) has shown that in *Aspergillus* the ratio of *nia* to *cnx* mutants is influenced by the nitrogen source on which selection is made. It is clear, however, that a number of different Mo-co gene mutants have been isolated in higher plants on the basis of chlorate resistance and it is likely that all the *cnx* loci described in *Aspergillus* will eventually be shown to have their counterpart in the plant genome.

Nitrate uptake (Doddema and Telkamp, 1979) and regulatory gene mutations (Ryan *et al.*, 1983; Nelson *et al.*, 1984) are poorly represented, if at all, amongst the nitrate assimilation mutants so far identified. It is unfortunate that the large number of chlorate resistant clones which retain NR activity (Müller and Grafe, 1978; Murphy and Imbrie, 1981; Marton *et al.*, 1982a; Buchanan and Wray, 1982) have not been examined since, as discussed above, it could be that some of these lines carry either regulatory or nitrate uptake mutations. Buchanan and Wray (1982) have isolated a cell line, 041, on the basis of chlorate resistance which retains NR activity. Growth kinetics on nitrate as sole nitrogen source are different from the wild type as are nitrate and NR levels, although growth of 041 and wild-type cells on either glutamine or ammonium is the same (Qureshi *et al.*, 1982; J. A. Qureshi and J. L. Wray, unpublished). It is clear that nitrate assimilation is altered in the 041 cell line although whether this is due to uptake or regulatory perturbations is uncertain.

With only a few exceptions mutations have been induced with base substitution mutagens. It may be that other types of mutagenesis would yield other types of mutations, such as regulatory mutations. Insertion mutagenesis is a possible technique which might be carried out either with plant controlling elements or even with the T-DNA region of the *Agrobacterium tumefaciens* Ti plasmid. Deletion mutants might be present amongst NR-minus progeny regenerated from tissue culture (Larkin and Scowcroft, 1981).

One disadvantage of selecting mutants in cell culture is that it may not be possible to regenerate whole, fertile plants. This makes it impossible to study the mutation by conventional genetic analysis. This is unfortunate since it has been argued that the most satisfactory evidence that a mutational event has in fact occurred relies on the regeneration of fertile plants from the cell lines and transmission of the variant trait in sexual crosses and that until this has been carried out the isolates should be described as "phenotypic variants" or "cell lines" rather than as mutants (Maliga, 1976). Inability to regenerate mutants also precludes the possibility of studying the effect of the mutation on the physiology and biochemistry of the intact plant and limits their usefulness in somatic hybridization and plant transformation.

The evidence summarized in Table 1 suggests that it might be more difficult to regenerate *cnx* than *nia* mutant cell lines to fertile plants but it is at present not clear whether this is directly related to the *cnx* genotype.

IV. Applied Aspects

A. Somatic Hybridisation

Glimelius *et al.* (1978) were the first to show that somatic hybrids resulting from PEG fusion of protoplasts derived from *N. tabacum* L. var. Gatersleben NR-minus *nia* and *cnx* cell lines could be recovered on the basis of their growth on medium containing nitrate as sole nitrogen source. The possibilities of reversion or of cross-feeding were excluded as possible explanations for the occurrence of nitrate-utilising colonies. In reconstruction experiments one wild-type colony amongst 4×10^{-4} *nia* colonies could be selected on nitrate medium at nearly 100% efficiency (Pental *et al.*, 1982). Reversion of Nia30 occurs at rates of about 10^{-6} (R. Grafe, pers. comm.). These results suggest that the NR-minus phenotype might be useful as a selection marker in the identification of somatic hybrids and perhaps also in plant cell transformation studies.

Complementation in somatic hybrids via protoplast fusion has been very useful in determining allelism of NR mutant cell lines, and intraspecific, interspecific and intergeneric complementation has been reported. Biasini and Marton (1985) have described a rapid assay for genetic complementation of NR deficiency via bulk protoplast fusion. Several workers have reported the use of *cnx* mutant protoplasts as one partner in somatic hybrid selection schemes (Wallin *et al.*, 1979; Glimelius *et al.*, 1981; Glimelius and Bonnett, 1981; Hein *et al.*, 1983). Hamill *et al.* (1983) have constructed an NR-minus streptomycin resistant double mutant of *N. tabacum*. On selective medium containing nitrate as sole nitrogen source and also streptomycin it is likely that only nuclear/cytoplasmic recombinants could grow as green colonies when the NR-minus, SR-plus protoplasts are fused with wild-type protoplasts of other species. Thus the double mutant might be useful in constructing new nuclear/cytoplasmic gene combinations.

Although it was hoped that somatic hybridization by protoplast fusion would lead to unique hybrid plants impossible to obtain by conventional sexual hybridization, limited success has been achieved only with closely related species (Schieder and Vasil, 1980). This is due largely to somatic incompatibility problems between distantly related or unrelated genomes and is manifested in the failure of hybrid development and in the elimination of chromosomes from both cell hybrids and regenerants. Attempts have therefore been made to bypass somatic incompatibility barriers between distantly related plants by the transfer of limited amounts of genetic information. As Shephard *et al.* (1983) have argued, introgression of genes from diverse alien species could significantly expand germplasm pools for such characters as pest or stress resistance provided that the introduced genes were expressed and were capable of being manipulated by breeding techniques.

Gupta *et al.* (1982) have described the transfer of nuclear genes from *Physalis minima* and *Datura innoxia* to *N. tabacum*. When X-irradiated, mitotically inactive protoplasts of *P. minima* or *D. innoxia* were fused with Cnx68 mutant protoplasts, cell lines were isolated able to grow on nitrate as sole nitrogen source. Restoration of NR and xanthine dehydrogenase activity to the Cnx68 mutant is probably due to transfer and expression of the equivalent genes from the X-irradiated fusion partner. Whilst this procedure might have application for introducing desirable characters from diverse genomes, it is clearly a relatively crude method of gene transfer which does not allow the investigator total control over the genes being transferred. Much more precise procedures for gene transfer exist and some of these are briefly discussed below.

B. Cloning of Nitrate Assimilation Genes

Molecular cloning of a higher plant NR apoprotein gene or of a Mo-co gene has not yet been accomplished although there is intense activity in this area. The approaches which might be used to clone Mo-co genes are more limited than those for apoprotein gene cloning since we have no information on any higher plant Mo-co gene-product. This means that techniques for the isolation of genes which depend either on the use of antibody probes or of synthetic oligonucleotide probes cannot be employed.

Several general approaches to either apoprotein or Mo-co gene cloning are available. One possibility is the use of "gene-tagging". This has been used to clone an NR gene from the non-nitrogen fixing cyanobacterium *Anacystis nidulans.* Eight mutants representing three loci designated *nar*A, *nar*B and *nar*C were isolated by insertion mutagenesis of *A. nidulans* with the transposon Tn901, which encodes ampicillin resistance (Kuhlmeier *et al.,* 1984). Whether these are apoprotein gene or Mo-co gene loci is unknown but the mutants could not be repaired by growth on medium supplemented with 1 mM molybdate. The *nar*B gene was cloned in the following way. The DNA of a *nar*B mutant allele, Nar6, was restricted with *Eco*R1 which does not cleave in the transposon sequence. The *Eco*R1 fragments were then cloned into the *E. coli* vector pACYC184 and, after transformation, selection was made for ampicillin-resistant colonies. The plasmid from one such colony, designated pNRT63, was able to transform the Nar6 allele to the wild-type phenotype and contained a 20 kb *Eco*R1 insert. The transposon was located on a 9.2 kb *Sal*1 fragment. pNRT63 was then used as a probe against a wild-type genomic cosmid library. Several cosmids were isolated which were able to transform the Nar6 mutant to the wild-type phenotype. One of these, pNR63, was found to contain a *Sal*1 fragment with the expected size of 5.1 kb, i. e. the size of the pNRT631 *Sal*1 fragment minus the size of the transposon. Cloning of the 5.1 kb *Sal*1 fragment into pACYC184 gave pNR631, which also transformed the Nar6 allele to the wild-type phenotype. The insert of pNR631 hybridised to a unique chromosomal fragment, suggesting that the *nar*B gene is represented by a single-copy.

A possible candidate "tag" for higher plants is the controlling element *Mu*1 found only in Robertson's *Mutator* maize lines (Robertson, 1978; Bennetzen, 1984). *Mu*1 produces mutations at most loci investigated at rates near 10^{-4}, 30—50 fold higher than in non-*Mutator* lines (Robertson, 1978). Alcohol dehydrogenase mutants have been induced by insertion of *Mu*1 into the *adh* gene (Strommer *et al.*, 1982; Freeling *et al.*, 1982) and it is not unreasonable to suppose that NR-minus mutants could also be generated. DNA isolated from such mutants could be probed with the cloned *Mu*1 sequence and plant genomic sequences carrying a *Mu*1 insertion isolated. However, since *Mutator* lines carry 15—40 copies of *Mu*1 per diploid genome (Bennetzen, 1984) *Mu*1 will be inserted in numerous other places besides the NR structural gene or a Mo-co gene. This would considerably complicate the identification of the genomic NR-related sequence carrying a *Mu*1 insertion. Further complication is introduced by the fact that *Mu*1 insertions are unstable. Recently, however, the *a*1 locus of *Zea mays* has been cloned using *Mu*1 and other controlling elements, *Spm*-18 and *En*1, as gene tags (O'Reilly *et al.,* 1985). "Tagging" with the T-DNA of the Ti plasmid of *Agrobacterium tumefaciens* might be useful in *Nicotiana* species and other dicotyledonous plants.

A further general approach is by functional complementation of plant

NR-minus mutants but this requires an appropriate system for the transfer and expression of plant DNA sequences and for the identification of transformants. Since most of the plant NR-minus mutants are in *Nicotiana* species, a system based on the Ti plasmid of *Agrobacterium tumefaciens* might be suitable (reviewed in Schell and van Montagu, 1983). A plant genomic library might be constructed in T-DNA from which oncogenic functions have been deleted (Zambryski *et al.,* 1983) and which carries an antibiotic resistance marker (Herrera-Estrella *et al.,* 1983; Fraley *et al.,* 1983). Initial screening for transformants would be on the basis of antibiotic resistance followed by identification of those transformants which were able to utilize nitrate as sole nitrogen source. Due to the nature of this system, complementing DNA sequences would be integrated into the host chromosome and appropriate procedures would be needed to rescue them.

Over the past three years transformation systems have been developed for *Aspergillus nidulans* (Ballance *et al.,* 1983; Tilburn *et al.,* 1984; Johnstone *et al.,* 1985) which make use of a cloned gene on appropriate vectors to functionally complement the corresponding mutant *A. nidulans* recipient to the wild-type phenotype. Transformation is by integration but spontaneous excision of integrated plasmids is frequent enough to allow their re-isolation in *E. coli* and has permitted the molecular cloning of the *A. nidulans* developmental gene, *brl*A (Johnstone *et al.,* 1985). As indicated above, a wealth of nitrate assimilation mutants is available in *A. nidulans* and it is likely that one or other of these transformation systems will shortly allow the cloning of *A. nidulans* nitrate assimilation genes.

Cloning of Mo-co genes from *E. coli* and from *N. crassa* has in fact already been accomplished by functional complementation of equivalent *E. coli* Mo-co mutants. The Mo-co is highly conserved between eukaryotes and prokaryotes and the *chl*A, *chl*B, *chl*D, *chl*E and *chl*G loci have been implicated in the synthesis of the *E. coli* Mo-co (Amy, 1981). Giordano *et al.* (1981) first cloned the *E. coli chl*A gene and, more recently, Kleinhofs *et al.* (1983) and Taylor *et al.* (1983) have shown that the *chl*A locus is divisible into two components, one of which was designated *chl*M (and which could complement the *chl*A mutant, SA493) and *chl*N (which could complement the *chl*A mutant, JP382). A 1.9 kb *Bcl*1 fragment in pBR322 complemented the mutant SA493 when inserted in either orientation suggesting that the entire gene was present and was active from its own promoter. *In vivo* transcription-translation of the 1.9 kb *Bcl*1 fragment in maxicells and minicells identified the *chl*M gene product as a ca. 15,000 mol. wt. polypeptide (Kleinhofs *et al.,* 1983; Taylor *et al.,* 1983).

Kleinhofs *et al.* (1983) chose to clone the *chl*A gene since they believed it was equivalent to the mutated gene in the *N. tabacum cnx*A allele, Cnx68, (Müller and Grafe, 1978) and the *Hyoscyamus muticus* mutant MA-2 (Strauss *et al.,* 1981). Extracts from *E. coli* wild type and *chl*B, *chl*C, *chl*E and *chl*G, but not *chl*A, mutants were able to reconstitute NR activity in each of the plant mutant extracts. Dunn-Coleman (1984b) has recently argued that the *chl*D locus rather than the *chl*A locus is equivalent in function to the *N. tabacum cnx*A locus since *E. coli chl*D mutants, like *N.*

*tabacum chl*A mutants, can be repaired by growth on unphysiologically high levels of molybdate (Glaser and DeMoss, 1971).

Several fungal genes have been cloned on the basis of their ability to functionally complement equivalent *E. coli* mutations. From *N. crassa* these include the *qa*2 gene encoding catabolic dehydrogenase (Vapnek *et al.*, 1977), the *pyr*4 gene encoding orotidine 5′-phosphate carboxylase (Buxton and Radford, 1983), and the trifunctional *trp*1 gene (Schechtman and Yanofsky, 1983). Kinghorn and Hawkins (1982) have cloned the synthetic dehydrogenase function of the *A. nidulans arom* cluster. Expression of these genes in *E. coli* is likely to be due to fortuitous recognition of accompanying fungal promoters. It is unlikely that this approach will be of use as a general procedure for fungal gene cloning.

However, Dunn-Coleman (1984b) has taken a similar approach to clone a *N. crassa* Mo-co gene. He constructed an *N. crassa* genomic library in pRK9, a derivative of pBR332, and used it to transform two *E. coli chl*D Mo-co mutations (*chl*D and *chl*D::Mu) to wild type. Three independent transformants examined had plasmids carrying an identical 4.2 kb insert. Southern blot analysis with one of the transforming plasmids, pMoCo, 1:4, showed hybridization to a single band of *N. crassa* genomic DNA. When pMo-co, 1:4 was used to transform various *E. coli* nitrate reductase mutants (*chl*A, *chl*B, *chl*C, *chl*D, *chl*E, *chl*G and *chl*M) the pMo-Co plasmid was capable of restoring NR activity to only the *chl*D mutant. Mo-co from *chl*D; pMo-co cell free extracts was capable of reconstituting more NADPH-NR activity from extracts of the *N. crassa* mutants *nit*-1, *nit*-7 and *nit*-8 than was cell-free extract from the original *E. coli chl*D mutant. The *chl*D mutant of *E. coli* can be repaired by growth on unphysiologically high levels of molybdate like mutants of the *nit*-9 locus of *N. crassa* (*nit*-9 A, *nit*-9 B and *nit*-9 C) (Dunn-Coleman, 1984a). This suggests that the 4.9 kb insert of pMp-co, 1:4 might carry sequences equivalent to the *nit*-9 locus of *N. crassa.*

Other approaches which might be used for apoprotein gene cloning exist. Differential screening of a cDNA library prepared from polyA$^+$ RNA enriched for NR mRNA might be employed. Colonies showing a stronger hybridization signal with a 'nitrate-induced' cDNA probe than an 'uninduced' cDNA probe would be selected as putative NR cDNA clones. Identity might be confirmed by hybrid-select-translation (Maniatis *et al.*, 1982) and immune precipitation with NR antiserum. A problem with this approach is that the abundance of the NR mRNA is likely to be very low (0.005—0.01 % of the total polyA$^+$ RNA) and its large size (greater than 3 kb) suggests it will be unstable. It is, however, encouraging that molecular cloning of oat phytochrome cDNA clones has recently been accomplished using an equivalent approach (Hershey *et al.*, 1984) since its apoprotein is approximately the same size as that of NR (Vierstra and Quail, 1983) and its mRNA constitutes only 0.005 % of the total polyA$^+$ RNA (Gottman and Schafer, 1983). Hurlburt and Garrett (1985) have recently isolated eight nitrate-inducible DNA sequences in *Aspergillus* using this differential approach.

Two techniques have recently been applied to the identification of cDNA clones derived from mRNA species present in low abundance. One of these, which may be used if monospecific antiserum to NR is available, is the immunological screening of expression libraries constructed in either plasmid, for example pUC8 (Helfman *et al.*, 1983), or phage cloning vehicles. Several workers have in fact reported the production of polyclonal antisera to higher plant NR (Graf *et al.*, 1975; Smarrelli and Campbell, 1981; Kuo *et al.*, 1981) whilst Notton *et al.*, (1985) have prepared monoclonal antibodies to the spinach enzyme.

An example of a phage cloning vehicle is the lambdoid expression vector, λ gtll (Young and Davies, 1983a). This vector permits the insertion of cDNA or genomic DNA into a unique *Eco*R1 cleavage site located within *lacZ*, 53bp upstream of the β-galactosidase termination codon. Phage containing inserts can be induced to produce an inactive β-galactosidase protein fused to antigenic protein specified by the foreign DNA, with the *lac* inducer IPTG. Screening with specific antibody probes allows the detection of antigen in populations of up to 10^6 lysogens per 82 mm nitrocellulose filter and thus this system should be ideal for the identification of NR cDNA clones. The yeast RNA polymerase II gene (Young and Davies, 1983b), human β-glucocerebrosidase gene (Ginns *et al.*, 1984) and the gene encoding the γ-subunit of bovine retinal GTPase (Yatsunami *et al.*, 1985) have been cloned using this approach.

The other technique requires a knowledge of the partial or complete amino acid sequence of the NR protein. A mixture of synthetic oligonucleotides which represents all possible coding combinations for a small portion of the amino acid sequence can be used directly as a probe for the detection of unique genes in Southern blot filter hybridization and in colony and bacteriophage library screening (Wallace *et al.*, 1981; Suggs *et al.*, 1981). Alternatively the oligonucleotides can be used indirectly as primers for cDNA synthesis (Sood *et al.*, 1981).

The abundance of NiR mRNA is likely to be of the same order as that of the NR mRNA, suggesting that immunological screening of expression libraries or the use of synthetic oligonucleotide probes might be the best approach for cloning the plant NiR apoprotein gene. NiR has been purified from several plant species (Hucklesby *et al.*, 1976; Gupta and Beevers, 1984; Small and Gray, 1984) and antibodies produced. Cloning strategies for plant nitrate uptake genes are very limited since the protein products of the genes have not been identified and cloning by functional complementation is confounded by the very few and poorly characterised nitrate uptake mutants available in higher plants and *Aspergillus*.

An approach which may be utilised here, and which may also be applicable to NR and NiR, is the use of equivalent cloned genes from other organisms as heterologous DNA probes. Clearly for this approach to be successful there has to be considerable nucleotide homology between the DNA sequences involved. Functional- or amino acid sequence-homology of the gene products is not sufficient. Shah *et al.* (1983) have isolated actin genes from genomic libraries of two highly divergent plants, maize and soybean, using a *Dictyostelium* actin gene as a heterologous probe. The nitrate assimilation genes of *Aspergillus* are linked and ordered in the sequence *crn*A *nii*A *nia*D (Tomsett and Cove, 1979; Brownlee and Arst, 1983). Some genomic sequences which functionally complement *Aspergillus nia*D or *nii*A mutants might be expected to carry the *crn*A gene. If not it would be relatively simple to fish-out the *crn*A sequence from an *Aspergillus* genomic library using either the cloned *nii*A or *nia*D gene as a probe.

Cloned nitrate assimilation genes will allow a path to be cut through the jungle of physiological data and permit studies on gene structure, organisation, and regulation in response to environmental parameters. The possibility that the heterologous DNA probe approach may permit the identification of a plant gene equivalent to the *Aspergillus* regulatory gene, *nir*A, is particularly exciting.

C. Plant Gene Transfer Systems

Gene transfer systems based on the Ti plasmid of *Agrobacterium tumefaciens,* from which oncogenic functions have been deleted, have been described recently by several groups. These systems allow the transformation of either protoplasts (De Block *et al.,* 1984; An *et al.,* 1985) or leaf disc cells (Horsch *et al.,* 1985). Transformants are identified on the basis of antibiotic resistance markers carried on the T-region. Regenerated plants express these resistance genes and transmit them to their progeny in a Mendelian fashion. Vectors of this type allow the transformation of plant species which can be infected by *Agrobacterium* (tobacco, petunia and tomato in the cases above).

A novel gene transfer vehicle based on an *E. coli* plasmid and containing an antibotic resistance gene marker under the control of the CMV gene VI expression signal has recently been described (Paszowski *et al.,* 1984). This allows *Agrobacterium*-independent direct gene transfer into plants and transmission of the antibiotic gene to the sexual offspring in a Mendelian fashion. Systems of these types should allow the transfer of NR structural genes and Mo-co genes to plants. One might also envisage NR apoprotein genes or Mo-co genes, instead of antibiotic resistance genes, as selectable markers in vector-mediated gene transfer to either *nia*- or *cnx*-type recipients. Transformants would regain the ability to utilise nitrate as sole nitrogen source and could be selected as nitrate-utilizers.

The Ti plasmid system has been extremely useful in the study of regulated plant genes and in promoter mapping. A bacterial chloramphenicol acetyltransferase gene has been linked to the 5′ portion of *Pisum sativum* ribulose 1,5-bisphosphate carboxylase small subunit gene and introduced into *N. tabacum* via the Ti system. The chimaeric gene is light inducible (Herrera-Estrella *et al.,* 1984). Introduction of derivatives of the gene into *Petunia* cells indicated that a conserved 33-base pair sequence close to the TATA box of the gene is sufficient to confer light inducibility (Morelli *et al.,* 1985). Studies of this type should allow the identification of nitrate regulatory sequences when cloned NR and NiR apoprotein genes become available.

D. Whole Plant Studies

Studies at the physiological and biochemical level show that nitrate uptake and nitrate flux, as well as NR activity, correlate to some extent with accumulation of reduced nitrogen by the plant (Reed and Hageman, 1980a, b; Shaner and Boyer, 1976). Partitioning of nitrate between storage and metabolic pools (Ferrari *et al.*, 1973) and between different plant parts, and the ability of the plant to mobilise stored nitrate and redistribute reduced nitrogen to the grain (Dalling *et al.*, 1975) must also be related in some way to accumulation of grain reduced nitrogen. However, since the ability of the plant to acquire nitrate from the environment influences both nitrate flux and also NR activity, it is clear that nitrate uptake is a major point of control in the assimilatory process.

Estimates show that more than two-thirds of the edible dry matter and half the protein produced in the world are contributed by the cereals (Evans, 1975). The increased use of fertiliser nitrogen has been a significant factor in increasing productivity of these crops. However, concerns about the high cost of energy required to produce fertiliser nitrogen and the effect of nitrate pollution on the quality of the environment (Foster *et al.*, 1982; Magee, 1982; Wilkinson and Greene, 1982) calls for more efficient management and utilization of fertiliser nitrogen. These objectives can be approached at the plant level by improving the efficiency of nitrate utilization through an understanding of the molecular processes involved in the acquisition and assimilation of nitrate nitrogen and the factors influencing these processes in the plant. Our present inability to analyse nitrate uptake in molecular terms is particularly disappointing.

Grain yield, protein yield, grain reduced nitrogen or plant reduced nitrogen have been shown to be related to NR levels in a wide variety of crop-plants and some workers, for example Johnson *et al.* (1976), have suggested that NR levels might be useful as a predictive indicator of grain yield. However, the correlation values are often low and grain yield and plant reduced nitrogen were little affected in NR-deficient mutants of barley (Oh *et al.*, 1980; Warner and Kleinhofs, 1981) possessing less than 10% of wild-type NR activity. The ability to introduce multiple copies of cloned NR apoprotein genes into crop plants would allow the effect of NR gene dosage to be analyzed in the same genetic background and perhaps finally resolve this question. Of relevance here is the observation of Müller (1983) that in tobacco the level of NR activity is independent of the number of *nia*$^+$ genes, indicating complete compensation of gene dosage effects by regulatory mechanisms.

An alternate approach might be to introduce NR apoprotein genes which have more efficient promoters or which are kinetically more efficient either as the result of natural variation or due to site-directed mutagenesis. In this respect the NR of *Erythrina senegalensis* is of some interest. Nitrate reductase levels in this tropical leguminous tree are some 100-fold those of most crop plants (Stewart and Orebamjo, 1979). Characterisation of the cloned gene would allow the reasons for this to be determined.

Field-grown soybeans reportedly obtain only 25 % of their nitrogen from atmospheric dinitrogen, the remainder being derived from soil reserves of nitrate and ammonium (Hardy *et al.*, 1968). Nitrate retards nodulation and nodule development (Munns, 1968), causes nodule senescence (Chen and Phillips, 1977) and reduces dinitrogen fixation. Understanding how nitrate brings about these effects is important for devising strategies for maximum dinitrogen fixation. The responses could be due to nitrate itself or due to the products of the reduction of nitrate by the plant or bacteroid NR. No decrease in inhibitory effects of nitrate were seen when legumes were inoculated with NR deficient mutants of *Rhizobium* suggesting that bacteroid NR was not involved (Gibson and Pagan, 1977; Manhart and Wong, 1980). Attempts to distinguish between the inhibitory effects of nitrate and of its reduction products by treating soybeans with tungstate were complicated by the fact that both NR activity and nitrate uptake were inhibited (Harper and Nicholas, 1978).

Feenstra *et al.* (1982) have used the E1 mutant of pea, which possesses 20 % of the wild-type activity but has normal nitrate uptake, in an attempt to distinguish between the two possibilities. Whilst ammonium ions inhibited acetylene reduction to the same extent in both the wild type and the E 1 mutant the inhibitory effect of nitrate was much reduced suggesting that the effects are due in part to the product(s) of nitrate reduction. This is unlikely to be nitrite (Streeter, 1985).

More recently, Jacobsen and Feenstra (1984) have isolated a mutant from M_2 populations of pea seedlings on the basis of its ability, unlike wild-type plants, to nodulate abundantly in the presence of 15 mM nitrate. This line is unlike other 'highly nodulating' lines carrying the nod_1 and nod_2 genes which produce abundant nodules only in the absence of nitrate. The mutant character is monogenic and recessive and the new gene has been designated nod_3. The biochemical basis of the mutation is unknown. The results of studies with a pea line carrying both nod_1 and nod_2 shows that nodule mass and seed yield are positively correlated and suggests that the mutant character of nod_3 may be of economic significance.

The ureides allantoin and allantoic acid are important transport forms of nitrogen in some legumes such as *Phaseolus vulgaris* and soybean where they can make up to 80 % of the total sap nitrogen in nodulated plants (McClure and Israel, 1979). Synthesis of ureides within the nodules is believed to be via purine breakdown and involves the xanthine dehydrogenase-catalysed conversion of xanthine to uric acid (Triplett, 1985) and its further metabolism to allantoin and allantoic acid by uricase and allantoinase respectively (Thomas and Schrader, 1981; Reynolds *et al.*, 1982). Use of Mo-co defective mutants of soybean would be invaluable in confirming this proposed pathway and in determining whether nodule xanthine dehydrogenase is a plant or *Rhizobium* encoded function. If the former, then these mutants should provide a much better tool than the xanthine dehydrogenase inhibitor, allopurinol, (Fujihara and Yamaguchi, 1978) for studying the long-term role of ureides in the nitrogen economy of the plant and the adaptation in nitrogen metabolism and transport which might occur as a result of the inability to synthesise them. However, Mo-co mutants have not so far been reported from ureide-transporting legumes. (Chapter 6 in this volume deals further with genetic approaches to nodulation and nitrogen fixation in legumes.)

Acknowledgements

I thank all those colleagues who were prepared to provide me with unpublished details of their work and, further, Dr. C. Deane-Drummond, Dr. R.-R. Mendel and Dr. A. J. Müller for useful discussions. This review was written whilst I was in receipt of research grant AG 49/23 from the United Kingdom Agricultural and Food Research Council. I am indebted to Margaret Wilson for typing the manuscript.

V. References

Åberg, B., 1974: On the mechanism of the toxic action of chlorates and some related substances upon young wheat plants. Kungl. Lantbrukshogkolans Ann. **15,** 37—107.

Amy, N. K., 1981: Identification of the molybdenum cofactor in chlorate-resistant mutants of *Escherichia coli.* J. Bacteriol. **148,** 274—282.

Amy, N. K., Rajagopalan, K. V., 1979: Characterization of molybdenum cofactor from *Escherichia coli.* J. Bacteriol. **140,** 114—124.

An, G., Watson, B. D., Stachel, S., Gordon, M. P., Nester, E. W., 1985: New cloning vehicles for transformation of higher plants. EMBO J. **4,** 277—284.

Arst, H. N., Cove, D. J., 1973: Nitrogen metabolite repression in *Aspergillus nidulans.* Mol. Gen. Genet. **126,** 111—141.

Arst, H. N., MacDonald, D. W., Cove, D. J., 1970: Molybdate metabolism in *Aspergillus nidulans.* I. Mutations affecting nitrate reductase and/or xanthine dehydrogenase. Mol. Gen. Genet. **108,** 129—145.

Arst, H. N., Tollervey, D. W, Sealy-Lewis, H. M., 1982: A possible regulatory gene for the molybdenum-containing cofactor in *A. nidulans.* J. Gen. Microbiol. **128,** 1083—1093.

Ashton, N. W., Cove, D. J., 1977: The isolation and preliminary characterisation of auxotrophic and analogue resistant mutants of the moss, *Physcomitrella patens.* Mol. Gen. Genet. **154,** 87—95.

Ballance, D. J., Buxton, F. P., Turner, G., 1983: Transformation of *Aspergillus nidulans* by the orotidine-5′-phosphate decarboxylase gene of *Neurospora crassa.* Biochem. Biophys. Res. Commun. **112,** 284—289.

Beadle, G. W., Tatum, E. L., 1945: Neurospora II. Methods of producing and detecting mutations concerned with nutritional requirements. Am. J. Bot. **32,** 678—686.

Beevers, L., Hageman, R. H., 1980: Nitrate and nitrite reduction. In: Miflin, B. J. (ed.), Amino Acids and Derivatives (The Biochemistry of Plants, Vol. 5), pp. 115—168. New York: Academic Press.

Bennetzen, J. L., 1984: Transposable element *Mu*1 is found in multiple copies only in Robertson's *Mutator* maize lines. J. Mol. Appl. Genet. **2,** 519—524.

Biasini, G., Marton, L., 1985: A rapid assay for genetic complementation of nitrate reductase deficiency via bulk protoplast fusion. Mol. Gen. Genet. **198,** 353—355.

Birkett, J. A., Rowlands, R. T., 1981: Chlorate resistance and nitrate assimilation in industrial strains of *Penicillium chrysogenum.* J. Gen. Microbiol. **123,** 281—285.

Braaksma, F. J., Feenstra, W. J., 1982a: Isolation and characterization of nitrate reductase-deficient mutants of *Arabidopsis thaliana.* Theor. Appl. Genet. **64,** 83—90.

Braaksma, F. J., Feenstra, W. J., 1982b: Nitrate reduction in the wild type and a nitrate reductase deficient mutant of *Arabidopsis thaliana.* Physiol. Plant. **54,** 351—360.

Braaksma, F. J., Feenstra, W. J., 1982c: Reverse mutants of the nitrate reductase-deficient mutant B25 of *Arabidopsis thaliana.* Theor. Appl. Genet. **61,** 263—271.

Bray, R. C., 1980: The reaction and the structures of molybdenum centres in enzymes. In: Meister, A. (ed.), Advances in Enzymology, Vol. 51, pp. 107—166. New York: Academic Press.

Bright, S. W. J., Norbury, P. B., Franklin, J., Kirk, D. W., Wray, J. L., 1983: A conditional-lethal *cnx*-type nitrate reductase-deficient barley mutant. Mol. Gen. Genet. **189,** 240—244.

Brown, J., Small, I. S., Wray, J. L., 1981: Age-dependent conversion of nitrate reductase to cytochrome c reductase species in barley leaf extract. Phytochemistry **20,** 389—398.

Brownlee, A. G., Arst, H. N., Jr., 1983: Nitrate uptake in *Aspergillus nidulans* and involvement of the third gene of the nitrate assimilation gene cluster. J. Bacteriol. **155,** 1138—1146.

Buchanan, R. J., Wray, J. L., 1982: Isolation of molybdenum cofactor-defective lines of *Nicotiana tabacum.* Mol. Gen. Genet. **188,** 228—234.

Buxton, F. R., Radford, A., 1983: Cloning of the structural gene for orotidine 5′-phosphate carboxylase of *Neurospora crassa* by expression in *Escherichia coli.* Mol. Gen. Genet. **190,** 403—405.

Campbell, A. M., Campbillo-Campbell, A. D., Villaret, D. B., 1985: Molybdate reduction by *Escherichia coli* K-12 and its *chl* mutants. Proc. Nat. Acad. Sci. U.S.A. **87,** 227—231.

Campbell, J. McA., Wray, J. L., 1983: Purification of barley nitrate reductase and demonstration of nicked subunits. Phytochemistry **22,** 2375—2382.

Campbell, J. McA., Small, I. S., Wray, J. L., 1984: Purification of an NADH cytochrome c reductase from cell-free extracts of barley. Phytochemistry **23,** 1391—1395.

Campbell, W. H., 1976: Separation of soybean leaf nitrate reductases by affinity chromatography. Plant Sci. Lett. **7,** 239—247.

Campbell, W. H., 1978: Isolation of NAD (P) H: nitrate reductase from the scutellum of maize. Z. Pflanzenphysiol. **88,** 357—361.

Chen, P. C., Phillips, D. A., 1977: Induction of root nodule senescence by combined nitrogen in *Pisum sativum* L. Plant Physiol. **59,** 440—442.

Clement, C. R., Hopper, M. J., Jones, L. P. H., 1978: The uptake of nitrate by *Lolium perenne* from flowing nutrient solution. I. Effect of NO_3^- concentration. J. Exp. Bot. **29,** 453—464.

Coddington, A., 1976: Biochemical studies on the *nit* mutants of *Neurospora crassa.* Mol. Gen. Genet. **145,** 195—206.

Cove, D. J., 1963: Studies of the biochemistry and genetics of inorganic nitrogen metabolism in the fungus *Aspergillus nidulans.* Ph. D. dissertation, University of Cambridge, England.

Cove, D. J., 1976a: Chlorate toxicity in *Aspergillus nidulans.* Studies of mutants altered in nitrate assimilation. Mol. Gen. Genet. **146,** 147—159.

Cove, D. J., 1976b: Chlorate toxicity in *Aspergillus nidulans:* The selection and characterisation of chlorate-resistant mutants. Heredity **36,** 191—203.

Cove, D. J., 1979: Genetic studies of nitrate assimilation in *Aspergillus nidulans.* Biol. Rev. **54,** 291—327.

Cove, D. J., Pateman, J. A., 1969: Autoregulation of the synthesis of nitrate reductase in *Aspergillus nidulans.* J. Bacteriol. **97,** 1374—1378.

Cramer, S. P., Wahl, R., Rajagopalan, K. V., 1981: Molybdenum sites of sulphite oxidase and xanthine dehydrogenase. A comparison by EXAFS. J. Am. Chem. Soc. **102,** 7721—7727.

Dailey, F. A., Kuo, T., Warner, R. L., 1982a: Pyridine nucleotide specificity of barley nitrate reductase. Plant Physiol. **69,** 1196—1199.

Dailey, F. A., Warner, R. L., Somers, D. A., Kleinhofs, A., 1982b: Characteristics of a nitrate reductase in a barley mutant deficient in NADH nitrate reductase. Plant Physiol. **69,** 1200—1204.

Dalling, M. J., Halloran, J. M., Wilson, H. J., 1975: The relation between nitrate reductase activity and grain nitrogen productivity in wheat. Aust. J. Agric. Res. **26,** 1—10.

Deane-Drummond, C. E., 1982: Mechanisms for nitrate uptake into barley (*Hordeum vulgare* cv. Fergus) seedlings grown at controlled nitrate concentrations in the nutrient medium. Plant Sci. Lett. **24,** 79—89.

Deane-Drummond, C. E., 1985: Regulation of nitrate uptake into *Chara corallina* cells via NH_4^+ stimulation of NO_3^- efflux. Plant Cell Environ. **8,** 105—116.

Deane-Drummond, C. E., 1986: A substrate cycling model for nitrate uptake by *Pisum sativum* L. seedlings — a key to sensitivity of response of net flux to substrate and effectors? Plant and Soil, **91,** 307—311.

Deane-Drummond, C. E., Glass, A. D. M., 1983: Short term studies of nitrate uptake into barley plants using ion-specific electrodes and $^{36}ClO_3$. I. Control of net uptake by NO_3 efflux. Plant Physiol. **73,** 100—104.

De Block, M., Herrera-Estrella, L., van Montague, M., Schell, J., Zambryski, P., 1984: Expression of foreign genes in regenerated plants and in their progeny. EMBO J. **3,** 1681—1689.

Dirks, R., Negrutiu, I., Sidorov, V., Jacobs, M., 1985: Complementation analysis by somatic hybridisation and genetic crosses of nitrate reductase-deficient mutants of *Nicotiana plumbaginifolia.* Evidence for a new category of *cnx* mutants. Mol. Gen. Genet. **201,** 339—343.

Doddema, H., Telkamp, G. P., 1979: Uptake of nitrate by mutants of *Arabidopsis thaliana* disturbed in uptake or reduction of nitrate. Physiol. Plant. **45,** 332—338.

Dunn-Coleman, N. S., 1984a: Biochemical characterisation of the molybdenum cofactor mutants of *Neurospora crassa: In vivo* and *in vitro* reconstitution of NADPH-nitrate reductase activity. Curr. Genet. **8,** 581—588.

Dunn-Coleman, N. S., 1984b: Cloning and preliminary characterization of a molybdenum cofactor gene of *Neurospora crassa.* Curr. Genet. **8,** 589—595.

Dunn-Coleman, N. S., Smarelli, J., Jr., Garrett, R. H., 1984: Nitrate assimilation in eukaryotic cells. Int. Rev. Cytol. **92,** 1—50.

Evans, H. J., Nason, A., 1953: Pyridine nucleotide-nitrate reductase from extracts of higher plants. Plant Physiol. **28,** 233—254.

Evans, L. T., 1975: Crop and world food supply, crop evolution, and the origins of crop physiology. In: Evans, L. T. (ed.), Crop Physiololgy, pp. 1—22. Cambridge: Cambridge University Press.

Evola, S. V., 1983 a: Chlorate-resistant variants of *Nicotiana tabacum* L. I. Selection *in vitro* and phenotypic characterization of cell lines and regenerated plants. Mol. Gen. Genet. **189,** 447—454.

Evola, S. V., 1983 b: Chlorate-resistant variants of *Nicotiana tabacum* L. II. Parasexual genetic characterization. Mol. Gen. Genet. **189,** 455—457.

Fankhauser, H., Bucher, F., King, P. J., 1984: Isolation of biochemical mutants using haploid mesophyll protoplasts of *Hyoscyamus muticus* IV. Biochemical characterisation of nitrate non-utilising clones. Planta **160,** 415—421.

Feenstra, W. J., Jacobsen, E., 1980: Isolation of a nitrate reductase deficient mutant of *Pisum sativum* by means of selection for chlorate resistance. Theor. Appl. Genet. **58,** 39—42.

Feenstra, W. J., Jacobsen, E., van Swaay, A. C. P. M., De Visser, A. J. C., 1982: Effect of nitrate on acetylene reduction in a nitrate reductase deficient mutant of pea (*Pisum sativum* L.). Z. Pflanzenphysiol. **105,** 471—474.

Ferrari, T. E., Yoder, O. C., Filner, P., 1973: Anaerobic nitrite production by plant cells and tissues: Evidence for two nitrate pools. Plant Physiol. **51,** 423—431.

Fido, R. J., Notton, B. A., 1984: Spinach nitrate reductase: further purification and removal of 'nicked' sub-units by affinity chromatography. Plant Sci. Lett. **37,** 87—91.

Fido, R. J., Hewitt, E. J., Notton, B. A., Jones, O. T. G., Narulhaq-Boyce, A., 1979: Haem of spinach nitrate reductase: low temperature spectrum and mid-point potential. FEBS Lett. **99,** 180—182.

Foster, S. S. D., Cripps, A. C., Smith-Carington, A., 1982: Nitrate leaching to groundwater. Philos. Trans. R. Soc. London, Ser. B **296,** 477—489.

Fraley, R. T., Rogers, S. G., Horsch, R. B., Sanders, P. R., Flick, J. S., Adams, S. P., Bitter, M. I., Brand, I. A., Fink, C. L., Fry, J. S., Gallupi, G. R., Goldberg, S. B., Hoffman, N. I., Woo, S. C., 1983: Expression of bacterial genes in plant cells. Proc. Nat. Acad. Sci., U.S.A. **80,** 4803—4807.

Freeling, M., Cheng, D. S.-K., Alleman, M., 1982: Mutant alleles that are altered in quantitative, organ-specific behaviour. Dev. Genet. **3,** 179—196.

Fujihara, S., Yamaguchi, M., 1978: Effects of allopurinol (4'-hydroxypyrazolo (3,4-d) pyrimidine) on the metabolism of allantoin in soybean plants. Plant Physiol. **62,** 134—138.

Garrett, R. H., Nason, A., 1969: Further purification and properties of *Neurospora* nitrate reductase. J. Biol. Chem. **244,** 2870—2882.

Gebhardt, C., Schnebli, V., King, P. J., 1981: Isolation of biochemical mutants using haploid mesophyll protoplasts of *Hyoscyamus muticus.* II. Auxotrophic and temperature-sensitive clones. Planta **153,** 81—89.

Gewitz, H.-S., Piefke, J., Vennesland, B., 1981: Purification and characterization of demolybdonitrate reductase (NADH cytochrome c oxidoreductase) of *Chlorella vulgaris.* J. Biol. Chem. **256,** 11527—11531.

Gibson, A. H., Pagan, J. D., 1977: Nitrate effects on the nodulation of legumes inoculated with NR-deficient mutants of *Rhizobium.* Planta **134,** 17—22.

Ginns, E. I., Choudary, P. V., Martin, B. M., Winfield, S., Stubblefield, B., Mayor, J., Merkle-Lehman, D., Murray, G. J., Bowers, L. A., Barranger, J. A., 1984: Isolation of cDNA clones for human β-glucocerebrosidase using the λgtll expression system. Biochem. Biophys. Res. Commun. **123,** 574—580.

Giordano, G., Fimmel, A. L., Powell, L. M., Haddock, B. A., Barr, G. C., 1981: Cloning and expression of the *chl*A gene of *Escherichia coli* K12. FEMS Microbiol. Lett. **12,** 61—64.

Glaser, J. H., DeMoss, J. A., 1971: Phenotypic restoration by molybdate of nitrate reductase activity in *chl*D mutants of *Escherichia coli*. J. Bacteriol. **108,** 854—860.

Glass, A. D. M., Thompson, R. G., Bordelean, L., 1985: Regulation of nitrate influx in barley. Studies using $^{13}NO_3^-$. Plant Physiol. **77,** 379—381.

Glimelius, K., Bonnett, H. T., 1981: Somatic hybridization in *Nicotiana:* Restoration of photoautotrophy to an albino mutant with defective plastids. Planta **153,** 497—503.

Glimelius, K., Eriksson, T., Grafe, R., Müller, A. J., 1978: Somatic hybridization of nitrate reductase-deficient mutants of *Nicotiana tabacum* by protoplast fusion. Physiol. Plant. **44,** 273—277.

Glimelius, K., Chen, K., Bonnett, H. T., 1981: Somatic hybridization in *Nicotiana:* Segregation of organellar traits among hybrid and cybrid plants. Planta **153,** 504—510.

Gottman, K., Schafer, E., 1983: Analysis of phytochrome kinetics in light-grown *Avena sativa* L. seedlings. Planta **157,** 392—400.

Graf, L. A., Notton, B. A., Hewitt, E. J., 1975: Serological estimation of spinach nitrate reductase. Phytochemistry **14,** 1241—1243.

Grafe, R., Müller, A. J., 1983: Complementation analysis of nitrate reductase deficient mutants of *Nicotiana tabacum* by somatic hybridization. Theor. Appl. Genet. **66,** 127—130.

Gray, J. C., Jung, S. D., Wildman, S. G., Sheen, S. J., 1974: Origin of *Nicotiana tabacum* detected by polypeptide composition of fraction I protein. Nature (London) **252,** 266—267.

Grossman, A. R., Barlett, S. G., Schmidt, G. W., Mullet, J. E., Chua, N.-H., 1982: Optimal conditions for post-translational uptake of proteins by isolated chloroplasts. *In vitro* synthesis and transport of plastocyanin, ferredoxin-$NADP^+$ oxidoreductase and fructose-1,6-bisphophatase. J. Biol. Chem. **257,** 1558—1563.

Guerrero, M. G., Vega, J. M., Losada, M., 1981: The assimilatory nitrate-reducing system and its regulation. Annu. Rev. Plant Physiol. **32,** 169—204.

Gupta, S. C., Beevers. L., 1984: Synthesis and degradation of nitrite reductase in pea leaves. Plant Physiol. **75,** 251—252.

Gupta, P. P., Gupta, M., Schieder, O., 1982: Correction of nitrate reductase defect in auxotrophic plant cells, through protoplast-mediated intergenic gene transfer. Mol. Gen. Genet. **188,** 378—383.

Hamano, T., Oji, Y., Okamoto, S., Mitsuhashi, Y., Matsuki, Y., 1984a: Purification and characterisation of thiol proteinase as a nitrate reductase-inactivating factor from leaves of *Hordeum distichum* L. Plant Cell Physiol. **25,** 419—427.

Hamano, T., Oji, Y., Mitsuhashi, Y., Matsuki, Y., Okamoto, S., 1984b: Action of thiol proteinase on nitrate reductase in leaves of *Hordeum distichum* L. Plant Cell Physiol. **254,** 1469—1475.

Hamill, J. D., Pental, D., Cocking, E. C., Müller, A. J., 1983: Production of a nitrate reductase deficient streptomycin resistant mutant of *Nicotiana tabacum* for somatic hybridization studies. Heredity **50,** 197—200.

Hardy, R. W. F., Holsten, R. D., Jackson, E. K., Burns, R. C., 1968: The aceylene-ethylene assay for N_2 fixation: Laboratory and field evaluation. Plant Physiol. **43,** 1185—1207.

Harper, J. E., 1981a: Effect of chlorate, nitrogen source and light on chlorate toxicity and nitrate reductase activity in soybean leaves. Physiol. Plant **53,** 505—510.

Harper, J. E., 1981b: Evolution of nitrogen oxide(s) during *in vivo* nitrate reductase assay of soybean leaves. Plant Physiol. **68,** 1488—1493.

Harper, J. E., Nicholas, D. J. D., 1978: Nitrogen metabolism of soybeans. I. Effect of tungstate on nitrate utilization, nodulation and growth. Plant Physiol. **62,** 662—664.

Hartley, M. J., 1969: Mutation in Aspergillus. Ph. D. Dissertation, University of Cambridge, England.

Hartley, M. J., 1970: Contrasting complementation patterns in *Aspergillus nidulans.* Genet. Res. **16,** 123—125.

Heath-Pagliuso, S., Huffaker, R. C., Allard, R. W., 1984: Inheritance of nitrite reductase and regulation of nitrate reductase, nitrite reductase and glutamine synthetase isoenzymes. Plant Physiol. **76,** 353—358.

Hein, T., Przewozny, T., Schieder, O., 1983: Culture and selection of somatic hybrids using an auxotrophic cell line. Theor. Appl. Genet. **64,** 119—122.

Helfman, D. M., Feramisco, J. R., Fiddes, J. C., Thomas, G. P., Hughes, S. H., 1983: Identification of clones that encode chicken tropomyosin by direct immunological screening of a cDNA expression library. Proc. Nat. Acad. Sci., U.S.A. **80,** 31—35.

Herrera-Estrella, L., De Block, M., Messens, E., Hernalsteens, J. P., van Montague, M., Schell, J., 1983: Chimaeric genes as dominant selectable markers in plant cells. EMBO J. **2,** 987—995.

Herrera-Estrella, L., Van den Broeck, R., Maenhaut, R., Van Montagu, M., Schell, J., Timko, M., Cashmore, A., 1984: Light-inducible and chloroplast-associated expression of a chimaeric gene introduced into *Nicotiana tabacum* using a Ti plasmid vector. Nature (London) **310,** 115—120.

Hershey, H. P., Colbert, J. T., Lissemore, J. L., Burker, R. F., Quail, P. H., 1984: Molecular cloning of cDNA of *Avena* phytochrome. Proc. Nat. Acad. Sci., U.S.A. **81,** 2332—2336.

Hewitt, E. J., Notton, B. A., 1980: Nitrate reductase systems in eukaryotic and prokaryotic organisms. In: Coughlan, M. (ed.), Molybdenum and Molybdenum-containing Enzymes, pp. 273—325. Oxford: Pergamon Press.

Hewitt, E. J., Notton, B. A., Rucklidge, G. J., 1977: Formation of nitrate reductase by recombination of apoprotein fraction from molybdenum deficient plants with a molybdenum containing complex. J. Less-Common Metals **54,** 537—545.

Highfield, P. E., Ellis, R. J., 1978: Synthesis and transport of the small subunit of chloroplast ribulose bisphosphate carboxylase. Nature (London) **271,** 420—424.

Hofstra, J. J., 1977: Chlorate toxicity and nitrate reductase activity in tomato plants. Physiol. Plant **41,** 65—69.

Horner, R. D., 1983: Purification and comparison of *nit*-1 and wild-type nitrate reductases of *Neurospora crassa.* Biochim. Biophys. Acta **744,** 7—15.

Horsch, R. B., Fry, J. E., Hoffmann, N. L., Eichholtz, D., Rogers, S. G., Fraley, R. T., 1985: A simple and general method for transferring genes into plants. Science **227,** 1229—1231.

Hucklesby, P. D., James, D. M., Banwell, M. J., Hewitt, E. J., 1976: Properties of nitrite reductase from *Cucurbita pepo.* Phytochemistry **15,** 599—603.

Huffaker, R. C., 1982: Biochemistry and physiology of leaf proteins. In: Boulter, D., Parthier, B. (eds.), Encyclopaedia of Plant Physiology, New Series, Vol. 14A, pp. 370—400. New York — Berlin — Heidelberg: Springer-Verlag.

Hulburt, B. K., Garrett, R. H., 1985: Cloning and regulation of nitrate-inducible genes in *Neurospora crassa.* Abst. UCLA Symp. "Molecular Genetics of Filamentous Fungi", p. 1597. New York: Alan R. Liss.

Ida, S., 1977: Purification to homogeneity of spinach nitrite reductase by ferredoxin-sepharose affinity chromatography. J. Biochem. **82,** 915—918.

Jackson, W. A., Flesher, D., Hageman, R. H., 1973: Nitrate uptake by dark-grown corn seedlings: some characteristics of apparent induction. Plant Physiol. **51,** 120—127.

Jackson, W. A., Kurik, K. D., Volk, R. J., 1976: Nitrate uptake during recovery from nitrate deficiency. Plant Physiol. **36,** 174—181.

Jacobsen, E., Feenstra, W. J., 1984: A new pea mutant with efficient nodulation in the presence of nitrate. Plant Sci. Lett. **33,** 337—344.

Jacobsen, E., Braaksma, F. J., Feenstra, W. J., 1984: Determination of xanthine dehydrogenase activity in nitrate reductase deficient mutants of *Pisum sativum* and *Arabidopsis thaliana.* Z. Pflanzenphysiol. **113,** 183—188.

Jacq, C., Lederer, F., 1974: Cytochrome b_2 from Bakers yeast (L-lactate dehydrogenase). A double-headed enzyme. Eur. J. Biochem. **41,** 311—320.

Johnson, C. B., Whittington, W. J., Blackwood, G. C., 1976: Nitrate reductase as a possible predictive test of crop yield. Nature (London) **262,** 133—134.

Johnson, J. L., 1980: The molybdenum cofactor common to nitrate reductase, xanthine dehydrogenase and sulfite oxidase. In: Coughlan, M. (ed.), Molybdenum and Molybdenum-containing Enzymes, pp. 345—383. Oxford: Pergamon Press.

Johnson, J. L., Rajagopalan, K. V., 1977: Tryptic cleavage of rat liver sulfite oxidase: Isolation and characterization of molybdenum and heme domains. J. Biol. Chem. **252,** 2017—2025.

Johnson, J. L., Rajagopalan, K. V., 1982: Structural and metabolic relationship between the molybdenum cofactor and urothione. Proc. Nat. Acad. Sci., U.S.A. **79,** 6856—6860.

Johnson, J. L., Hainline, B. E., Rajagopalan, K. V., 1980: Characterization of the molybdenum cofactor of sulfite oxidase, xanthine oxidase and nitrate reductase. Identification of a pteridine as a structural component. J. Biol. Chem. **255,** 1783—1786.

Johnson, J. L., Hainline, B. E., Rajagopalan, K. V., Arison, B. H., 1984: Pterin component of the molybdenum cofactor. Structural characterization of two fluorescent derivatives. J. Biol. Chem. **259,** 5414—5422.

Johnstone, I. J., Hughes, S. G., Clutterbuck, A. J., 1985: Cloning an *Aspergillus nidulans* developmental gene by transformation. EMBO J. **4,** 1307—1311.

Jolly, S. O., Campbell, W. H., Tolbert, N. E., 1976: NADPH and NADH-nitrate reductase from soybean leaves. Arch. Biochem. Biophys. **174,** 431—439.

Jones, P. W., Ní Mhuimhneacháin, M., 1985: The activity and stability of wheat nitrate reductase *in vitro.* Phytochemistry **24,** 385—392.

Kakefuda, G., Duke, S. H., Duke, S. O., 1983: Differential light induction of nitrate reductases in greening and photobleached soybean seedlings. Plant Physiol. **73,** 56—60.

Ketchum, P. A., Cambier, H. Y., Frazier, W. A. III., Madansky, C. H., Nason, A., 1970: *In vitro* assembly of *Neurospora* mutant assimilatory nitrate reductase from protein subunits of a *Neurospora* mutant and the xanthine oxidising and aldehyde oxidase systems of higher plants. Proc. Nat. Acad. Sci., U.S.A. **66,** 1016—1023.

King, J., Khanna, V., 1980: A nitrate reductase-less variant isolated from suspension cultures of *Datura innoxia* (Mill). Plant Physiol. **66,** 632—636.

Kinghorn, J. R., Hawkins, A. R., 1982: Cloning and expression in *Escherichia coli* of the *Aspergillus nidulans arom* cluster gene. Mol. Gen. Genet. **186,** 145—154.

Kleinhofs, A., Warner, R. L., Muehlbauer, F. J., Nilan, R. A., 1978: Induction and selection of specific gene mutations in *Hordeum* and *Pisum.* Mutat. Res. **51,** 29—35.

Kleinhofs, A., Kuo, T., Warner, R. L., 1980: Characterization of nitrate reductase-deficient barley mutants. Mol. Gen. Genet. **177,** 421—425.

Kleinhofs, A., Taylor, J., Kuo, T. M., Somers, D. A., Warner, R. L., 1983: Nitrate reductase genes as selectable markers for plant cell transformation. In: Lurquin, P. F., Kleinhofs, A. (eds.), Genetic Engineering in Eukaryotes, pp. 215—231. New York: Plenum.

Kleinhofs, A., Warner, R. L., Narayanan, K. R., 1985: Current progress towards an understanding of the genetics and molecular biology of nitrate reductase in higher plants. In: Miflin, B. J. (ed.), Oxford Surveys of Plant Molecular and Cell Biology, Vol. 2, pp. 91—121. Oxford: Oxford University Press.

Kramer, S., Hageman, R. V., Rajagopalan, K. V., 1984: *In vitro* reconstitution of nitrate reductase activity of the *Neurospora crassa* mutant *nit*-1: specific incorporation of molybdopterin. Arch. Biochem. Biophys. **233,** 821—829.

Kuhlmeier, C. J., Logtenberg, T., Stoorvogel, W., van Heughten, H. A. A., Borrias, W. E., van Arkel, G. A., 1984: Cloning of nitrate reductase genes from the cyanobacterium *Anacystis nidulans.* J. Bacteriol. **159,** 36—41.

Kuo, T., Kleinhofs, A., Warner, R. L., 1980: Purification and partial characterisation of nitrate reductase from barley leaves. Plant Sci. Lett. **17,** 371—381.

Kuo, T., Kleinhofs, A., Somers, D., Warner, R. L., 1981: Antigenicity of nitrate reductase deficient mutants in *Hordeum vulgare* L. Mol. Gen. Genet. **181,** 20—23.

Kuo, T. M., Somers, D. A., Kleinhofs, A., Warner, R. L., 1982: NADH-nitrate reductase in barley leaves. Identification and amino-acid composition of subunit protein. Biochem. Acta **708,** 75—81.

Kuo, T. M., Kleinhofs, A., Somers, D. A., Warner, R. L., 1984: Nitrate reductase-deficient mutants in barley: enzyme stability and peptide mapping. Phytochemistry **23,** 229—232.

Laan, P. H., Van der, Oostindier-Braaksma, F. J., Feenstra, W. J., 1971: Chlorate resistant mutants of *Arabidopsis thaliana.* Arabidopsis Inf. Serv. **8,** 22.

Lahav, E., Harper, J. E., Hageman, R. H., 1976: Improved soybean growth in urea with pH buffered by a carboxy resin. Crop Sci. **16,** 325—328.

Lancaster, J. R., Vega, J. M., Kamin, H., Orme-Johnson, N. R., Orme-Johnson, W. H., Kreuger, R. J., Siegel, L. M., 1979: Identification of the iron-sulphur centre of spinach ferredoxin-nitrite reductase as a tetranuclear centre and preliminary EPR studies of mechanism. J. Biol. Chem. **254,** 1268—1272.

Larkin, P. J., Scowcroft, W. R., 1981: Somaclonal variation — a novel source of variability from cell cultures for plant improvement. Theor. Appl. Genet. **60,** 197—214.

Lazar, G. B., Fankhauser, H., Potrykus, I., 1983: Complementation analysis of a nitrate reductase deficient *Hyoscyamus muticus* cell line by somatic hybridization. Mol. Gen. Genet. **189,** 359—364.

Lé, K. H. D., Lederer, F., 1983: On the presence of a heme-binding domain homologous to cytochrome b_5 in *Neurospora crassa* assimilatory nitrate reductase. EMBO J. **2,** 1909—1914.

Lee, K. Y., Pan, S.-S., Erickson, P., Nason, A., 1974: Involvement of molybdenum and iron in the *in vitro* assembly of assimilatory nitrate reductase utilising mutant *nit*1. J. Biol. Chem. **249,** 3941—3952.

Lijlestrom, S., Åberg, B., 1966: Studies on the mechanism of chlorate toxicity. Kungl. Lantbrukshogkolans Ann. **32,** 93—107.

MacDonald, D. W., Cove, D. J., 1974: Studies on temperature-sensitive mutants affecting the assimilatory nitrate reductase of *Aspergillus nidulans.* Eur. J. Biochem. **47,** 107—110.

MacDonald, D. W., Cove, D. J., Coddington, A., 1974: Cytochrome c reductase from wild type and mutant strains of *Aspergillus nidulans*. Mol. Gen. Genet. **128,** 187—199.

Magee, P. N., 1982: Nitrogen as a potential health hazard. Philos. Trans. R. Soc. London, Ser. B **296,** 543—550.

Maliga, P., 1976: Isolation of mutants from cultured cells. In: Dudits, D., Farkas, G. L., Maliga, P. (eds.), Cell Genetics in Higher Plants, pp. 59—76. Budapest: Akademiai Kiado.

Manhart, J. R., Wong, P. P., 1980: Nitrate effects on nitrogen fixation (acetylene reduction). Plant Physiol. **65,** 502—505.

Maretski, A., De La Cruz, A., 1967: Nitrate reductase in sugarcane tissues. Plant Cell Physiol. **7,** 605—611.

Marton, L., Dung, T. M., Mendel, R. R., Maliga, P., 1982a: Nitrate reductase deficient cell lines from haploid protoplast cultures of *Nicotiana plumbaginifolia*. Mol. Gen. Genet. **182,** 301—304.

Marton, L., Sidorov, V., Biasini, G., Maliga, P., 1982b: Complementation in somatic hybrids indicates four types of nitrate reductase deficient lines in *Nicotiana plumbaginifolia*. Mol. Gen. Genet. **187,** 1—3.

Marton, L., Biasini, G., Maliga, P., 1985: Co-segregation of nitrate reductase activity and normal regeneration ability in selfed sibs of *Nicotiana plumbaginifolia* somatic hybrids, heterozygotes for nitrate reductase deficiency. Theor. Appl. Genet. **70,** 340—344

McClure, P. R., Israel, D. W., 1979: Transport of nitrogen in the xylem of soybean plants. Plant Physiol. **64,** 411—416.

Mendel, R.-R., 1983: Release of molybdenum co-factor from nitrate reductase and xanthine oxidase by heat treatment. Phytochemistry **22,** 817—819.

Mendel, R. R., Müller, A. J., 1976: A common genetic determinant of xanthine dehydrogenase and nitrate reductase in *Nicotiana tabacum*. Biochem. Physiol. Pflanzen. **170,** 538—541.

Mendel, R. R., Müller, A. J., 1979: Nitrate reductase-deficient mutant cell lines of *Nicotiana tabacum*. Further biochemical characterization. Mol. Gen. Genet. **177,** 145—153.

Mendel, R.-R., Müller, A. J., 1980: Comparative characterisation of nitrate reductase from wild-type and molybdenum cofactor-defective cell cultures of *Nicotiana tabacum*. Plant Sci. Lett. **18,** 277—288.

Mendel, R.-R., Müller, A. J., 1985: Repair *in vitro* of nitrate reductase-deficient tobacco mutants (*cnx*A) by molybdate and by molybdenum cofactor. Planta **163,** 370—375.

Mendel, R.-R., Alikulov, Z. A., Lvov, N. P., Müller, A. J., 1981: Presence of the molybdenum-cofactor in nitrate reductase-deficient mutant cell lines of *Nicotiana tabacum*. Mol. Gen. Genet. **181,** 395—399.

Mendel, R. R., Alikulov, Z. A., Müller, A. J., 1982: Molybdenum cofactor in nitrate reductase-deficient tobacco mutants. III. Induction of cofactor synthesis by nitrate. Phytochemistry **27,** 95—101.

Mendel, R.-R., Buchanan, R. J., Wray, J. L., 1984: Characterization of a new type of molybdenum cofactor-mutant in cell cultures of *Nicotiana tabacum*. Mol. Gen. Genet. **195,** 186—189.

Mendel, R.-R., Kirk, D. W., Wray, J. L., 1985: Assay of molybdenum cofactor of barley. Phytochemistry **24,** 1631—1634.

Mendel, R. R., Marton, L., Müller, A. J., 1986: Comparative biochemical characterization of mutants at the nitrate reductase/molybdenum cofactor loci *cnx*A, *cnx*B and *cnx*C of two *Nicotiana* species. Plant Sci. **43,** 125—129.

Meriwether, L. S., Marzluff, W. F., Hodgson, W. G., 1966: Molybdenum-thiol complexes as models for molybdenum bound in enzymes. Nature (London) **212,** 465—467.

Miller, B. L., Huffaker, R. C., 1981: Partial purification and characterization of endoproteinases from senescing barley leaves. Plant Physiol. **68,** 930—936.

Minotti, P. L., Williams, D. C., Jackson, W. A., 1968: Nitrate uptake and reduction as affected by calcium and potassium. Soil Sci. Soc. Am. **32,** 692—698.

Morelli, G., Nagy, F., Fraley, R. T., Rogers, S. C., Chua, N.-H., 1985: A short conserved sequence is involved in the light-inducibility of a gene encoding ribulose 1,5-bisphosphate carboxylase small subunit of pea. Nature (London) **315,** 200—204.

Müller, A. J., 1983: Genetic analysis of nitrate reductase-deficient tobacco plants regenerated from mutant cells. Evidence for duplicate structural genes. Mol. Gen. Genet. **192,** 275—281.

Müller, A. J., Grafe, R., 1978: Isolation and characterization of cell lines of *Nicotiana tabacum* lacking nitrate reductase. Mol. Gen. Genet. **161,** 67—76.

Mulvaney, C. S., Hageman, R. H., 1984: Acetaldehyde oxime, a product formed during *in vivo* nitrate reductase assay of soybean leaves. Plant Physiol. **76,** 118—124.

Munns, D. N., 1968: Nodulation of *Medicago sativa* in solution culture. III. Effects of nitrate on root hairs and infection. Plant and Soil **29,** 33—47.

Murphy, J. M., Imbrie, C. W., 1981: Induction and characterization of chlorate-resistant strains of *Rosa damascena* cultured cells. Plant Physiol. **67,** 910—916.

Nakagawa, H., Yonemura, Y., Yamamoto, H., Sato, T., Ogura, N., Sato, R., 1985: Spinach nitrate reductase. Purification, molecular weight and subunit composition. Plant Physiol. **77,** 124—128.

Narayanan, K. R., Somers, D. A., Kleinhofs, A., Warner, R. L., 1983: Nature of Cytochrome c reductase deficient mutants in barley. Mol. Gen. Genet. **190,** 222—226.

Narayanan, K. R., Müller, A. J., Kleinhofs, A., Warner, R. L., 1984: *In vitro* reconstitution of NADH: nitrate reductase in nitrate reductase deficient mutants of barley. Mol. Gen. Genet. **197,** 358—362.

Naslin, L., Spyridakis, A., Labeyrie, F., 1973: A study of several bonds hypersensitive to proteases in a complex flavohaemoenzyme, yeast cytochrome b_2. Modification of their reactivity with ligand-induced conformational transitions. Eur. J. Biochem. **34,** 268—283.

Nason, A., Lee, K.-Y., Pan, S.-S., Ketchum, P. A., Lamberti, A., Davies, J., 1971: *In vitro* formation of assimilatory reduced nicotinamide adenine dinucleotide phosphate: nitrate reductase from a *Neurospora* mutant and a component of molybdenum-enzymes. Proc. Nat. Acad. Sci., U.S.A. **68,** 3242—3246.

Negrutiu, I., Dirks, R., Jacobs, M. L., 1983: Regeneration of fully nitrate reductase-deficient mutants from protoplast culture of *Nicotiana plumbaginifolia* (Viviani). Theor. Appl. Genet. **66,** 341—347.

Nelson, R. S., Ryan, S. A., Harper, J. E., 1983: Soybean mutants lacking constitutive nitrate reductase activity. I. Selection and initial plant characterization. Plant Physiol. **72,** 503—509.

Nelson, R. S., Streit, L., Harper, J. E., 1984: Biochemical characterization of nitrate and nitrite reduction in the wild type and a nitrate reductase mutant of soybean. Physiol. Plant **61,** 384—390.

Notton, B. A., Hewitt, E. J., 1971 a: Incorporation of radioactive molybdenum into protein during nitrate reductase formation and effect of molybdenum on nitrate reductase and diaphorase activities of spinach (*Spinacea oleracea* L). Plant Cell Physiol. **12,** 465—477.

Notton, B. A., Hewitt, E. J., 1971 b: The role of tungsten in the inhibition of nitrate reductase activity in spinach *(Spinacea oleracea)* leaves. Biochem. Biophys. Res. Commun. **44,** 702—710.

Notton, B. A., Hewitt, E. J., 1979: Structure and properties of higher plant nitrate reductase especially *Spinacea oleracea* L. In: Hewitt, E. J., Cutting, C. V. (eds.), Nitrogen Assimilation in Plants, pp. 227—244. New York: Academic Press.

Notton, B. A., Fido, R. J., Hewitt, E. J., 1977: The presence of a functional haem in a higher plant nitrate reductase. Plant Sci. Lett. **8,** 165—170.

Notton, B. A., Fido, R. J., Galfre, G., 1985: Monoclonal antibodies to a higher plant nitrate reductase: differential inhibition of enzyme activity. Planta **165,** 114—119.

Oh, J. Y., Warner, R. L., Kleinhofs, A., 1980: Effect of nitrate reductase deficiency upon growth yield and protein in barley. Crop Sci. **20,** 487—490.

Oostindier-Braaksma, F. J., 1970: A chlorate resistant mutant of *Arabidopsis thaliana.* Arabidopsis Inf. Serv. **7,** 24.

Oostindier-Braaksma, F. J., Feenstra, W. J., 1972: Chlorate resistant mutants of *Arabidopsis thaliana* II. Arabidopsis Inf. Serv. **7,** 9—10.

Oostindier-Braaksma, F. J., Feenstra, W. J., 1973: Isolation and characterization of chlorate-resistant mutants of *Arabidopsis thaliana.* Mutat. Res. **19,** 175—185.

O'Reilly, C., Shepherd, N. S., Pereira, A., Schwarz-Sommer, Z., Bertram, I., Robertson, D. S., Peterson, P. A., Saedler, H., 1985: Molecular cloning of the *a*1 locus of *Zea mays* using the transposable elements *En* and *Mu*1. EMBO J. **4,** 877—882.

Orihuel-Iranzo, B., Campbell, W. H., 1980: Development of NAD (P) H: and NADH: nitrate reductase activities in soybean cotyledons. Plant Physiol. **65,** 595—599.

Pan, S-S., Nason, A., 1978: Purification and characterization of homogeneous assimilatory reduced nicotinamide adenine dinucleotide phosphate-nitrate reductase from *Neurospora crassa.* Biochim. Biophys. Acta **523,** 297—313.

Paszowski, J., Shillito, R. D., Saul, M., Mandak, V., Hohn, T., Hohn, P., Potrykus, I., 1984: Direct gene transfer to plants. EMBO J. **3,** 2712—2722.

Pateman, J. A., Cove, D. J., Rever, B. M., Roberts, D. B., 1964: A common co-factor for nitrate reductase and xanthine dehydrogenase which also regulates the synthesis of nitrate reductase. Nature (London) **201,** 58—60.

Pateman, J. A., Rever, B. M., Cove, D. J., 1967: Genetic and biochemical studies of nitrate reduction in *Aspergillus nidulans.* Biochem. J. **104,** 103—111.

Pental, D., Cooper-Bland, S., Harding, K., Cocking, E. C., Müller. A. J., 1982: Cultural studies on nitrate reductase deficient *Nicotiana tabacum* mutant protoplasts. Z. Pflanzenphysiol. **105,** 219—227.

Pienkos, P. T., Shah, V. K., Brill, W. J., 1977: Molybdenum co-factors from molybdo-enzymes and *in vitro* reconstitution of nitrogenase and nitrate reductase. Proc. Nat. Acad. Sci., U.S.A. **74,** 5468—5471.

Qureshi, J. A., Buchanan, R. J., Wray, J. L., 1982: Isolation and partial characterization of a chlorate-resistant cell line of *Nicotiana tabacum* possessing nitrate reductase activity. Proc. Internat. Symp. "Nitrate Assimilation — Molecular and Genetic Aspects", Gatersleben, DDR, P. 22.

Rao, K. P., Rains, D. W., 1976: Nitrate absorption by barley. I. Kinetics and energetics. Plant Physiol. **57,** 55—58.

Redinbaugh, M. G., Campbell, W. H., 1981: Purification and characterization of NAD(P)H: nitrate reductase and NADH: nitrate reductase from corn roots. Plant Physiol. **68,** 115—120.

Redinbaugh, M. G., Campbell, W. H., 1983: Purification of squash NADH: nitrate reductase by zinc chelate affinity chromatography. Plant Physiol. **71,** 205—207.

Redinbaugh, M. G., Campbell, W. H., 1985: Quaternary structure and composition of squash NADH: nitrate reductase. J. Biol. Chem. **260,** 3380—3385.

Reed, A. J., Hageman, R. H., 1980a: Relationship between nitrate uptake, flux, and reduction and the accumulation of reduced nitrogen in maize (*Zea mays* L). I. Genotypic variation. Plant Physiol. **66,** 1179—1183.

Reed, A. J., Hageman, R. H., 1980b: Relationship between nitrate uptake, flux, and reduction and the accumulation of reduced nitrogen in maize (*Zea mays* L.). Plant Physiol. **66,** 1184—1189.

Relimpio, A. Ma., Aparicio. P. J., Paneque, A., Losada, M., 1971: Specific protection against inhibitors of the NADH-nitrate reductase complex from spinach. FEBS Lett. **17,** 226—230.

Rever, B. M., 1966: Biochemical and genetical studies of inorganic nitrogen metabolism in *Aspergillus nidulans.* Ph. D. Dissertation, University of Cambridge, England.

Reynolds, P. H. S., Boland, M. J., Blevins, D. G., Schubert, K. R., Randall, D. D., 1982: Enzymes of ureide biogenesis in developing soybean nodules. Plant Physiol. **69,** 1334—1338.

Robertson, D. S., 1978: Characterization of a mutator system in maize. Mutant Res. **51,** 21—28.

Robin, P., Streit, L., Campbell, W. H., Harper, J. E., 1985: Immunochemical characterization of nitrate reductase forms from wild-type (cv. Williams) and nr_1 mutant soybean. Plant Physiol. **77,** 232—236.

Ruiz-Herrera, J., Showe, M. K., DeMoss, J. A., 1969: Nitrate reductase complex of *Escherichia coli* K-12: Isolation and characterization of mutants unable to reduce nitrate. J. Bacteriol. **97,** 1291—1297.

Ryan, S. A., Nelson, R. S., Harper, J. E., 1983: Soybean mutants lacking constitutive nitrate reductase activity. I. Nitrogen assimilation, chlorate resistance and inheritance. Plant Physiol. **72,** 510—514.

Savage, A. D., King, J., Gamborg, O. L., 1979: Recovery of a pantothenate auxotroph from a cell suspension culture of *Datura innoxia.* Mill. Plant. Sci. Lett. **16,** 367—376.

Schechtman, M. G., Yanofsky, C., 1983: Structure of the trifunctional *trp*-1 gene from *Neurospora crassa* and its aberrant expression in *Escherichia coli.* J. Mol. Appl. Genet. **2,** 83—99.

Schell, J., van Montagu, M. 1983: The Ti plasmids as natural and as practical gene vectors for higher plants. Biotechnology **1,** 175—180.

Schieder, O., Vasil, I. K., 1980: Protoplast fusion and somatic hybridisation. In: Vasil, I. K. (ed.), Recent Advances in Plant Cell and Tissue Culture. Int. Rev. Cytol. Suppl. **118,** pp. 21—46. New York: Academic Press.

Schrader, L. E., Ritenour, G. L., Eilrich, G. L., Hageman, R. H., 1968: Some characteristics of nitrate reductase from higher plants. Plant Physiol. **43,** 930—940.

Serra, J. L., Ibarlucea, J. M., Arizmerdi, J. M., Llama, M. J., 1982: Purification and properties of the assimilatory nitrite reductase from barley *Hordeum vulgare* leaves. Biochem. J. **201,** 167—170.

Shah, D. M., Hightower, R. C., Meagher, R. B., 1983: Genes encoding actin in higher

plants: Intron positions are highly conserved but the coding sequences are not. J. Mol. Appl. Genet. **2,** 111—126.
Shah, V. K., Brill, W. J., 1977: Isolation of an iron-molybdenum cofactor from nitrogenase. Proc. Nat. Acad. Sci., U.S.A. **74,** 3249—3252.
Shaner, D. L., Boyer, J. S., 1976: Nitrate reductase activity in maize (*Zea mays* L.) leaves. I. Regulation by nitrate flux. Plant Physiol. **58,** 499—504.
Shen, T. C., Funkhauser, E. A., Guerrero, M. G., 1976: NADH- and NAD(P)H-nitrate reductases in rice seedlings. Plant Physiol. **58,** 292—297.
Shepard, J. F., Bidney, D., Barsby, T., Kemble, R., 1983: Genetic transfer in plants through interspecific protoplast fusion. Science **219,** 683—688.
Shumny, V. K., Tokarev, B. I., 1982: Genetic control of nitrate reductase activity in barley. Proc. IVth Intl. Barley Symp.
Small, I. S., Gray, J. C., 1984: Synthesis of wheat leaf nitrite reductase *de novo* following induction with nitrite. Eur. J. Biochem. **145,** 291—297.
Small, I. S., Wray, J. L., 1980: NADH nitrate reductase and related NADH cytochrome c reductase species in barley. Phytochemistry **19,** 387—394.
Smarrelli, J., Campbell, W. H., 1981: Immunological approach to structural comparisons of assimilatory nitrate reductases. Plant Physiol. **68,** 1226—1230.
Solomonson, L. P., Vennesland, B., 1972: Nitrate reductase and chlorate toxicity in *Chlorella vulgaris* Beyerinck. Plant Physiol. **50,** 421—424.
Solomonson, L. P., Barber, M. J., Howard, W. D., Johnson, J. L., Rajagopalan, K. V., 1984: Electron paramagnetic resonance studies on the molybdenum center of assimilatory NADH: nitrate reductase from *Chlorella vulgaris.* J. Biol. Chem. **259,** 849—853.
Somers, D. A., Kuo, T., Kleinhofs, A., Warner, R. L., 1982: Barley nitrate reductase contains a functional cytochrome b_{557}. Plant Sci. Lett. **24,** 261—265.
Somers, D. A., Kuo, T. M., Kleinhofs, A., Warner, R. L., 1983: Nitrate reductase deficient mutants in barley. Plant Physiol. **71,** 145—149.
Sood, A. K., Pereira, D., Weissman, S. M., 1981: Isolation and partial nucleotide sequence of a cDNA clone for human histocompatibility antigen H_2A-B by use of an oligodeoxynucleotide primer. Proc. Nat. Acad. Sci., U.S.A. **78,** 616—620.
Sorger, G. J., Giles, N. H., 1965: Genetic control of nitrate reductase in *Neurospora crassa.* Genetics **52,** 777—788.
Southerland, W. H., Rajagopalan, K. V., 1978: Domain interactions in oxidised and reduced forms of rat liver sulfite oxidase. J. Biol. Chem. **253,** 8753—8758.
Srivastava, H. S., 1980: Regulation of nitrate reductase activity in higher plants. Phytochemistry **19,** 725—733.
Steffen, A., Schieder, O., 1984: Biochemical and genetical characterization of nitrate reductase deficient mutants of *Petunia.* Plant Cell Rep. **3,** 134—137.
Stewart, G. R., Orebamjo, T. O., 1979: Some unusual characteristics of nitrate reductase in *Erythrina senegalensis* DC. New Phytol. **83,** 311—319.
Stewart, V., MacGregor, C. H., 1982: Nitrate reductase in *Escherichia coli* K-12: Involvement of *chl*C, *chl*E and *chl*G loci. J. Bacteriol. **151,** 788—799.
Strauss, A., Bucher, F., King, P. J., 1981: Isolation of biochemical mutants using haploid mesophyll protoplasts of *Hyoscyamus muticus.* I. A NO_3^- non-utilising clone. Planta, **153,** 75—80.
Streeter, J. G., 1985: Nitrate inhibition of legume nodule growth and activity. I. Long term studies with a continuous supply of nitrate. Plant Physiol. **77,** 321—324.
Streit, L., Nelson, R. S., Harper, J. E., 1985: Nitrate reductases from wild-type and nr_1-mutant soybean (*Glycine max* (L.) Merr.) leaves. Plant Physiol. **78,** 80—84.
Strommer, J. N., Hake, S., Bennetzen, J. L., Taylor, W. C., Freeling, M., 1982: Regu-

latory mutants of the maize *Adh*1 gene caused by DNA mutations. Nature (London) **300,** 542—544.

Suggs, S. Y., Wallace, R. B., Hirose, T., Kawashima, E. H., Itakura, K., 1981: Use of synthetic oligonucleotides as hybridization probes: Isolation of cloned cDNA sequences for β_2-microglobulin. Proc. Nat. Acad. Sci., U.S.A. **78,** 6613—6617.

Taylor, J. L., Bedbrook, J. R., Grant, F. J., Kleinhofs, A., 1983: Reconstitution of plant nitrate reductase by *Escherichia coli* extracts and the molecular cloning of the *chl*A gene of *Escherichia coli* K12. J. Mol. Appl. Genet. **2,** 261—271.

Thomas, R. J., Schrader, L. E., 1981: Ureide metabolism in higher plants. Phytochemistry **20,** 361—371.

Thompson, S. T., Kass, K. H., Stellwagen, E., 1975: Blue dextran-Sepharose: An affinity column for the dinucleotide fold in proteins. Proc. Nat. Acad. Sci., U.S.A. **72,** 669—672.

Tilburn, J., Scazzochio, C., Taylor, G. T., Zabicky-Zissman, J. H., Lockington, R. A., Davies, W. R., 1984: Transformation by integration in *Aspergillus nidulans.* Gene **26,** 205—221.

Tokarev, B. I., Shumny, U. K., 1977: Clarification of barley mutants with lowered nitrate reductase activity after treatment of the grain with ethylmethanesulphonate. Genetika **13,** 2097—2103.

Tomsett, A. B., Cove, D. J., 1979: Deletion mapping of the *nii*A *nia*D gene region of *Aspergillus nidulans.* Genet. Res. **34,** 19—32.

Tomsett, A. B., Garrett, R. H., 1980: The isolation and characterization of mutants defective in nitrate assimilation in *Neurospora crassa.* Genetics **95,** 649—660.

Tomsett, A. B., Dunn-Coleman, N. S., Garrett, R. H., 1981: The regulation of nitrate assimilation in *Neurospora crassa:* The isolation and genetic analysis of *nmr*-1 mutants. Mol. Gen. Genet. **182,** 229—233.

Triplett, E. W., 1985: Intercellular nodule localization and nodule-specificity of xanthine dehydrogenase in soybean. Plant Physiol. **77,** 1004—1009.

Vapnek, D., Hautala, J. A., Jacobson, J. W., Giles, N. H., Kushner, S. R., 1977: Expression in *Escherichia coli* K-12 of the structural gene for catabolic dehydrogenase of *Neurospora crassa.* Proc. Nat. Acad. Sci. USA **74,** 3508—3512.

Vega, J. M., 1972: NADH-nitrate reductase de Chlorella. Nouvelle contribution a l'etude de ses properties. Physiol. Veg. **10,** 637—652.

Vega, J. M., Kamin, H., 1977: Spinach nitrite reductase. Purification and properties of a siroheme-containing iron-sulphur centre. J. Biol. Chem. **252,** 896—909.

Vierstra, R. D., Quail, P. H., 1983: Purification and initial characterization of 124 K dalton phytochrome from *Avena.* Biochem. **22,** 2498—2505.

Wahl, R. C., Hageman, R. V., Rajagopalan, K. V., 1984: The relationship of Mo, molybdopterin, and the cyanolyzable sulphur in the Mo cofactor. Arch. Biochem. Biophys. **230,** 264—273.

Wallace, W., Oaks. A., 1984: Role of proteinases in the regulation of nitrate reductase. In: Dalling, M. J. (ed.), Plant Proteases. Boca Raton: CRC Press.

Wallace, R. B., Johnson, N. J., Hirose, T., Mijake, M., Kawashima, E. H., Itakura, K., 1981: The use of synthetic nucleotides as hybridization probes. II. Hybridization of oligonucleotides of mixed sequence to rabbit β-globin DNA. Nuc. Acids Res. **9,** 879—894.

Wallin, A., Glimelius, K., Eriksson, T., 1979: Formation of hybrid cells by transfer of nuclei via fusion of miniprotoplasts from cell lines of nitrate reductase deficient tobacco. Z. Pflanzenphysiol. **91,** 89—94.

Warner, R. L., Kleinhofs, A., 1981: Nitrate utilization by nitrate reductase-deficient barley mutants. Plant Physiol. **67,** 740—743.

Warner, R. L., Lin, C. J., Kleinhofs, A., 1977: Nitrate reductase-deficient mutants in barley. Nature (London) **269,** 406—407.

Warner, R. L., Kleinhofs, A., Muehlbauer, F. J., 1982: Characterization of nitrate reductase-deficient mutants in pea. Crop Sci. **22,** 389—393.

Weaver, R. J., 1942: Some responses of the bean plant to chlorate and perchlorate ions. Plant Physiol. **17,** 123—128.

Wilkerson, J. O., Janick, P. A., Siegel, L. M., 1983: Siroheme-Fe_4S_4 interactions in spinach nitrite reductase (NiR). Fed. Proc. Fed. Am. Soc. Exp. Biol. **42,** 2060.

Wilkinson, W. B., Greene, L. A., 1982: The water industry and the nitrogen cycle. Philos. Trans. R. Soc. (London), Ser. B **296,** 459—475.

Wray, J. L., Filner, P., 1970: Structural and functional relationships of enzyme activities induced by nitrate in barley. Biochem. J. **119,** 715—725.

Wray, J. L., Kirk, D. W., 1981: Inhibition of NADH-nitrate reductase degradation in barley leaf extracts by leupeptin. Plant Sci. Lett. **23,** 207—213.

Wray, J. L., Steven, B., Kirk, D. W., Bright, S. W. J. 1985: A conditional-lethal molybdopterin-defective mutant of barley. Mol. Gen. Genet. **201,** 462—466.

Xuan, L. T., Grafe, R., Müller, A. J., 1983: Complementation of nitrate reductase deficient mutants in somatic hybrids between *Nicotiana* species. In: Potrykus, I., Harms, C. T., Hinnen, A., Hütter, R., King, P. J., Shillito, R. D. (eds.), Protoplasts 1983 (Poster Proceedings), pp. 76—77. Basel: Birkhäuser.

Yatsunami, K., Pandya, B. V., Oprian, D. D., Khorana, H. G., 1985: cDNA-derived amino acid sequence of the γ-subunit of GTPase from bovine rod outer segments. Proc. Nat. Acad. Sci., U.S.A. **82,** 1936—1940.

Young, R. A., Davies. R. W., 1983a: Efficient isolation of genes by using antibody probes. Proc. Nat. Acad. Sci., U.S.A. **80,** 1194—1198.

Young, R. A., Davies, R. W., 1983b: Yeast RNA polymerase II genes: Isolation with antibody probes. Science **222,** 778—782.

Zabala, G., Filner, P., 1980: Reduction of chlorate by nitrate reductase of tobacco XD cells. 15th Ann. Report MSU/DOE Plant Res. Lab., pp. 72—73.

Zambryski, P., Joos, H., Genetello, C., Leemans, J., van Montagu, M., Schell, J., 1983: Ti plasmid vector for the introduction of DNA into plant cells without alteration of their normal regeneration capacity. EMBO J. **2,** 2143—2150.

Chapter 6

Plant Genetic Approaches to Symbiotic Nodulation and Nitrogen Fixation in Legumes

Peter M. Gresshoff and Angela C. Delves

Botany Department, Australian National University, Canberra ACT 2600, Australia

With 6 Figures

Contents

I. Introduction

We happily accepted the invitation to write this chapter because we have experienced the development of new approaches to research into the genetics, microbiology and biochemistry of symbiotic nitrogen fixation. We were confronted with a broad range of advances in the biochemical genetics of *Rhizobium,* especially relating to the genetic elements controlling the key symbiotic phenotypes: nitrogen fixation (the *nif* and *fix* genes) and nodulation (controlled by the *nod* genes). Likewise there has been a major application of plant molecular biology to the analysis of nodule-specific or nodule-enhanced plant gene products (Lee *et al.,* 1983; Fuller *et al.,* 1983; Govers *et al.,* 1985; and Kaninakis and Verma, 1985). This major class of proteins or nodulins allows the molecular visualisation of the plant genome's contribution to the symbiosis (Verma *et al.,* 1985).

The first volume of Plant Gene Research deals with the recent advances in plant-bacterial interactions (see also references by Dazzo and Gardiol, 1984; Miflin and Cullimore, 1984; Rolfe and Shine, 1984; Verma and Nadler, 1984). However, we feel that the molecular analyses of the bacterial genome (see Djordjevic *et al.*, 1983; Fischer and Hennecke, 1984; Schmidt *et al.*, 1984; Schofield *et al.*, 1983, 1984; Masterson *et al.*, 1985) and the nodulin sequence analysis have largely outstripped the level of understanding at the phenotypic level of whole plant biology. Even our knowledge of cellular processes is limited, although it is possible now to use a well-defined bacterial mutant to probe the plant genotype. For example, one can use *lacZ* fusions to bacterial nodulation promoters to evaluate the excretion of plant signals during early nodulation (Olson *et al.*, 1985).

Numerous plant functions are involved in nodule initiation and nodule function. Some of these are under direct bacterial control, others are developmentally regulated by endogenous plant factors. Nearly every plant function is integrated in the symbiotic process. However, we know little about essential mechanisms such as growth regulator activity, translocation, cell wall biosynthesis, developmental gradients, plant genome structure, gas exchange and especially gene regulation. No wonder that we really know so little about symbiotic nitrogen fixation. Perhaps by focussing on two easily recognisable phenotypes and assayable characters of development, like nodule development and nitrogen fixation, one can study phenomena which relate to plants in general, rather than the nodulation response *per se*.

Nodulation in legumes by *Rhizobium* offers a unique experimental system to study plant development as it is possible to isolate plant mutants altered in the symbiotic process, without their fitness being affected (Carroll *et al.*, 1985 a, b; Gresshoff *et al.*, 1985 b). This is due to the ability of legumes to use alternative modes of nitrogen acquisition such as nitrate reduction. This flexibility of course aids the investigation of both plant and bacterial mutants. In contrast, studies of flower initiation, root development, phytohormone biochemistry and/or fruit ripening are experimentally more complex, particularly if one requires a genetic approach which is to lead into analytial biochemistry of the organism and its development.

Thimann in 1936 stated that it was 50 years since the original discoveries of the nitrogen fixing root nodule, and the first successful cultivation of *Rhizobium* outside the plant. Yet, he remarked, the three main questions of the nodulation phenomenon remained: (i) how does the bacterium invade?, (ii) how is the nodule tissue induced? and (iii) how is nitrogen fixation regulated in the nodule? Today, another 50 years on, the answers are still not available.

The great contribution of genetics to the biochemical analysis of a wide range of organisms is the result of the researcher's ability to use genetic conditionality (i. e. mutant versus wild type) to provide a structure-function relationship. This is more difficult to establish through biochemical or physiological experiments alone. There is now a need to use the same approach to study the mechanisms controlling symbiotic nitrogen fixation and by doing so help to improve the legume symbiotic potential (see Tudge, 1984). Dommergues (1978) stated: "Enhancing nitrogen fixation by plant breeding is one of the most promising possibilities . . . we would like to reemphasise the need to develop research on plant genetics towards the improvement of the relationship between plants and microorganisms". For all of the above reasons we will focus here on areas of symbiotic nodulation and nitrogen fixation relating specifically to plant genetics. Our account will be somewhat speculative and personal, giving our work some extra, and perhaps otherwise not necessary, emphasis.

II. A General Description of Legume Nodule Ontogeny

Not all legumes nodulate. However, most members of the Viciaceae and Phaseoleae form nitrogen-fixing root nodules in symbiosis with the soil bacterium *Rhizobium.* In general, temperate legumes (such as *Pisum sativum, Medicago sativa,* and *Trifolium* species) develop indeterminate nodules which are characterised by (i) a defined meristem during nodule growth, (ii) an open vascular system connecting the root's vascular system with the nodule meristem, (iii) vacuolated infected cells and (iv) asparagine/glutamine as the major translocation products of nitrogen fixation. These legumes tend to be infected by the bacteria belonging to the *Rhizobium* genus (as compared to the newly recognised *Bradyrhizobium* genus which contains bacterial species formerly incorporated in the Rhizobium group; see Jordan, 1982). Tropical legumes of the *Glycine, Vigna, Lupinus, Macroptilium* and *Arachis* genera develop nodules of the determinate type, i. e. those in which the cell divisions responsible for the *de novo* synthesis of bacteroid-containing plant tissue cease early in nodule ontogeny. This means that mature, nitrogen-fixing nodules in contrast to the indeterminate nodule type do not show meristematic regions. Figure 1 shows nodules on roots of pea and soybean. Soybean nodules are spherical and pea nodules cylindrical in shape. However, it is now clear that the plant genome controls the nodule morphology. This is best exemplified by the *Parasponia* symbiosis, in which the same bacterial strain induces two totally distinct nodule types on two different plant hosts. Nodule sizes vary considerably between species and even within the same plant. For example, on clover or peas we have detected occasional nodule clusters of over 10 mm in diameter. Such variation appears to be environmentally induced.

The mechanisms of infection of legumes by *Rhizobium* are varied and mainly involve entry of the bacterium by an infection thread through the extending tip region of a young root hair (Bauer, 1981; Dazzo and Gardiol, 1984; Rolfe and Shine, 1984) or through hydrolysis of middle lamellar material of epidermal cells. Variations on these themes are found in stem nodulation, as seen in *Sesbania* and *Aeschynomene* species (Legocki *et al.,* 1983; Tsien *et al.,* 1983).

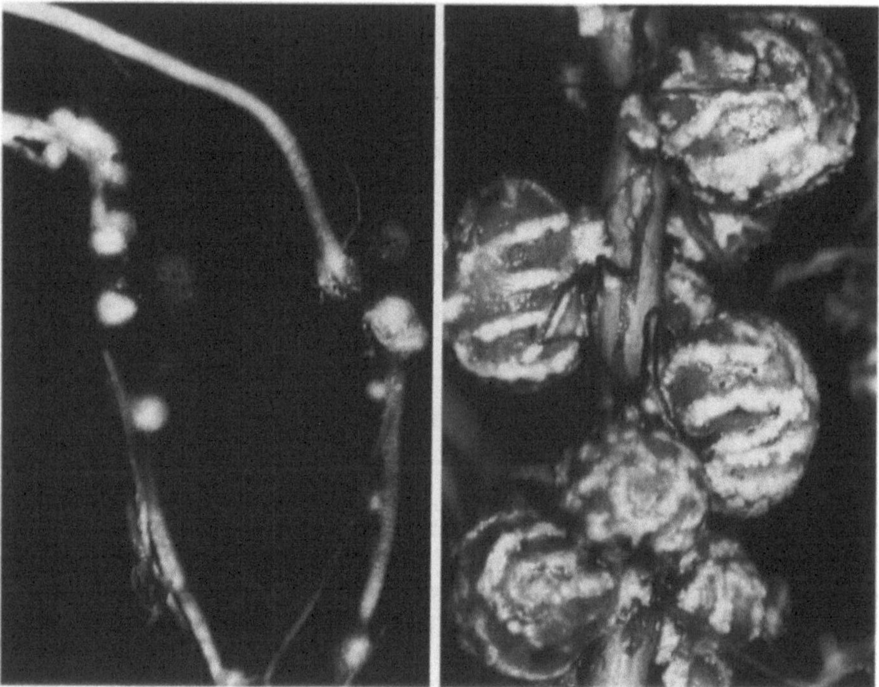

Fig. 1. Comparison of the nodule morphology of soybean and pea. The pea nodules (left side) are of effective and ineffective type caused by a mixed infection of two *Rhizobium* strains. The soybean nodules (right) are those formed on the taproot of a supernodulating mutant. Note the distinct pattern of lenticel development

The invasion by *Rhizobium* induces sustained cortical cell divisions. Recent analysis demonstrated that even prior to infection some cellular division zones exist in legume root tissue. For example, Collins (1983) found cortical "foci" located adjacent to the endodermis in infected and uninfected white clover roots. These were meristematic foci with up to 12 cells per cluster, which coincided with the site of nodule initiation. Microdensitometry of nodules showed that a large proportion of nodule cells have an increased DNA content. This led to the conclusion that the invading *Rhizobium* induced polyploidisation in cortical tissues and that this tissue was then stimulated to develop into the mature nodule. Gresshoff and Mohapatra (1981) reported studies with infected and uninfected (nitrate grown) white clover plants, from which cortical cell protoplasts were isolated by differential enzymatic digestion. These protoplasts were Feulgen-stained and air-dried prior to Feulgen microdensitometry. Cortical cells contained two distinct populations with DNA contents of 2C and 4C. There was no evidence for mitotic divisions or S-phase nuclei. The 4C cells thus were either tetraploid G_1 or diploid G_2 arrested. It was thus unnecessary to postulate the endoduplication induction by invading *Rhizobium*.

Appropriate target cells already existed. What is of significance is that legume plants do show G_2-arrest in cortical cells. The arrest is achieved through the translocation of a compound (7-methyl-nicotinate), which is found in cotyledons of beans and peas, but not in those of two non-legumes (Evans and Van't Hof, 1973). At present one can only speculate about the involvement of this mechanism in the nodulation process.

Calvert *et al.,* (1984) used serial sections of infected soybean roots to describe in detail the distribution of infection events. Many subepidermal (hypodermal) divisions were not associated with infection threads or even early infection events such as curling of the root hair. Many infection threads were found to abort without eliciting a nodule. The hypodermal infection foci were called "pseudoinfections" and could be induced only by ***Bradyrhizobium*** (i. e. the normal soybean microsymbiont). The induction did not require direct contact, as experimental separation of plant and bacterium by a millipore membrane resulted in the same phenomenon (Bauer, pers. comm.). Likewise it was seen that pseudoinfections could be induced by an exogenous application of cytokinins such as benzyladenine.

Figure 2 presents a model to explain the ontogeny of nodule development in soybean. The initial stages encompass the findings of Calvert *et al.* (1984), while the latter ones are highly speculative and are designed to include the observable structures of mature, nitrogen-fixing nodules. Of importance to the theme of this chapter is the presence of plant structures which also develop in other plant organs. Likewise it is suggested that *Rhizobium* as the microsymbiont is restricted by the plant in its growth (mainly through the plant-controlled milieu affecting oxygen and nutrient relations; Sinclair and Goudriaan, 1981; Witty *et al.,* 1985). Thus the plant is actively involved in the prevention of a parasitic invasion by *Rhizobium* (Werner *et al.,* 1984).

The nitrogen-fixing nodule is of benefit to the plant, because of its ability to maintain bacteroids, which convert atmospheric nitrogen gas to ammonia. Nitrogenase is made up of two major subunits, the iron protein (also called nitrogenase reductase, coded for by the bacterial gene *nifH;* molecular weight = 31 000, Scott *et al.,* 1983) and the molybdenum-iron protein (coded for by the *nifD* and *nifK* genes; molecular weight = 56 000 each; Weinman *et al.,* 1984; see Brill (1980) for general references). The nitrogenase complex requires ATP, electrons and redox potential to facilitate the reduction of nitrogen gas. The following reaction is thought to best describe the stoichiometry:

$$N_2 + 10H^+ + 16ATP + 8e^- \rightarrow H_2 + 2NH_4^+ + 16ADP + 16\ P_i$$

Hydrogen production by nitrogenase is an inescapable part of the reaction. So far all nitrogenase systems produce hydrogen and therefore this process is viewed as an energy loss to the bacterium and thus the plant. Several *Rhizobium* strains possess a hydrogen uptake *(hup)* system, which "re-cycles" hydrogen gas to form water and ATP (Schubert and Evans, 1976).

The *hup* system may be of value to the symbiosis, as H. J. Evans (pers. comm.) has recently shown that hup^+ strains of *Bradyrhizobium japonicum* give up to 10% better seed yields in symbiosis with soybean when compared with hup^- isolates. The ATP, electrons and redox potential are supplied through the respiratory chain of the bacteroid. There exists evidence that *Rhizobium* species possess several respiratory chains (Appleby, 1984).

The nitrogenase enzyme is extremely oxygen sensitive (Bergersen and Turner, 1967, 1978; Mohapatra and Gresshoff, 1984). Exposure of isolated bacteroids to atmospheric oxygen for just one minute eliminates most nitrogenase activity through enzyme instability rather than transcriptional control (Van den Bos *et al.,* 1983). The regulation of transcription of nitrogenase genes in *Rhizobium* bacteroids is different than that found in free-living bacteria such as *Klebsiella pneumoniae,* although there may be several similarities such as promoter sequences and *nifA*-like activation of transcription (see Elmerich, 1984) for a review of the regulation of

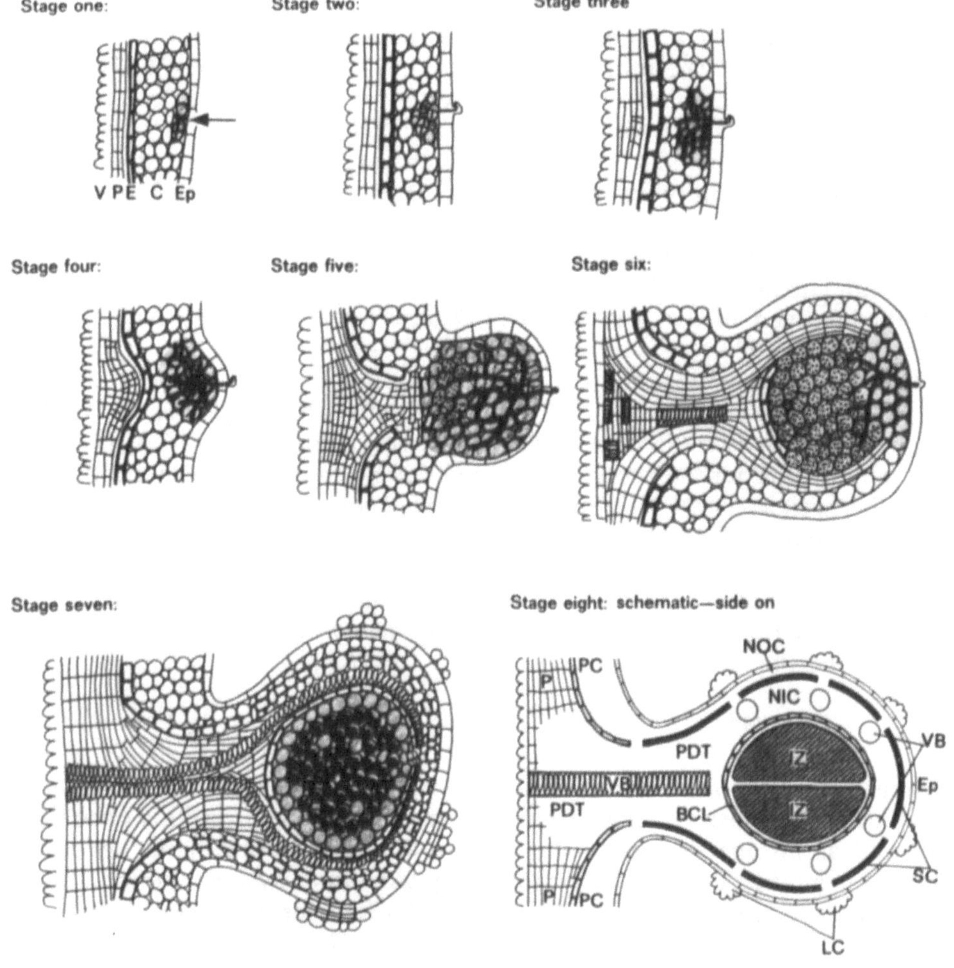

nitrogenase in free-living diazotrophs). Neither oxygen nor ammonia repressed transcription of the nifHDK message of *R. leguminosarum* bacteroids (Van den Bos *et al.,* 1983). Likewise *ntrC* mutants of *R. meliloti* were symbiotically active, although the development of the symbiosis was delayed (Ausubel *et al.,* 1985).

Oxygen is needed for nitrogen fixation, as illustrated by studies with isolated bacteroids under oxygen-free conditions (Bergersen and Turner, 1967). The legume host achieves a solution to the apparent paradox through the development of regulatory and restrictive structures in the nodule (Sinclair and Goudriaan, 1981; Ralston and Imsande, 1982). The iron-containing leghemoglobin protein plays a key role in the facilitated oxygen transport to the bacteroid (Appleby, 1984). Recent work by Minchin and collaborators demonstrated the presence of a variable oxygen diffusion barrier in several legume nodule types (Minchin *et al.,* 1983; Sheehy *et al.,* 1983; Witty *et al.,* 1984). Carroll (1985) also provided evidence, through a variety of experiments, which suggested the existence of the oxygen diffusion barrier in soybean nodules. This barrier may be activated as soon as there is an alteration to the efficient functioning of the nitrogenase system. It is even conceivable that a "feed-forward" control

---

Fig. 2. Model of the nodule ontogeny in the soybean-*Bradyrhizobium* symbiosis. *Stage one:* the bacterium (arrow) infects an epidermal cell and elicits subepidermal cell divisions. V = vascular bundle; P = pericycle (the site of lateral root formation); E = endodermis; C = cortex; Ep = epidermis. *Stage two:* after penetration of the infection thread of the root hair, cortical cells divide in advance of the ramifying infection thread. Such hypodermal divisions can occur without infection and are termed pseudoinfections. *Stage three:* increased cortical activity "signals" pericycle tissue to initiate divisions. *Stage four:* progressing cortical activity results in a visible lump. Further pericycle divisions distort and eventually rupture the endodermis. *Stage five:* bacteria leave the infection thread and start bacteroid development leading towards nitrogenase activity. Pericycle tissues push through the endodermis and proliferate towards invaded (shaded) zone. Nodule expansion may break through the epidermis (this is not shown on the diagram). *Stage six:* continued development of pericycle and possibly cortical cell material to develop vascular traces surrounding the infected zone. Tracheids at first develop in parallel to the root axis, often in unconnected cell clusters, but then "grow" together to form perpendicular vascular strands (indicated by the spiral tissue). The early boundary cell layer around the infected zone is noticeable as is the formation of an endodermis-like layer in the nodule cortex. *Stage seven:* the infected zone is fully nitrogen fixing and is characterised by infected and uninfected cells. The latter type harbours the enzymes related to ureide metabolism. Vascular bundles are connected and surround the infected tissue. Scleroid (thick-walled) cells develop in a nearly continuous layer in the nodule cortex. Gaps in this layer are associated with the position of the vascular bundles, which also govern the location of the lenticels (LC). *Stage eight:* as stage seven, but indicating some possible cell lineages. P = pericycle; PC = plant cortex; LC = lenticel; SC = scleroid cell; IZ = infected zone; VB = vascular bundle; EP = epidermis (casparian strip); NOC = nodule outer cortex; NIC = nodule inner cortex; PDT = pericycle derived tissue; BCL = boundary cell layer.

loop is in action, as the plant itself does not appear to have a chance to monitor the altered nitrogen output of the bacteroid within the one to two minutes that are taken to facilitate the barrier's action. Anatomically this barrier resides in the cortex of the nodule as confirmed by direct measurements using microelectrodes (Tjepkema and Yocum, 1974; Witty *et al.*, 1985). The cortex contains a layer of scleroid-type cells which is not continuous and shows gaps above the position of the vascular traces. Lenticels, being parenchyma-like outgrowths on the nodule surface, also follow the vascular traces (see Fig. 1 for well-developed lenticels on the surface of the soybean nodules). Pankhurst and Sprent (1975 a, b) found that desiccation or waterlogging inhibited nitrogenase activity of nodules. This loss of activity was correlated with the anatomical collapse of the lenticel structure. Similarly, the same authors found that elevated oxygen concentrations reversed the waterlogging-induced decrease in nitrogenase. Dark-induced and nitrate-induced losses of nodule nitrogenase activity were also recovered after nodules were incubated at elevated oxygen levels (Carroll, 1985).

In general it is arguable that legume nodules regulate bacteroid nitrogenase activity not through the controlled supply of carbohydrates (say malate or succinate), but through the provision of oxygen, which is needed for oxidative phosphorylation. Additional barriers to oxygen transport into the nodule interior are (i) the lack of airspaces in the inner cortex (Bergersen and Goodchild, 1973), (ii) the bacteroid-boundary cell layer (see Fig. 2) and (iii) plant respiration and cytoplasmic oxidases which lower the oxygen concentration close to the bacteroid surface. Bacteroids scavenge low amounts of free oxygen (being about 1—10 micromolar in most legumes) through the interaction of leghemoglobin and specific oxidases which accept oxygen from the hemoglobin molecule. This protein is an example of a molecular symbiosis. The globin moiety is a product of plant genes which are directly related to globin genes found in animals (Appleby, 1985; Verma *et al.*, 1985). The heme molecule is the presumed product of the bacterium and is excreted into the plant cytoplasm during bacteroid development. However, there is no evidence for heme export into the peribacteroid space or for heme transporter proteins in the bacterial and peribacteroid membranes. Part of the argument for the bacteroid origin of heme is the observation that heme biosynthesis was regulated by microaerobic conditions similar to those expected in the developing nodule (Nadler and Avissar, 1977; Keithley and Nadler, 1983). Furthermore, there is an apparent correlation between bacteroid content and heme content in soybean nodules (D. Day, pers. comm.) and genetic data from *Rhizobium meliloti* mutants, which lack the first enzyme of the heme biosynthetic pathway (being delta-aminolevulinic acid (ALA) synthetase), provided evidence for the bacterial origin of the heme moiety. The mutants produced white nodules while their revertants gave red nodules (Leong *et al.*, 1982; Ditta *et al.*, 1983). Such an argument is, however, not sufficient, as one finds that, for instance, leucine-requiring auxotrophs of *R. trifolii* also form white nodules and that their revertants are symbiotically effective (Bassam and Gresshoff, 1986). Yet bacterial leucine biosynthesis is not directly essential for leghemoglobin biosynthesis. The classical assumptions of this process were questioned recently through the isolation of an ALA-synthetase-deficient mutant of *Bradyrhizobium japonicum* which still nodulated and gave leghemoglobin-containing nodules (Guerinot and Chelm, 1985). Whether ALA was crossfed from the plant to the bacteroid for further processing, or whether an alternative pathway existed in the bacterium, or whether the plant after all supplied the heme, is not clear.

Robertson *et al.* (1984) using immunogold labelling localised leghemoglobin in the cytoplasm and not in the peribacteroid space of pea and lupin nodules. Verma *et al.* (1985) obtained similar results with soybean nodules. How oxygen is transported across the peribacteroid space is not clear. Simple diffusion could suffice, especially since the volume of the peribacteroid vesicle may be enlarged in electron-micrographs due to fixation artifacts (G. D. Price, pers. comm.). In relation to this point and the next, it is clear that there are various aspects of symbiotic nitrogen fixation where the final analysis of the mechanism has not been completed. In other words, gray areas remain in mechanisms which could have been studied by existing technologies of biochemistry and microbiology. For example, the succinate-uptake mutants of Ronson (see next paragraph) still induce leghemoglobin-containing nodules. However, bacterial ALA synthetase uses glycine and succinate as substrates. So where does the succinate-uptake mutant, which is really struggling to get enough dicarboxylic acids to maintain nitrogenase, obtain the succinate to permit an excessive excretion of heme to the plant host? It is these shortcomings or gaps in the knowledge of the basic biology of nitrogen fixation that hopefully can be closed by the application of biochemical genetics of the plant and the bacterium.

Genetic studies by Ronson *et al.* (1981, 1984) substantiated the biochemical and physiological findings of the importance of dicarboxylic acid transport to the nitrogen-fixing bacteroid. Thus bacterial mutants, which lacked the succinate-malate-fumarate uptake system were capable of nodulation and bacteroid development, but not of nitrogen fixation. The *dctA* gene has been cloned and sequenced. Hydropathic analysis of the amino-acid sequence showed 12 transmembrane helices consistent with a membrane-associated molecule. Analysis of the carbon requirement of isolated bacteroids also implicated the critical role of succinate-malate. Often it was claimed that succinate increased nitrogenase activity but this was a restrictive interpretation, as it depended on the external oxygen concentration. Investigations with isolated bacteroids from soybean and *Parasponia rigida* (a non-legume nodulated effectively by certain *Bradyrhizobium* strains) showed that maximal nitrogenase activity did not change through the addition of succinate, but that instead the oxygen sensitivity profile was altered. This meant that in the presence of higher oxygen levels during incubation, succinate increased detectable activities (McNeil *et al.*, 1984; Carroll, 1985; Sandeman and Gresshoff, 1985). McNeil *et al.* (1984) showed that optimum nitrogenase activities over the first 100 min. at six levels of oxygen between 0.1 % and 1.5 % in the gas phase did not differ between the additions of 50 mM sucrose, 50 mM glucose or 50 mM succinate. These optimum activities were identical to the "no carbon-source" control, suggesting that soybean bacteroids had large endogenous supplies of available carbon source. If bacteroids were starved of carbon *in planta* by preincubation of the plant in the dark for 2.5 days and assayed after anaerobic isolation for initial nitrogenase activity, then the addition of 50 mM pyruvate or succinate (even as low as 1.5 mM) resulted in substantial increases of nitrogenase activity. Dicarboxylic acids other than succinate may be of physiological importance. For example, the carbon compound most important in the support of nitrogen fixation may be malate, which is produced by phospho-enol-pyruvate carboxylase (PEPC) in the nodule cytoplasm without "taxing" the already "stressed" nodule mitochondria. PEPC is a major enzyme found in nodule cytosols (Gadal, 1983). A more speculative suggestion was made by Kahn *et al.* (1985), who proposed that glutamate may be a key compound transported into the *Rhizobium* cell. It may be of value to test this hypothesis with *dct* mutants or with isolated bacteroid systems as described above.

Ammonia produced by nitrogenase diffuses out of the bacteroid. There is no direct evidence for an ammonia export system under bacteroid conditions, as the plant cytoplasm contains high amounts of plant-derived glutamine synthase which assimilates the ammonia with a low K_m to form glutamine, which in turn serves as a translocation product or is used for transamination reactions (Streeter, 1977; Duke and Henson, 1985). In tropical legumes the major translocation products are the ureides (Schubert, 1981; Hanks *et al.,* 1983). The level of ureides in the xylem, especially if expressed as relative ureides, has been taken as a measure of the amount of nitrogen fixed (Herridge, 1982). McNeil and LaRue (1984) demonstrated that ureides were also produced by nitrate-grown and non-nodulated soybean plants. Additionally it should be considered that the determination of the relative ureide level assumes that nitrate uptake is constant between comparative samplings. This may be so within the same cultivar monitored at the same developmental stage, but may differ significantly between different cultivars. The conclusion is that we do not even know how to measure nitrogen fixation in the field accurately!

III. The *Parasponia-Bradyrhizobium* Symbiosis

As stated above, *Rhizobium* and *Bradyrhizobium* species form a variety of nodule types on a range of legumes. This was taken in the past as a basic characteristic of *Rhizobium.* One notable exception to the legume specificity of *Rhizobium* is the nodulation of plants within the *Parasponia* genus by many *Bradyrhizobium* isolates. This was first discovered by Trinick (1973), although the plant species was misidentified as the related species *Trema.* The *Trema* genus has not yet been reported to nodulate, although it belongs to the *Ulmaceae,* of which several species are nodulated by the bacterium *Frankia* (Lalonde, 1979). Since then numerous studies have utilised the *Parasponia* system and progress has been reviewed by Gresshoff *et al.* (1984). A few distinct features will be highlighted here. The association is as efficient as the legume symbiosis in terms of nitrogen fixation, although nodule number per plant in *Parasponia* is increased, and the amount of infected tissue per nodule is decreased when compared to the legume (such as siratro, *Macroptilium atropurpureum*) control grown under identical conditions with the identical bacterial inoculant. The *Parasponia* nodule seems similar to a modified lateral root as it has a central vascular cylinder. Nodule growth starts in the pericycle and not the cortex. The infected zone contains infected and uninfected cells with the hemoglobin being localised in the cytoplasm of the infected cell. The amino-acid sequence of the *P. andersonii* hemoglobin shows very close homology to that of the soybean Lb_c hemoglobin, suggesting a common evolutionary origin (Appleby *et al.*, 1983; Kortt *et al.*, 1985). The preliminary *Parasponia* hemoglobin sequences referred to by Gresshoff *et al.* (1984) appear to be artifacts and erroneous. Bacteria in the *Parasponia* nodules apparently fix nitrogen, while still contained in the modified infection thread. This structure has been termed the fixation thread (Price *et al.*, 1984).

The infected zone is surrounded by a layer of cells which show extensive tannin accumulation, especially in the vacuole (Price *et al.*, 1984). This deposition of phenolic compounds may be related to the recognition by the plant of the microsymbiont as a potential pathogen (compare to later discussion of the gene-for-gene concept).

Isolated bacteroids of *Bradyrhizobium* strain ANU289 obtained from nodules of *P. rigida* showed similar oxygen sensitivities as siratro-derived bacteroids of the same strain (Sandeman and Gresshoff, 1985). *In vitro* derepression studies of strain ANU289 indicated the further similarities to other *Bradyrhizobium* strains (Mohapatra and Gresshoff, 1984). This similarity was further confirmed by molecular analysis of the nitrogenase genes, which share sequences as well as the genomic arrangement with other bacteria which infect soybean or peanut but not *Parasponia* (Scott *et al.*, 1983; Weinman *et al.*, 1984). This non-legume symbiosis demonstrates clearly that one specific *Bradyrhizobium* strain can be flexible enough to invade a root, induce a morphologically different nodule, and maintain a functional symbiosis in two distinctly different plant hosts. It also suggests that the two hosts are capable of supplying a relatively similar biochemical and physiological niche. Hence it is possible that many other non-legume plants are capable of infection and nodulation by *Rhizobium*, although these may not have been recognised as yet. The taxonomists of the last two centuries have seldom included root characteristics in their botanical writings. The question arises now as to the nature of the plant genes which facilitate infection and nodule formation in this symbiotic system. Are they similar to legume genes? Do other plants possess similar genes? How does the bacterium nodulate the two hosts? Are different bacterial genes employed? What is the function of symbiotically activated genes in non-symbiotic tissue and/or species?

IV. Biochemical and Molecular Analysis of Plant Functions

There are numerous plant gene products involved in nodulation and nitrogen fixation. Many should also have "normal" plant functions, which are utilised in this alternative developmental pathway. Others will be more or less specific. Recent molecular analyses have focussed on the presence of nodule-specific proteins called nodulins (Bisseling *et al.*, 1983, 1985; Fuller and Verma, 1984; Fuller *et al.*, 1983; Miflin and Cullimore, 1984; Verma and Nadler, 1984; Govers *et al.*, 1985; Verma *et al.*, 1985). The definition of the term nodulin may also have to include nodule-amplified proteins. At present the concept also seems to focus on the root as a control tissue. This is not a necessity, as some symbiotic functions may utilise genes which are only expressed in other tissues at another time in the plant's ontogeny.

The major approach has been the production of nodule-specific antisera which were used to visualise peptides on two-dimensional gels. Antisera against nodule proteins were titrated with excess root proteins, thereby precipitating the "common" antibodies. The remaining uncomplexed antiserum contained antibodies which were predominantly specific for nodule proteins. If used in a Western blot, these sera would recognise the nodule-specific peptides and this was visualised by a second reaction in which the now-complexed antibody was recognised by a second antibody which carried a means of visualisation. Usually this was a radioactive iodine tag or a covalently-linked enzyme such as peroxidase. The presence of the antibody-protein interaction can thus be recognised by either autoradiography or colorimetric methods. In a similar fashion, it was possible to prepare cDNA from nodule RNA and to use such molecules in the differential screening experiments. To do so, cDNA libraries from root and nodule tissues of similar age were screened for differential hybridisation against nodule-derived cDNA probes. Relevant nodule-specific clones were then analysed, rechecked for specificity, sequenced and used to isolate genomic sequences of homology. This procedure allowed the isolation of three characterised nodulin genes. These are the leghemoglobin family (Lee *et al.*, 1983), glutamine synthetase (Cullimore *et al.*, 1983) and nodulin 35 (being the enzyme uricase, which is essential in the ureide biosynthetic pathway; Legocki and Verma, 1979; Bergmann *et al.*, 1983).

Major advances have been achieved in the pea and soybean system (for reviews see Bisseling *et al.*, 1983, 1985, as well as Verma *et al.*, 1985). In pea, Bisseling and collaborators found several (up to eight) nodulins which were coordinately expressed with leghemoglobin. This occurred whether the nodule was induced by a bacterium capable or incapable of nitrogen fixation. Similarly the *Rhizobium* chromosomal (as compared to plasmid) replicon was required for complete nodulin expression, as *Agrobacterium tumefaciens* transconjugants showed only one nodulin. Similar results were obtained in soybean and in alfalfa *(Medicago sativa)*. The composite data of Bisseling *et al.* (1985), Lang-Unnasch and Ausubel (1985) and Mohapatra and Pühler (1986) argued for a cascade-type of activation during nodule development. A key event may be the success of the invasion within the first 24 hours after infection. This apparent escape from the host defences is then followed by the plant regulation of further infection and nodule proliferation (autoregulation) and finally the nodule control over nitrogen fixation through the control of oxygen diffusion to the encased bacteroid. Thus the plant successfully maintains control over the infection through a series of coordinated developmental steps.

In our laboratory we tried to test whether this coordination of symbiotic steps is maintained if the symbiosis is restricted or inhibited by external factors. Schuller *et al.* (1986) applied 10 mM nitrate to established symbiotic soybean plants and followed a number of symbiotic parameters. They found that nitrate inhibited nitrogenase activity (up to 50 % inhibition after 2 days), but that all nodule parameters such as leghemoglobin, glutamine synthetase, ureide export, uricase, xanthine dehydrogenase, and isolated bacteroid nitrogenase activity were maintained at control levels, until general nodule senescence set in after about 14 days, when most relevant peptides and their activity disappeared. It is thus likely that the symbiotically important plant functions, once induced in a coordinate way, are maintained and that no further gene regulation is possible other than through general abscission of the organ. These findings fit to those related to cascade-type inductions and eventually may tell us more about determination and differentiation in plants.

After the significant findings of bacterial genetics leading to the definition of the nodulation genes (Schofield *et al.*, 1983, 1984; Kondorosi *et al.*, 1984; Schmidt *et al.*, 1984) and the recognition of plant exudates (Halverson and Stacey, 1985; Olson *et al.*, 1985; Mulligan and Long, 1985) and lectin binding features, several researchers have turned their attention to the plant's direct contribution in the symbiosis. Although the bacterium harbours the causative factors (*nod* genes, *nif* genes and perhaps heme biosynthesis genes), the plant has retained control over the key processes. Such considerations are progressively becoming important for those who desire to improve commercially the legume-*Rhizobium* symbiosis (see also a general article on this subject by Barton and Brill, 1983; a comprehensive review dealing with the potential of *Rhizobium* improvement was prepared by Hodgson and Stacey, 1986).

V. Gene-for-Gene Aspects of Nodulation

Pathogenic interactions between plants and microbes are of two main types — either successful invasion and multiplication of the pathogen at the expense of the plant, or successful resistance of the plant at the expense of the pathogen, preventing its invasion and increase. Situated in neither of these camps is the legume/*Rhizobium* symbiosis, in which both plant and bacterium benefit from their association, but which nevertheless has many features in common with the usual plant/pathogen interaction (Vance and Johnson, 1983). These features include the attachment of bacteria to the host plant, their successful penetration, the response of the host to this invasion, a change in plant metabolism and obvious morphological changes as a result of the infection.

As in any form of interaction between two organisms there must be a genetically controlled recognition sequence, which determines whether or not the symbiosis or pathogen attack will be successful, or elicit a negative interaction such as a hypersensitive response, phytoalexin accumulation or similar (Albersheim and Anderson-Prouty, 1975; Daly, 1984; Sequira, 1984). This interaction is a function of genes present in both bacterium and plant. Whereas considerable information is available on the genes which condition successful nodulation in *Rhizobium,* much less, as stated in the introduction, is known about the plant genes. It has been estimated that at least 10 plant genes are involved in nodule development alone (Holl and LaRue, 1976). A similar number may also be controlling the infection process as well as the functioning of the nitrogen fixation processes, including ammonia assimilation, carbohydrate supply and oxygen diffusion. Whatever this number, it is large and thus gives potential for genetic variation within the plant. This in turn has phylogenetically resulted in the specificity of infection and nitrogenase expression seen in efficient symbioses. The exact nature and mechanics of this specificity, and the genetic components of its control are not well understood, particularly in relation to the plant. In other successful pathogen attacks a gene-for-gene relationship has been demonstrated (Flor, 1955, 1971; Day, 1974), which describes the response of the host plant to various gene products from the pathogen, but does not, in most cases, explain the way in which a host plant recognises infective or non-infective pathogens. With the *Rhizobium*-legume symbiosis involving fast-growing *Rhizobium* strains the genes coding for the ability to recognise a host plant are located on large indigenous plasmids (Djordjevic *et al.,* 1983, and many others). These symbiosis (Sym) plasmids can be removed (cured) from the strain, rendering it unable to interact with the legume host. When transferred to a different *Rhizobium* species, the presence of the Sym plasmid allows initiation of nodulation but not nitrogen fixation in the new host (Rolfe and Shine, 1984), thus indicating that at this point specificity is controlled by genes on the chromosomal rather than the Sym plasmid replicon. In *Bradyrhizobium* the symbiotic genes are not located on plasmids (Fischer and Hennecke, 1984; Masterson *et al.,* 1985). The use of large deletions spanning the symbiotic region, and replacement analysis, however, have aided complex genetic analysis, so that the absence of symbiotic plasmids is no longer a problem to further genetic investigation of *Bradyrhizobium.*

Whatever systems are involved, they must include an interchange of signals which enable more specialised pathogens to avoid being identified by the host defence systems as foreign, and allow them to continue their development within the plant tissue. An incompatible combination of pathogen or host results in a reaction which effectively prevents any further pathogen development. That plants can respond to microorganisms has been known from the beginning of this century; the term "hypersensitive response" was coined for the reaction, in which cells in a resistant host adjacent to the infection site rapidly become discoloured, granular and necrotic (Klement, 1982). There are many examples of this type of disease resistance to be found in cereals against rusts and mildews, in potato against *Phytophthera infestans*, and in other plant species against bacteria, viruses and nematodes (Misaghi, 1982). In all cases the response is similar. In comparisons of the mode of host response to virulent and avirulent pathogens, no difference was detected in the way in which the epidermis was initially penetrated. After infection, however, resistant plant varieties showed a loss of turgour, browning of localised tissues, and cellular death. These changes in the cell were associated with loss of permeability of the plasma membrane, increased respiration, accumulation and oxidation of phenolic compounds and the production of phytoalexins (Keen and Kennedy, 1974; Heath, 1980). The infected cell and a few surrounding ones died, and the invader was engulfed by necrotic tissue which blocked successful proliferation of the pathogen. This mechanism appears commonly in a range of plants infected by a range of microorganisms, and thus may indicate a common resistance mechanism in plants to perceived pathogens (Heath, 1981). Halverson and Stacey (1986) reviewed the general area of plant-microbe interaction and discussed in detail aspects of the molecular signalling between host and potential pathogen. (See also Chapter 8 in this volume).

Recent studies have concentrated on the host-pathogen interaction in early stages of infection, with particular reference to host plant responses (Fraser, 1982; Misaghi, 1982). Avirulent strains of *Pseudomonas solanacearum* which attached only weakly to cell walls of tobacco and were agglutinated by a hydroxyproline-rich glycoprotein extracted from potato tubers, lacked the O-antigen oligosaccharide of the lipopolysaccharide (Keen and Williams, 1971; Billing, 1982). Electron-microscopic analysis showed that the pathogen, when infiltrated into tobacco leaves, was closely attached to plant cell walls and at a later stage became enveloped in a fibrillar material. This did not happen with virulent strains of the pathogen, which multiplied in the intercellular spaces. Clearly the plant recognised the avirulent strain but not the virulent one.

With the *P. solanacearum* system the recognition is known to be associated with lectins in the plant, which bind to the lipopolysaccharide (LPS) of the bacterial cell wall. Purified LPS binds to plant walls in the same way as do avirulent bacteria. LPS of both virulent and avirulent strains behave in a similar way. The only difference is that virulent bacteria produce extra slime (mainly in the form of exopolysaccharides, EPS). This additional external "dressing" may prevent the recognition of the active LPS factors by the plant lectin and the concomitant hypersensitive response (Billing, 1982).

The legume-*Rhizobium* symbiosis, as stated above, requires a number of precise interactions and different responses from the plant to develop the functioning nodule. Interaction is probable prior to attachment (cf. pseudoinfections in soybean), during attachment, infection thread growth and penetration into neighbouring cells, induction of cell proliferation to initiate nodule growth, bacteroid release from the infection thread, and during the bacteroid development-bacteroid maintenance phase. *Rhizobium* mutants have been isolated which have altered EPS production. Their symbiotic phenotypes vary, but in general one finds less efficient symbioses, less nodules, and stronger rejection of the bacterium (Johansen *et al.*, 1984; Müller *et al.*, 1985). The EPS of *Rhizobium* may be important as a protective barrier when inside the root environment, but may also be necessary to elicit the correct response from the plant to complete the nodulation sequence. Mutants altered in their EPS biosynthesis in general are recognised as pathogens and cause a hypersensitive response thereby preventing further growth. For example, the nodules induced by the *R. meliloti* EPS-deficient mutants decribed by Müller *et al.* (1985) were empty of bacteria. The nodules were tumour-like and showed obvious signs of initial bacterial infection as the nodule surface showed small infection pockets. These were clearly enclosed and aborted infections. The cell wall surrounding the infection pocket showed thickenings and clear secondary cell wall deposition. Of interest is that the correct set of nodulins were still induced, including hemoglobin (Mohapatra and Pühler, 1986), suggesting that the nodulation induction signal was only required in the early stages of the interaction.

Analysis of soybean nodules formed by an ineffective *Bradyrhizobium* strain (61-A-24) revealed an increase of the phytoalexin glyceollin, which occurred in the root or nodules formed by wild-type bacteria (Werner *et al.*, 1985). The ineffective nodules lacked hemoglobin and had an unstable peribacteroid membrane. Nodules were numerous and small compared to wild type. Another *fix*$^-$ mutant lacked these symbiotic phenotypes. The data suggest that in some *Rhizobium* mutants the latter stages of symbiotic communication are defective, resulting in the host's recognition of the invader and the accompanying phytoalexin effect.

VI. Existing Plant Variation in Symbiotic Nitrogen Fixation

It has long been recognised that the plant host genotype influenced the major symbiotic characteristics (Nutman, 1946, 1949, 1953; Holl, 1983). This was despite the failure of plant breeders and agronomists to recognise the importance of nodulation and nitrogen fixation to plant productivity. For example, in a recent paper Nelson and Bernard (1984) investigated the production and performance of hybrid soybean. They described heterotic interaction for characters such as yield, maturity date, lodging, height, harvest index, percentage oil and protein, seed quality and weight, but not symbiotic characters.

A major input into the recognition of plant cultivar effects came from multistrain — multicultivar interaction trials (Pacovsky *et al.*, 1984). Certain cultivars do better with certain *Rhizobium* strains. For example, Rennie and Kemp (1984) compared the nitrogen fixation abilities of two bean *(Phaseolus vulgaris)* cultivars by ^{15}N isotope dilution technique. They found that cultivar Aurora had superior nitrogenase activity at 12°C compared to cultivar Kentwood (average 119 kg N/ha versus 84 kg N/ha at 10 kg N/ha supplied nitrogen). This difference was environmentally influenced, so that culture at 15°C negated the superiority of Aurora. The essential finding was that Aurora was superior because of a longer vegetative phase supporting nitrogen fixation. Similar studies with soybean (Rennie and Dubetz, 1984) also demonstrated differences (e. g. cultivar Maple Amber fixed 91 kg N/ha while cultivar Maple Presto fixed 60 kg N/ha). The strong environmental influence may be explained by the apparent multigenic nature of this type of variation involving complex gene and strain interactions.

Differential nodulation patterns and nitrogen fixation rates were seen in many legume species. Harper and Gibson (1984) compared the nitrate sensitivity of soybean, lupin, chickpea, siratro, pea, subterranean clover, lablab bean and barrel medic in hydroponic solution. Clear differences emerged, with soybean, lablab bean and barrel medic being more sensitive to nitrate in terms of nodulation than the remainder of the tested species. Similar variation in nitrate sensitivity has been reported by Herridge and Betts (1985), who tested over 480 soybean cultivars and found continuous variation for symbiotic parameters such as nodulation index and ureide content. Their work claimed that two cultivars of Korean origin had increased nodulation and apparent nitrogen fixation on nitrate-containing soils, when compared to a control cultivar Bragg. Whether it is possible to use such variation in breeding programmes is not known at the moment and will depend on the genetic nature of the variation.

Skot (1983) found cultivar differences in peas, so that at 63 days after planting the plant genotype accounted for 75 % of the variation in dry weight, while the *Rhizobium* inoculum accounted for 63 % of the variation in nitrogen content and 70 % of the variation in dry weight/total N ratio.

Plant control over nitrogen fixation was demonstrated by Thomas *et al.* (1983), who inoculated soybean variety Wells with two different *Bradyrhizobium japonicum* mutants which had an enhanced ability for nitrogen fixation. Although improvements were seen by three weeks (most likely caused by increased early nodulation and perhaps competition), by five weeks there was no difference in N content of the xylem sap, acetylene reduction, leaf allantoinase activity and shoot ureide content.

Sen and Weaver (1984) studied the differential nitrogen fixation rates observed when the same cowpea *Bradyrhizobium* strain was tested on peanut and cowpea. Bacteroids isolated from both hosts had similar rates of oxygen consumption and acetylene reduction. Although the number of bacteroids per unit nodule volume was higher in cowpea than in peanut the nitrogenase activity as measured on whole plants showed that peanut was more active. This suggested a partial oxygen deficiency within the more "crowded" cowpea nodule and the conclusion was supported by the fact that increased partial pressures of oxygen increased the observable nitrogenase activity in cowpea plants.

Phillips and associates showed that the plant genotype affected hydrogen evolution by *R. leguminosarum* induced nodules on pea (Bedmar and Phillips, 1983; Bedmar *et al.*, 1983). A shoot factor in pea regulated the degree of hydrogen evolution (Bedmar and Phillips, 1984). Similarly, alfalfa cultivars were selected from existing material to produce lines with consistently higher nitrogen fixation rates. This increase was reflected in improved fitness in the absence and presence of exogenously supplied ammonium nitrate. Thus, performance of selected line HP32 exceeded that of HP for dry weight and total reduced nitrogen by 42 % and 45 % in ammonium nitrate-dependent plants and by 69 % and 81 % in plants dependent on nitrogen fixation alone (Phillips *et al.*, 1985). In field conditions, such differences were reduced to forage matter increases of 10 % and total forage nitrogen increases of about 22 %. Whilst agronomically significant, and again demonstrating the importance of the plant genotype in the joint symbiotic phenotype, the application of existing variability to biochemical analysis was hindered by the general inability to explain the causes for the improved line. This again goes back to the possible multigenic nature of such phenotypic differences, which will result in multiple biochemical changes.

Leghemoglobins of *Medicago sativa, Lathyrus, Lupinus polyphyllus* and subterranean clover were evaluated by cellulose acetate electrophoresis (Holl *et al.*, 1983). Although different profiles were observed for different species, no qualitative differences were found in almost 1000 plants of different cultivars of alfalfa. Only one plant of subterranean clover line PS-1003 failed to produce the major component of LbI, even if inoculated with different *Rhizobium* strains. This distribution was not affected by the nodule age.

The consistency of the Lb profile within one species may stem from the restricted gene pool found in many commercial crop species. Smartt (1984) reviewed the gene pools of grain legumes and concluded that utilisation of the plant genetic resources, unlike many other of our resources, does not bring about exhaustion, when they have been collected and conserved. Delannay *et al.* (1983) found that the northern and southern gene pools of soybean have maintained remarkable qualitative uniformity since 1951, showing that plant breeding techniques had used the same material repeatedly.

Bowen and Kennedy (1961) studied heritable variation in *Centrosema pubescens* and found that commercially available seeds contained stable sparsely nodulating and profusely nodulating lines. With most *Rhizobium* strains the sparsely nodulating lines were inferior in symbiotic effectiveness.

Jones (1962) described considerable genetic variation in nodulation parameters in white clover. Total nodule weight per plant was highly correlated with eventual harvest weight of the plant. Even other nodule-related parameters like bacteroid viability were strongly dependent on the host plant (Sutton and Paterson, 1983).

VII. Existing Single Locus Variation for Nodulation-Nitrogen Fixation

A relatively large number of naturally occurring plant mutants have been found which show alterations in their symbiotic phenotype. These differ from the varietal differences described above in that their genetic basis can be elucidated. In many cases the mutant phenotype was caused by a single, identifiable, genetic alteration.

Five nodulation-nitrogen fixation mutants of soybean were described prior to the present studies using induced mutagenesis (Carroll *et al.*, 1985a, b). The mutants are labelled rj_1, Rj_2, Rj_3, Rj_4, and Rj_5. Only one of these (mutant rj_1) was studied in any detail (Williams and Lynch, 1954; Clark, 1957; Tanner and Anderson, 1963; Weber, 1966; Devine and Weber, 1977). It was restricted in its nodulation of soybean by ***Bradyrhizobium*** and was controlled by a recessive allele. All others resulted in ineffective (non-nitrogen-fixing) nodules, were dominant and showed strong strain dependence. For example, mutant Rj_5 was recently found in cultivar Evans, which was homozygous dominant for Rj_5 and gave nod^+ and fix^- symbioses with strain USDA205. Cultivar Peking, in contrast, was rj_5rj_5 (Devine, 1985). Mutations Rj_2 and Rj_3 gave small white nodules and were independent dominants with particular strains belonging to the Cl and 122 serogroups of ***Bradyrhizobium*** (Caldwell, 1966; Vest, 1970). Mutant Rj_4, which gave normal nodulation in early phases of development, then led to totally ineffective symbioses with strain USDA61 (Vest and Caldwell, 1972).

Mutant rj_1, which was characterised by an inability to form nodules with most ***Bradyrhizobium*** strains at normal inoculation densities, mapped on linkage group 11 of soybean, being 40 map units distant from the F locus, which controls fasciated stem formation in soybean (Devine *et al.*, 1983). The rj_1 mutation was bred into many commercial cultivars and serves as a convenient non-nodulation control for nitrogen balance experiments, and in the determination of nitrogen fixation in the field using relative abundance and isotope dilution techniques. The mutant was found to be conditional in its phenotype as Clark (1957) showed that strains such as USDA76 were able to nodulate the mutant soybean sparsely. At first it was thought that the ability to produce the compound rhizobotoxin was correlated with the ability to suppress the mutant phenotype. However, recent work by LaFavre and Eaglesham (1984) disproved this. The same authors also showed that increased inoculation densities of ***Bradyrhizobium*** increased the degree of nodulation. For example, strain USDA76 at 2×10^{11} cells per pot gave about 48 nodules per plant (compared to 146 on Rj_1Rj_1) and about 1.5 nodules per plant at 10^9 cells per pot on rj_1rj_1. The correlation with rhizobotoxin production broke down as strains SM35 and IRj2133b (both being unable to produce the toxin) gave 30 and 21 nodules per plant respectively at high inoculation densities. Other strains like I-110 nodulated poorly (about 320 nodules per plant on wildtype compared to about 2.5 nodules per plant on the mutant). Other strains like IRj2147 and IRj2179, despite their ability to produce rhizobotoxin, gave no nodulation even at high inoculation densities. The apparent blockage in mutant rj_1 is very early in the nodulation process, as no infections were observed in normal nodulation trials. The genotype of the root controlled the phenotype and it was claimed that a diffusable rhizosphere factor (acting as an inhibitor) was involved (Elkan, 1961), although this has been disputed (Eskew and Schrader, 1977). Despite the presence of this genetic material, little biochemical analysis was carried out on the rj_1 system.

Other naturally occurring mutants were decribed in *Trifolium pratense* (5 genes, Nutman, 1969), *Trifolium incarnatum* (controlling a *fix*$^-$ phenotype, Smith and Knight, 1984), *Medicago sativa* (two clones with early senescence, two other clones with tumour-like nodules with infection threads or bacteroids, one with nodulation resistance, Viands *et al.*, 1979; Peterson and Barnes, 1981), *Arachis hypogaea* (non-nodulation, Gorbet and Burton, 1979; Nambiar *et al.*, 1983), and *Pisum sativum* (Gelin and Blixt, 1964; Lie, 1971; Holl, 1975; Ohlendorf, 1983a, b). The general nature of these mutants was described in the above references and in reviews by LaRue *et al.* (1985) and Verma and Nadler (1984). A general conclusion that can be drawn from the entire spectrum of non-nodulating and non-nitrogen-fixing plant mutants is that (a) they all exhibit strain specificity, (b) either dominant or recessive types exist, (c) they exist in a large range of legumes, (d) they were not numerous enough to have allowed an extensive biochemical analysis and (e) they seldom were the product of a directed search but rather appeared through serendipity.

VIII. Induced Mutation in Symbiotic Characters

Major advances in biology have stemmed from the tight coupling of biochemical and genetic approaches. The studies on *Neurospora* by Beadle and Tatum, the *Escherichia coli* lactose operon concept as initially proposed by Jacob and Monod, the analysis of development in bacteriophages T4 and lambda, the genetics and biochemistry of eye colour, immunology, and the basis of development in insects are just a short listing. The reason for this synergism of approaches can be found in the elegant nature by which a point mutation or small deletion precisely alters a phenotype, which then is determined by a variety of biochemical technologies. Using cross-feeding, genetic interaction, multiple mutant analysis, and precursor accumulation data it was possible to develop, for instance, an integrated idea of the development of bacteriophage assembly or tryptophan biosynthesis. The same approach should be of value in the analysis of symbiotic nodulation and nitrogen fixation.

The last two years have seen an escalation in the genetic analysis of the legume. The most useful procedure was the direct selection for obvious symbiotic phenotypes, such as non-nodulation or nitrate tolerance. Induced mutagenesis has now been applied successfully to the soybean *(Glycine max)*, pea *(Pisum sativum)* and chickpea *(Cicer arietinum)*. The following section will discuss recent advances in those three systems.

A. Pea and Chickpea Mutants

The essential features of these mutants were summarised by LaRue *et al.* (1985) and Gresshoff *et al.* (1985). Davies *et al.* (1985) used gamma-ray irradiation of chickpea seeds, raised the first generation (termed the M_1), and then harvested the M_2 seeds. These were planted out in large numbers and screened for their inability to fix nitrogen. This was easily monitored, as non-nodulated or non-fixing plants in the absence of exogenously supplied nitrogen turned yellow, because of nitrogen starvation. Such yellow plants were repotted and raised to maturity through the addition of nitrate fertiliser. If the plant failed to respond to nitrate application, it could be presumed that a general fitness gene, and not a symbiotic function, was mutated. It was thus possible to eliminate chlorophyll deficiencies and related deleterious mutations from the screen. Seeds from selected M_3 plants were then tested for the presumed mutant phenotype. The chickpea mutants isolated by Davies *et al.* (1985) included one non-nodulation and four non-fixation mutants. All fell into separate complementation groups. Two of the *fix*$^-$ mutants were temperature sensitive, as effective nodules were formed at 24°C but not at 29°C. Wild-type plants formed effective nodules at both temperatures but not at 34°C. Mutant alleles rn_4 and rn_5 determined the formation of ineffective nodules. One of these showed nodule growth alterations, as nodules were small and white, while the other (mutant PM796) had normal size nodules which were abnormally pale, i. e. they were free of leghemoglobin. All mutant alleles in chickpea were recessive. The non-nodulation mutant controlled by the rn_1 allele conferred resistance to nodulation at all tested temperatures and with all tested strains (Davies, 1985).

In *P. sativum* several research groups have taken advantage of the elegant genetic system that this plant offers. The selection strategy in all cases was the same as that described above. Kneen and LaRue (1984a, b) described a non-nodulation mutant of the pea variety "Sparkle". Since then they have isolated at least 30 further non-nodulation mutants using a range of mutagens and have defined perhaps up to 10 separate complementation groups (LaRue *et al.*, 1985; LaRue, pers. comm.). Perhaps a word of warning should be included here regarding complementation analysis. To achieve reliable data it is of importance to be certain that the cross has indeed occurred. Thus the use of flanking markers or at least of other donor pollen markers is essential. If such line differences have not yet been bred into the relevant mutant material, then one requires at least a several-fold repetition of the observation, before assigning markers to complementation groups. Failure to complement may result from an incomplete cross so that the presumed hybrid seed actually stemmed from a self-fertilisation. This would be easily confirmed in the subsequent generation. This word of caution does not apply so much to research groups that are familiar with plant breeding techniques, but to those that have only recently entered the area. For example, all 15 mutants isolated by Messager (1985; see later detailed description) fall into separate complementation groups.

Feenstra and Jacobsen (1985) described the isolation of two non-nodulation mutants in pea cultivar "Rondo". One mutant (K5) formed occasional nodules on lateral roots, which were capable of efficient nitrogen fixation. Root hair curling and infection thread formation were noted in both the mutant and wild type, but occurred to a larger extent both in degree and frequency over the root length of the wild-type plant. The phenotype of mutant K5 was shown through grafting experiments to be root-controlled. Another mutant isolated by Feenstra and Jacobsen was mutant K24, which also was monogenic, but showed nodulation resistance under all experimental conditions. Jacobsen *et al.* (1986) claimed that the K24 mutation was both root- and shoot-controlled as illustrated by grafting experiments with a nitrate-tolerant pea mutant. On the data supplied so far, however, we feel that such interpretation may be premature and K24 might fit into the same category of root-controlled non-nodulation mutants as previously described in pea (Kneen and LaRue, 1984 a, b).

Messager (1985) reported a large range of pea mutants with altered symbiotic properties. A total of 7 *nod*$^-$ independent mutants were described. All appeared to be monogenically inherited, with all of them being recessive except for mutant F.4.218, which was a dominant. Grafting showed that non-nodulation was root-controlled. Eight mutants with ineffective symbioses were described. All were monogenic recessives. Ineffectiveness like non-nodulation was root-controlled. One of the *fix*$^-$ mutants (F261) showed conditional effectiveness, as bacterial strains from diverse soils were isolated, which suppressed the symbiotic defect. Such conditionality was earlier pointed out in the naturally occurring soybean mutants, and was described in pea with the "Afghanistan" line (Lie, 1971; Degenhardt *et al.,* 1976) which failed to nodulate with most European *R. leguminosarum* strains, but was nodulated effectively by isolates TOM and HIM (Winarno and Lie, 1979). Messager (1985) furthermore described two allelic mutants that nodulated in the presence of nitrate. They found nodulation to be elevated by a factor of 3 to 6 compared to wild-type plants grown under the same conditions. The term "hypernodulation" was used to describe the phenomenon. The hypernodulation characteristic was associated with fasciation in the flower of the pea plant. This might argue for some pleiotropic effects, perhaps involving some growth regulator balance, or a major deletion or genome rearrangement, which also affects this character. Jacobsen and Feenstra (1984) and Jacobsen (1984) described a similar pea mutant which nodulated profusely in the presence of supplied nitrate. No fasciation of the flower or stem was reported, although a reduced ability to form a wild-type root system, even in the absence of nodules, was noted. The mutant allele *nod*$_3$ was recessive and was inherited in a monogenic fashion. Grafting led Jacobsen to conclude that the *nod*$_3$ phenotype was root-controlled. This conclusion may be entirely correct but again, with the present set of supplied data, one has to query whether the graft was made too late, and why the control plant was the non-nodulation mutant K5 rather than the wild type. The hypernodulation mutant of Messager (1985) was shoot-controlled, which coincides with observations made by Delves *et al.* (1986) and reported by Gresshoff *et al.* (1985). The nitrate tolerant pea mutant of Jacobsen developed more than 250 nodules per plant in the absence or presence of 15 mM nitrate, which in control plants lowered the nodule number from about 59 to 17 per plant. The nod_3 plants likewise showed increased nitrogenase activity, as measured by acetylene reduction, with about 5 micromoles ethylene being produced per hour per plant with or without nitrate being present, but with nitrate lowering the activity from 2 to 0.2 micromoles per hour per plant. Nodule fresh weight per plant was nitrate inhibited (125 mg versus 26 mg), but not in the mutant (589 mg versus 682 mg). The increase of nodule weight per plant on nitrate underlined the ability of this mutant to develop a symbiosis in the presence of the otherwise inhibitory levels of nitrate, and was caused by an increase in total plant weight under nitrate. The latter point, and direct measurements of nitrate reductase activity and nitrate content, demonstrated that the nod_3 mutation did not affect nitrate metabolism *per se.*

The pea system clearly provides several experimental advantages. The genetic system is more defined than in any other legume, and there are fewer linkage groups. Second, the plant is easy to grow under a variety of conditions. Growth to maturity is fast. However, advantages relating directly to the analysis of symbiotic nitrogen fixation are minimal, as the symbiotic genetics of *R. leguminosarum* are not significantly ahead of *Bradyrhizobium japonicum.* This was mainly achieved through the transfer of "know-how" and gene-specific clones, which allowed rapid advance in the slow-growing *Bradyrhizobium* species. Likewise one finds that the knowledge of nodulin molecular biology, nodule biochemistry and the developmental biology is nearly identical. Soybean has a large physiological and biochemical data-base mainly because of the greater ease of harvesting large amounts of nodule material. Determinate nodules, furthermore, have the characteristic that the development of bacteroids or nodule components occurs synchronously. Thus 13 day old soybean nodules are relatively uniform in terms of their developmental stages, whereas pea nodules contain a range of symbiotic stages with varying contributions relative to the age of the nodule. The new researcher is advised to look at those advantages in a plant system which are inherent biological characteristics. Other disadvantages may only stem from a dearth of research work with that class of organism. Often such short-falls can be quickly made up through exchange of information and material.

B. Soybean Nodulation Mutants

The following are considerations in an induced mutagenesis programme. The starting material should be homogeneous and homozygous. The plant species must be self-fertile. Mutagenesis must achieve workable mutation frequencies. There must be a convenient selection procedure.

The studies with soybean as described by Carroll *et al.* (1984, 1985 a, b; Gresshoff *et al.,* 1985 a, b) chose not to select for co-dominant or dominant mutations in the M_1 generation, because the selection screen looked at root-related phenotypes. The target tissue of induced mutagenesis is the developing shoot meristem, which lies preformed in the dormant seed. There is also a preformed root meristem, which upon germination will develop into a chimeric root, provided that a stem cell in the meristem was mutated. However, only mutations induced in the shoot meristem will be passed on in the "germline" to the next generation. Thus any root characteristic observed in the first generation of mutagenesis (M_1) would be genetically dead and the M_1 was screened only for the degree of fitness and for chlorophyll deficient sectors; both parameters allowed the evaluation of the mutagenic treatment. Three types of mutagen, gamma-rays, sodium azide and ethyl methanesulphonate (EMS), were tried at varying doses. Gamma-rays have great experimental advantages as seeds can be dry and there are no carry-over effects to the experimenter. In our hands, both gamma irradiation and chemical mutagenesis with sodium azide gave increased mutation frequencies for chlorophyll deficiency above controls, but these were only one-tenth of those observed with EMS. For optimal EMS treatments, large lots of seeds were pregerminated in flywire bags in well-aerated phosphate buffer at 28°C. For soybean it was necessary to change the buffer as germination inhibitors were abundantly released during pregermination. After 8 hours preincubation, EMS at 0.5 % (v/v) was added. The EMS solution was drained from the seeds into an inactivation vessel after 6 h and the seeds were further washed and rinsed prior to wet planting in the soil. Direct planting seemed optimal as it was difficult to dry soybean seeds partially without severely affecting the fitness of the seed. EMS in the inactivation vessel was inactivated by chemical reaction. Detailed procedures of mutagenesis and safety procedures can be obtained from us upon written request.

Optimal mutagenesis gave about 2.8 % chlorophyll deficiency in the M_2 families. Carroll *et al.* (1985 a, b) collected individual M_2 families, as this allowed the later conclusion to be drawn that individual mutant isolates were the product of separate mutagenic events. Additionally, one could obtain a preliminary estimate of the mode of segregation. It was possible to separate the mutant and wild-type segregants in the M_2, then to raise the M_3 and observe directly the stability or segregation of the mutant character. It was thus feasible to evaluate the genetic nature of the individual mutant alleles by the third generation after mutagenesis.

To select mutant phenotypes 12 seeds of each family were planted into 20 litre pots. The growth medium was river sand watered with nitrogen-free growth medium plus *Bradyrhizobium japonicum* (strain CB1809 = USDA136) for selection of *nod*$^-$ and *fix*$^-$ plants, or medium containing about 5 mM nitrate for the selection of nitrate-tolerant-symbiotic (nts) mutants. An additional screen in flat trays was used to isolate soybean mutants lacking the constitutive nitrate reductase activity (Carroll and Gresshoff, 1986).

i) Constitutive Nitrate Reductase Deficiency

Nitrate was known to inhibit many phases of the symbiosis involving legumes and *Rhizobium* (Carroll and Gresshoff, 1983). McNeil (1982) found that different *Rhizobium* strains produced slight differences in the symbiotic efficiency of soybean plants grown on nitrate. A similar involvement of the *Rhizobium* was shown by Herridge *et al.* (1984), who demonstrated that increased inoculant numbers were needed, if high nitrate levels were present. However, since data by Gibson and Pagan (1977) suggested that the plant contributed to the nitrate effect in a major way, induced mutagenesis was used to isolate plant mutants which were not affected by exogenous nitrate. Two strategies were utilised: one was to isolate the phenotype directly (see later section VIII (iii)), and the other was to attempt to isolate a blockage in nitrate uptake, which could be detected as a nitrate reductase deficiency.

Similar mutants were isolated in *Arabidopsis thaliana,* although the selection employed chlorate resistance screening (Braaksma, 1982). However, soybean possesses at least two nitrate reductase activities (Campbell, 1976); one inducible, $NADH^+$ requiring, molybdenum-containing enzyme similar to that found in other plants, and a constitutive enzyme, which prefers $NADPH^+$ and seems not to contain molybdenum. Ryan *et al.* (1983) and Nelson *et al.* (1983) using chlorate selection isolated a plant mutant which lacked constitutive nitrate reductase activity in the soybean cultivar Williams. Chlorate acts as an analogue of nitrate and is thought to act through its conversion by nitrate reductase to the toxic chlorite ion. However, other interpretations of the chlorate effect also exist (Cove, 1976). The soybean mutant LNR2 (defined by allele nr_1) showed no impairment of growth on nitrate and had identical symbiotic characteristics when compared to the parental cultivar. It lacked the ability to release nitrogenous compounds, which after catalytic conversion can be detected as NO_x in the form of nitrite. The NO_x compound released during the *in vivo* assay was identified as acetaldehyde oxime (Mulvaney and Hageman, 1984). The correlation between constitutive nitrate reductase activity and NO_x production was tight, but conclusions were based on only the one allele.

Additional mutant isolates lacking this enzyme function are therefore of importance to confirm or disprove the correlation. Until then, the physiological function of the constitutive nitrate reductase of soybean will remain unclear. Is it involved as a safety valve to high nitrate (or nitrite) levels? Or is the measured activity an artifact of *in vitro* biochemistry and does the enzyme have an alternative substrate *in vivo?* Spreit *et al.* (1985) reported further complexities as the constitutive nitrate reductase activity was separated into two molecular species. Thus the question arises whether mutant LNR2 was a regulatory mutation affecting both peptides, or whether there were multiple mutations (or large deletions), or whether the two peptides were the product of the same gene. A mutant such as LNR2 should also be good starting material to select further mutations in the inducible nitrate reductase activity of soybean.

The approach of Carroll and Gresshoff (1986) was to isolate a constitutive nitrate reductase mutant by a direct screen rather than through chlorate resistance. M_2 plants were grown under stringent nitrate-free conditions (to prevent the induction of the inducible nitrate reductase) and discs from primary leaves were taken and assayed for nitrite production using a colour assay in multi-titre plates. Putative mutants were retested and grown to maturity to allow further characterisation of the mutant character. Two mutants were isolated using this procedure (nr345 and nr328). Both had lowered enzymatic activity and their phenotype was stable for at least 4 generations after isolation. Mutant nr328 had a different phenotype than mutant LNR2, as it showed necrotic lesions on leaves when grown on high (10 or 15 mM) nitrate (Whitmore-Smith, 1985). Symbiotically both mutants behaved similarly to wild-type Bragg plants (Carroll, 1985).

ii) Non-Nodulation Mutants of Soybean

Three clear-cut non-nodulation mutants were obtained from mutant screens of EMS-mutagenised Bragg seeds (Carroll *et al.,* 1986). The isolation procedure did not specifically focus on delayed nodulation mutants although some of these may have been obtained. For example, in a mutant screen for non-nodulation in cultivar Williams, six presumptive *nod*$^-$ mutants segregated in M_2 families, but upon retesting at maturity these plants nodulated. Such mutants may define further nodulation genes which are at present overlooked because the large screening programmes present difficulties even for larger research groups. The analysis of the bacterial nodulation genes has recently recognised the value of delayed nodulation phenotypes. By this fashion extra nodulation genes were identified on the Sym plasmid (Kondorosi *et al.,* 1985).

Carroll *et al.* (1986) isolated mutants nod49, nod139 and nod772. All grew well on nitrate media, indicating that the non-nodulation was not a result of a general fitness mutation. Mutant nod49 had no pseudoinfection or infection threads. Root hairs were not curled (W. D. Bauer and A. Mathews, pers. comm.). *Bradyrhizobium* strain USDA76, which effectively suppressed the rj_1 mutation in soybean, failed to do so with mutant nod49. However, other *Bradyrhizobium* strains erratically caused nodules on nod49 plants grown in Leonard jars (A. Mathews, pers. comm.). The nodules were large (up to 113 mg per nodule compared to 11 mg per nodule on wild-type plants), capable of efficient nitrogen fixation, and were primarily positioned in the lower portions of the root system rather than the crown region. Such increased nodule size was expected as the plant compensated for the decreased nodule number (see Singleton and Stockinger, 1983, for an illustration of compensation). This phenomenon, already observed by Nutman (1953), implies that the developing meristem of the nodule is controlled by systemic mechanisms, which recognise the development of other nodule meristems in the plant — see also the effects of the systemic signalling in split root systems as shown by Hinson (1975) and Carroll and Gresshoff (1983). Whether the meristem is a source of signal compounds or a sink is not known. Bacterial isolates from nodules of nod49 yielded only the original inoculum strains suggesting that some physiological, epigenetic phenomenon resulted in the successful invasion. Whether this dealt with a plant (i. e. somatic) or bacterial effect was unclear. In two separate field trials, mutant nod49 had remained non-nodulated. In mutants nod49 and nod139 the root genotype controlled the mutant phenotype (Delves *et al.,* 1986; Gresshoff *et al.,* 1985). Furthermore, higher inoculation densities increased the nodule number found on nod49 but not on plants of the parent cultivar Bragg grown in Leonard jars. The mutant thus behaved in many ways like the rj_1 allele in other soybean cultivars. Allelism tests are in progress at the time of writing.

It is possible that the non-nodulation mutants described here are altered in the production of root exudate substances. These substances interact with the *nodD* gene product to activate a range of nodulation genes in the bacterium (Halverson and Stacey, 1986). On the other hand it is feasible that the *nod*⁻ mutants are less sensitive to the relevant bacterial signal which elicits the infection series. Or the plant mutant may have a hyperactive defense response system which eliminates most bacterial interactions at the infection stage. The ability of increased bacterial populations to achieve some degree of nodulation on the otherwise non-nodulation mutant nod49 suggests that this mutant fits into the latter two classes. In part this is supported by data from plant co-culture (nod49 with Bragg), which showed that the mutant phenotype was maintained despite close proximity and general asepsis (Mathews, pers. comm.).

Halverson and Stacey (1984, 1985) found that a slow-to-nodulate mutant of *Bradyrhizobium japonicum* strain USDA110 could produce a wild-type nodulation profile if preincubated with soybean seed lectin or soybean root exudate. The tentative interpretation of this finding was that the mutant was slow to interact with this exudate factor *in vivo,* and thus some essential function could not be induced or suppressed. The lectin bound to the bacterial surface may not necessarily help in the binding to the plant cell surface (which may be a secondary property of bound lectin), but rather makes the bacterium appear as if it is a non-pathogen (i. e. it is camouflaged). Perhaps lectin bound to exopolysaccharides prevents the recognition by the plant of LPS or other factors. Clearly the use of both plant and bacterial mutants blocked in several stages of the infection process will help in the biochemical solution to these questions. Preliminary analysis of the root exudate of seedlings of nod49 showed the presence of identical amounts of soybean root lectin if compared to parent cultivar Bragg or supernodulation mutant nts382 (see later section).

Carroll (1985) was also able to isolate some soybean mutants of cultivar Bragg which were characterised by decreased nodule mass. These may fit into the delayed nodulation category described above or may represent a new phenotypic class in which the growth of the nodule is restricted because of altered plant defense systems. This material has not been tested further but could allow the biochemical analysis of nodule function and nodule growth rather than nodule initiation.

iii) Nitrate-tolerant-symbiotic and Supernodulation Mutants of Soybean

Legumes utilise nitrate or atmospheric nitrogen gas as nitrogen source. Usually they "prefer" soil nitrate, and symbiotic nitrogen fixation and nodulation tend to be suppressed in the presence of nitrate. However, plants tend to yield the same whatever the soil nitrate level provided sufficient *Rhizobium* inoculant is present and environmental conditions favour good symbiotic development. Thus the normal legume crop removes nitrate from the soil, despite its natural ability to fix high amounts of nitrogen from the atmosphere (Herridge *et al.,* 1984). The mechanism of nitrate inhibition of nodulation and nitrogen fixation has not been clarified. As a matter of fact, one may have to consider multiple mechanisms in view of recent findings in a number of plant species (Gibson and Harper, 1985). The isolation of nitrate-tolerant legume mutants, as reported by Carroll *et al.* (1984, 1985 a, b) and Gresshoff *et al.* (1985 a, b) for soybean, and Jacobsen and Feenstra (1984), Jacobsen (1984) and Messager (1985) in pea, will aid with further investigations.

Carroll *et al.* (1985 a, b) originally isolated 15 independent nts mutants from EMS mutagenised Bragg families. Of these 12 were further characterised. Table 1 shows a summary of their symbiotic properties in the presence and absence of nitrate. Nitrate tolerant mutants have an increased nodule number and mass under both culture conditions. Nitrogenase activity in the mutants was significantly elevated in the presence of nitrate. This was confirmed by total N determinations of different plant parts (D. Day, pers. comm.). Mutants fell into different phenotypic classes as judged by nodule number, nodule pattern, and plant fitness. Tentatively Gresshoff et al. (1985 b) have defined at least three complementation groups. Two mutants (nts733 and nts 1116) were inherited as Mendelian co-dominant alleles, while mutants nts382, nts501, nts2264, nts1007 and nts183 were inherited as monogenic recessives. The mutants were described by the term supernodulation and hypernodulation to indicate the degree of extra nodulation compared to wild-type plants. Thus mutants nts382 and nts1007 were clear supernodulation mutants, whereas mutant nts1116 and nts739 were more intermediate, hypernodulation types. Mutant nts382 grew on nitrate and had nitrate reductase activity. This observation, together with the increased nodule number in the absence of nitrate, led to the conclusion that the nts phenotype was due to supernodulation caused by a cessation or diminishing of the autoregulation response.

The autoregulation response was described by Pierce and Bauer (1983) on the tap root of soybean and by Kosslak and Bohlool (1984) on split roots of soybean. It involves the regulation of nodule formation by other nodulation events in a systemic fashion, so that ontogenetically younger tissue close to the growing root tip recognises early nodulation stages in more mature root tissue, leading to symbiotic arrest after the infection stage (Calvert *et al.,* 1984). Mutant nts382 and nts1007 were tested in plastic growth pouches and both showed an absence or lessening of the autoregulation response. Mutant nts382 was further analysed in a quantitative fashion, and Bauer (pers. comm.) demonstrated that the same mechanism of infection was seen in the mutant as in the wild type. The number of infection events (being the sum of all pseudoinfections and infection thread structures) was increased, and the region of infectability was maintained for a longer period of root development. Using growth pouches, different soybean cultivars showed different degrees of autoregulation. In other words, the distribution of nodules was extended further down the root in cultivar Bragg than in Williams. Preliminary results from split root experiments involving Bragg and several other cultivars confirmed the single root observations (Bohlool and Gresshoff, unpublished data).

Table 1. Summary of the symbiotic characteristics of soybean nts mutants

Soybean genotype	Nodule number		Nodule dry weight (mg)		Nitrogenase activity nmol C_2H_4 min $^{-1}$	
	$-NO_3$	$+NO_3$	$-NO_3$	$+NO_3$	$-NO_3$	$+NO_3$
			Experiment 1			
Bragg	26 (6)	19 (7)	34 (10)	5 (3)	71 (13)	1 (1)
nts382	576 (77)	1007 (154)	166 (9)	193 (20)	119 (35)	69 (11)
nts1007	334 (292)	991 (231)	92 (49)	179 (35)	98 (29)	88 (20)
nts183	351 (71)	712 (188)	101 (10)	141 (37)	62 (8)	66 (7)
nts2264	478 (99)	907 (61)	124 (21)	188 (25)	99 (10)	55 (8)
nts1116	101 (26)	74 (45)	66 (12)	30 (12)	85 (17)	23 (10)
nts733	457 (107)	797 (180)	115 (22)	172 (19)	88 (22)	54
			Experiment 2			
Bragg	15 (5)	15 (15)	47 (15)	4 (5)	112 (18)	3 (3)
nts501	253 (101)	235 (91)	115 (5)	55 (27)	173 (13)	20 (12)
nts2282	99 (67)	445 (82)	63 (20)	70 (29)	58 (21)	24 (13)
nts97	282 (88)	512 (152)	87 (2)	92 (39)	72 (17)	28 (11)
nts739	94 (56)	175 (84)	92 (21)	77 (24)	154 (52)	85 (41)
nts246	579 (157)	581 (162)	116 (43)	154 (45)	69 (21)	48 (8)
nts2062	123 (46)	299 (69)	120 (17)	144 (13)	169 (24)	108 (53)

Plants were inoculated with *Bradyrhizobium* strain USDA110 and harvested 26 days after planting. Each entry is the mean of 3 to 6 plants (nodule number and dry weight) or 2 plants (nitrogenase activity). SD are given in brackets. Carroll *et al.*, unpublished data.

Growth analysis of mutant nts382 in glasshouse conditions evidenced that in the absence of *Bradyrhizobium* the shoot to root ratio of the nts mutant versus the Bragg parent was altered during the first 50 days of culture. For example, the shoot-root ratio of Bragg at 13 days was 5.5 compared to 4.0 in nts382, but changed to 3.7 in Bragg versus 4.5 in nts382 after

50 days. Such uninoculated plants were grown on nitrate. Similar plants at 7 days showed that nts382 had more lateral roots (40±2 versus 29±2) and similar tap root length (167±8 mm versus 167±8 mm) when compared to Bragg. This was maintained at 14 days (84±5 mm versus 68±6 mm for lateral root number and 246±15 mm versus 227±12 mm for tap root length). (Day *et al.*, pers. comm.; Delves *et al.*, 1985.)

Not only were early nodulation initiation functions altered in mutant nts382, but also late, i. e. nitrogen fixation related characteristics were different. For example, specific nitrogenase activity per nodule mass was reduced to about 30% of wild-type activity, but nitrogenase activity per milligram bacteroid was equivalent (Delves *et al.*, 1986; Day, pers. comm.). Mutant nts382 contained fewer bacteroids per unit nodule tissue. Mutant nts1116, however, which showed hypernodulation instead of supernodulation, and which may be controlled by a different gene than nts382 and nts1007, showed intermediate amounts of bacteroid tissue to the parent. Yet that mutant also had lowered specific nitrogenase activity. The causes for these differences are totally unclear. The lowered bacteroid density was confirmed by ultrastructural investigations of 30 day old nts382 nodules, which were filled with only about one half of the bacteroids as wild type. Additionally, the infected cells were considerably smaller showing less

Fig. 3. Root system of a supernodulating soybean plant grown under field conditions. Mutant nts1007 was inoculated with ***Bradyrhizobium*** USDA110 and grown in a moderately high nitrate soil. Irrigation was applied and the root system is shown at maturity. Upto 3000 nodules can be obtained on such field-grown, well-inoculated mutant plants. Control plants (not shown) have severe nodule senescence and about 150 nodules per plant.

hypertrophism. Bacteroid number per peribacteroid vesicle was also reduced. Mutant nts382 nodules resembled immature Bragg nodules or Bragg nodules grown on nitrate. Occasional vacuoles were found in mutant infected cells of young nodules, which were never seen in fully matured wild-type cells (G. D. Price, pers. comm.).

Root systems of soil-grown mutant nts1007 (tentatively allelic to mutant nts382) showed strong supernodulation (Fig. 3), but the nodules had not senesced at maturity compared to control Bragg plants.

Growth analysis of supernodulation mutants indicated that excessive nodulation was associated with a reduction of shoot growth. Thus more intermediate hypernodulators like mutant nts1116 may be of greater agronomic importance than mutant nts382. Day (pers. comm.) found that lower bacterial inoculant densities also could result in decreased levels of supernodulation and that in those cases the reduction of shoot weight was not as pronounced.

In a small soil-grown growth trial Gresshoff *et al.* (1985b) found that mutant nts1007 grew nearly as well as the Bragg parent, although the root system was heavily supernodulated (see Fig. 3 for a plant from that trial). The major agronomic potential of the nitrate-tolerant plant mutants may be their ability to fix nitrogen for a prolonged season in the presence of soil nitrate. This nitrate could potentially be spared as the symbiotic dependence of the plant is increased. This is of significance in rotational cropping systems and possibly pastures, as nitrate needs to be available for the companion or rotational species (see Tudge, 1984).

Table 2. Summary of grafting data with supernodulation soybean mutants

Graft	Nodule number	Nodule mass (mg)	Tap root length (cm)	% root nodulated	Nitrogenase activity
Bragg control	31 (11)	43 (10)	29 (5)	upper33	1860 (360)
Bragg shoot/ nts root	76 (26)	45 (21)	29 (5)	upper33	1320 (780)
nts control	278 (144)	151 (43)	18 (5)	95	480 (60)
nts shoot/ Bragg root	213 (175)	109 (41)	23 (3)	95	1080 (300)

Data are means of 10 plants with the standard deviation in brackets. Nodule mass is dry weight and nitrogenase activity is expressed as nmol ethylene produced per hour per gram nodule dry weight. Nodule number and nodule mass values are expressed per gram dry weight of plant.

Reciprocal grafting of different mutant shoots onto wild-type root stock showed that mutant nts382 (a supernodulation mutant) and mutant nts1116 (a hypernodulation mutant, tentatively placed in a separate complementation group) controlled the root phenotype through a graft transmissible shoot factor (Table 2, and Delves *et al.*, 1986).

The combined data allowed the development of a general model to explain the autoregulation response (Fig. 4). The model predicts that some autoregulation mutants may be root-controlled. For this reason it is essential that the apparently root-controlled *nod*$_3$ mutant of pea is further verified. The model also incorporates the results of split root studies of Hinson (1975), Carroll and Gresshoff (1983) and Kosslak and Bohlool (1984) in Fig. 5, and is expanded to involve a temporal sequence leading to the establishment of nodulation and the autoregulatory sequence (Fig. 6).

Grafting eliminated flowers and cotyledons as the source tissue of shoot-derived compound. Similarly approach grafts, in which a wild-type and mutant shoot were grafted together onto the same root stock, demonstrated that the mutant state could be associated with the absence of a factor, as the root system always showed a wild-type nodulation response (J. Olsson, pers. comm.). The observation also allows the expansion of this technique, as one can now remove individual plant parts to see whether the inhibitory signal from the wild-type plant can be removed. Also two nts shoots could be grafted onto an nts root then various compounds could be applied in bioassays to only one shoot.

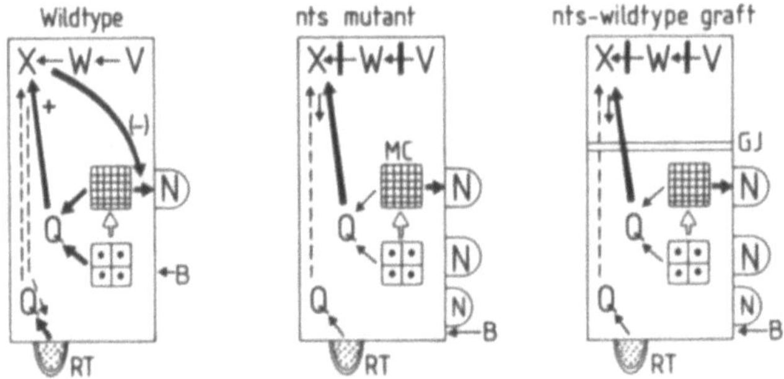

Fig. 4. A general model of autoregulation of nodulation in soybean. In wild-type plants the shoot produces a signal (X) after being stimulated to do so by the root-derived compound Q. Signal X acts as a inhibitor of symbiotic development at the continued cell division stage. Compound Q in turn is a product of nodule meristematic centers. In mutant nts382 and nts1116 two different steps of the shoot mechanism are disturbed. This results in a lesion of the regulatory loop as no X is translocated to the root. Hence symbiotic development can continue towards nodule formation. Basal levels of Q may also come from the other meristematic centers of the root such as the root tip or cambial regions. Thus nts mutants, which lack the ability to produce signal X may already be morphologically affected prior to infection and cortical cell division. Nts mutant-wild type grafts mimick the mutant situation, as no signal X is received by the root. Excessive nodulation may increase levels of compound Q, so that eventually the signal level is sufficient to produce some signal X, so that delayed inhibition of further nodulation occurs. RT = root tip, N = nodule, B = bacterial invasion, + = positive activation, − = inhibition. MC = meristematic centre, GJ = graft junction, W and V are unknown precursors of X.

In the natural situation, nodulation may be directly controlled by a phytohormone balance between shoot and root system. Lawn and Bushby (1982) studied the involvement of the root and shoot on nitrogen fixation in four *Vigna* species which varied in their symbiotic potential. The shoot had a pronounced effect on nodule fresh weight, while the root and the inoculant were associated with the specific nodule activity. This was also reported by Lawn and Brun (1974), who grafted between different soybean cultivars and illustrated that ontogenetic change in the shoot might influence the amount of photosynthate available for export to the root which influenced the amount of nodule tissue formed. Root genotype had little effect on the amount of nodule mass but did influence the ability of nodules to reduce acetylene. Whether the interpretations regarding the involvement of photosynthate as a primary signal were correct is uncertain, especially in view of the findings by Sheehy *et al.* (1980) and Williams *et al.* (1982), who showed that there were no short-term increases in symbiotic nitrogen fixation of soybean when whole plant photosynthesis was increased. Any increase in the symbiotic tissue stemmed from an increased root system which mirrored the growth increases in the shoot. Additionally, such larger shoots can be the sink for extra translocated nitrogen or they can be the source for translocatable compounds other than sugars which influence the plant's ability to regulate nodule development and growth.

The translocated compounds from shoots may include a range of phytohormones. Very little is known about the involvement of phytohormones in nodulation. Although *Rhizobium* synthesises cytokinins and auxins (Phillips and Torrey, 1970), this may not be critical as many other bacterial and fungal species produce such "common" metabolites as indole acetic acid or adenine derivatives. Zeatin riboside is thought to be a major form in which cytokinin moves from the root to the shoot (Summons *et al.*, 1981). Nodules were shown to be a sink for translocatable cytokinins (Badenoch-Jones *et al.*, 1983). Recently, Chen *et al.* (1985) reported that all meristematic tissues in a plant were able to produce cytokinins. There are suggestions that cytokinins move from the root tip to the developing nodule. Thimann (1936) found that nodules contained high concentrations of auxin. Indole acetic acid was recently unequivocally detected in culture supernatants of *Rhizobium*. Mutants which lacked the ability to curl plant root hairs still produced the same level of auxin as the parent strain (Badenoch-Jones *et al.*, 1982). Effective nodules of pea plants supplied at their apical meristem with labelled auxin (indole acetic acid) accumulated more radioactivity than ineffective nodules. Indole acetylaspartic acid appeared as the major metabolite (Badenoch-Jones *et al.*, 1983). The nodule tissue in all cases accumulated more auxin label than the corresponding root tissue.

The nts mutants may possibly have an altered auxin biosynthetic ability, so that the root system experiences increased cytokinin levels. This may prevent hypertrophic extension of infected cells and may lead to increased lateral root formation during early root growth, or increased nodulation. While such speculations may be premature, it is worthwhile to integrate these into models, as they aid experimental planning and hopefully stimulate discussion and feed-back. Of interest is the possible correlation between auxin and nitrate metabolism. Tanner and Anderson (1963) already suggested such links but focussed their thinking on the bacterial side. It is possible that exogenous nitrate increases the effective auxin levels in a particular tissue such as the root or the nodule. Such situations may be the natural situation for all non-legumes which do not nodulate. This may stimulate extensive root growth and may be involved in the suppression of root exudate substances which may help to produce microbial infestations of the root tissue. Likewise auxin may stimulate, either directly or through a secondary molecule such as ethylene, phenolic biosynthesis involved in the general anti-microbial mechanisms described previously.

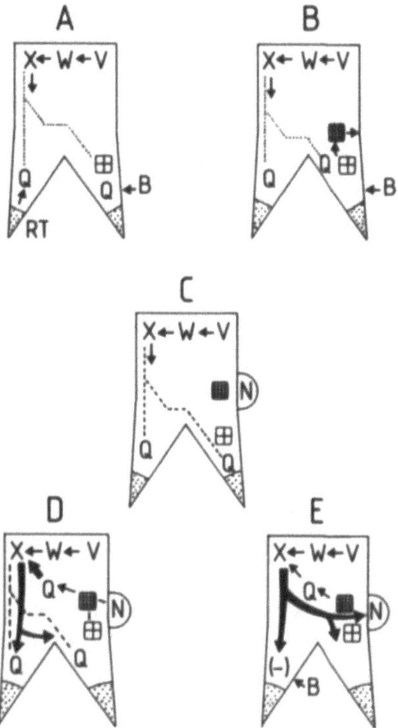

Fig. 5. The application of the model as shown in Fig. 4 to split root systems. At the time of infection *(A)* on one side of the root, basal levels of compound Q from the root tip stimulate basal levels of signal X release from the shoot. This level of X is insufficient to prevent nodule initiation. Thus *(B)* nodule foci form, which develop into nodules as shown in *C*. As sufficient meristematic centres develop in the cortex of the one root half, higher amounts of compound Q are transported to the shoot, resulting in the release of more signal X. This signal is transported to both root halves *(D)*, resulting in the suppression of new nodule development in the uninoculated half and further nodulation in the already nodulated half *(E)*. All symbols as in Fig. 4.

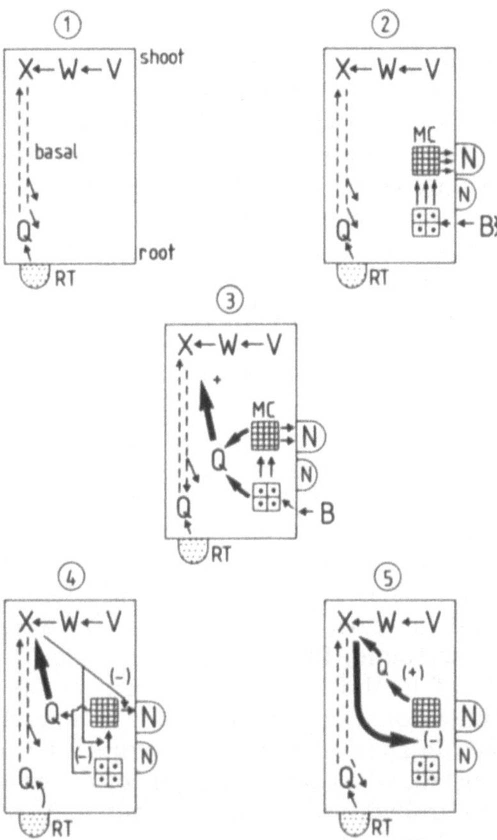

Fig. 6. A temporal analysis of autoregulation events as predicted by the model outlined in Fig. 4. *(1)* Prior to infection, the shoot biosynthetic system leading to the production of signal X is functioning at a basal level under basal stimulation of compound Q, which is produced in the root. *(2)* Bacterial infection leads to initial nodulation which comprises not only the formation of visible nodules but also of meristematic centres in the root cortex. Not all of these regions will develop into nodules. *(3)* These meristematic centres add to the level of compound Q, thereby increasing the release of signal X from the shoot. *(4)* This results in a level of X in the root which prevents any further symbiotic development and leads to symbiotic arrest of the meristematic regions, so that a continued supply of compound Q is produced; *(5)* thereby maintaining the inhibition of nodulation. Compound Q and signal X may be more than one molecular species. Possible alternative sites for mutation may be the inability to produce compound Q or the inability to release signal X from the shoot. Some of these may be lethal and thus would be nearly impossible to isolate without prior knowledge of their chemical requirements. Signal X is also proposed to act in the general growth control of uninoculated plants leading to altered shoot-root ratios and increased lateral root formation in its absence as illustrated by the mutant nts382. All symbols are as in Fig. 4.

Whether nitrate achieves this by a stimulation of auxin synthesis in the shoot, decreased breakdown in the root, an increased transport, or an increased "susceptibility" of root tissues to the auxin signals is not known (see Cross, 1985) for a review of auxin action).

Nitrate itself may be a "morphogen" and the nodulation response could be blocked in its presence. Thus most non-legumes are "locked" in a nitrate utilisation state and thus are "resistant" to nodulation by *Rhizobium*. For example, nitrate may control cell wall biochemistry and thus the release of specific oligosaccharides which may have morphogenetic effects, such as those described by Tran *et al.* (1985). The described nts mutants may be insensitive to the morphogenetic switch, which would explain the causes of the nitrate tolerance.

In a related study Beach and Gresshoff (1986) found that roots of red clover and siratro transformed by *Agrobacterium rhizogenes* were unable to form nodules when inoculated with the *Rhizobium*. Yet mixed infections of untransformed plants with *Rhizobium* and *Agrobacterium* at equal titres gave no inhibition. Perhaps the phytohormone changes associated with the transformed state prevented symbiotic development.

These speculations show how a few mutants can open up a range of different experimental approaches. More mutants will be isolated in a wider range of legumes and the existing material will be described in a large variety of ways. For example, Bisseling (pers. comm.) looked at *in vitro* translation products of total RNA isolated from 5 day old nts382 and Bragg roots by 2 dimensional electrophoresis, and found that the mutant root lacked one peptide spot of molecular weight of about 37000. Why such a difference existed, when the regulation of supernodulation in this mutant was shoot-controlled, seems unclear, but may suggest that some secondary molecular changes were already in place at this early stage of root development prior to nodulation.

IX. Conclusions

The nodulation of legumes is characterised by a wide variety of mechanisms showing immense ability for adaptation. Within this range of phenotypes there are, however, general principles. The analogy to leaves and photosynthesis comes to mind. Despite a seemingly unlimited range of form and environmental optimisation, most photosynthetic and gas exchange processes fall into two, possibly three, different categories. The same balance between diversity and stability is found in nodulation biology. The existence of mutants altered in the nodulation and nitrogen fixation capacities indicated that similar phenotypes, tissue control and genes of similar genetic function were detectable in a variety of legumes. This commonality spanned even the gap between the determinate and indeterminate nodule type. Fundamental functions necessary for nodulation control are similar between all legumes (and perhaps even non-legumes). Micro-evolution may have inserted numerous fine adjustments giving the range of symbioses seen today. Leghemoglobin amino-acid sequences can diverge by as much as 50 %, yet all molecules are still capable of fulfilling their symbiotic function of oxygen transport. Similar changes could have occurred in other genes. The commonality of the host defense systems is amazing, as is the similarity of the genes and mechanisms involved in other plant-bacterial interactions such as those described for *Agrobacterium* or pathogenic *Pseudomonas* species.

The biochemical analysis of the plant responses in symbiotic nodulation and nitrogen fixation has just started. Hopefully, with the new set of tools available to scientists, continued analysis will not be hampered, since, although seemingly important in crop productivity, not all discoveries will result in commercial advances. Despite the large data-base in photosynthesis, there have been no commercial applications of this knowledge. The study of plant development and plant biochemistry, however, goes beyond the commercial. Hopefully the coupling of plant breeding, plant cell culture, molecular genetics and classical biochemistry can help in the elucidation of the complex developmental processes in plants, which form the basis of all life on this planet.

Acknowledgements

The authors thank Dr. Bernard J. Carroll, Dr. David Day and Dr. Dean Price as well as our graduate students for extensive discussions and the provision of unpublished material. Special thanks also to the Genetics Department and Professor Alfred Pühler (University of Bielefeld), where the major part of the writing was completed. Dr. Wolfgang D. Bauer, Dr. Ben Bohlool and Dr. Ton Bisseling are thanked for providing additional information about the nts mutants. The Alexander von Humboldt Foundation and Agrigenetics Corporation are thanked for direct and indirect support.

X. References

Albersheim, P., Anderson-Prouty, A. J., 1975: Carbohydrates, proteins, cell surface and the biochemistry of pathogenesis. Annu. Rev. Plant Physiol. **26,** 31—52.

Appleby, C. A., 1984: Leghemoglobin and *Rhizobium* respiration. Annu. Rev. Plant Physiol. **35,** 443—478.

Appleby, C. A., 1985: Plant hemoglobin: properties, function and genetic origin. In: Ludden, P. W., Burris, J. E. (eds.), Nitrogen Fixation and CO_2 Metabolism, pp. 41—51. Amsterdam: Elsevier.

Appleby, C. A., Tjepkema, J. D., Trinick, M. J., 1983: Hemoglobin in the non-leguminous plant *Parasponia.* Possible genetic origins and function in nitrogen fixation. Science **220,** 951—953.

Ausubel, F., Buikema, W., Earl, C., Klingensmith, J., Nixon, B. T., Szeto, W., 1985: Organisation and regulation of *Rhizobium meliloti* and *Parasponia Bradyrhizobium* nitrogen fixation genes. In: Evans, H. J., Bottomley, P. J., Newton, W. E. (eds.), Nitrogen Fixation Research Progress, pp. 165—171. Dordrecht: Nijhoff Publ.

Badenoch-Jones, J., Summons, R., Djordjevic, M. A., Shine, J., Letham, D. S., Rolfe, B. G., 1982: Mass-spectrometric quantification of indole-3-acetic acid in *Rhizobium* culture supernatants: Relation to root hair curling and nodule initiation. Appl. Environ. Microbiol. **44,** 275—280.

Badenoch-Jones, J., Rolfe, B. G., Letham, D. S., 1983: Phytohormones, *Rhizobium* mutants and nodulation in legumes. III.: Auxin metabolism in effective and ineffective pea root nodules. Plant Physiol. **73,** 347—352.

Barton, K. A., Brill, W. J., 1983: Prospects in plant genetic engineering. Science **219,** 671—676.

Bassam, B., Gresshoff, P. M., 1986: Use of neomycin for preferential selection against *Rhizobium trifolii* in symbiosis with white clover *(Trifolium repens).* Aust. J. Biol. Sci. **37,** in press.

Bauer, W. D., 1981: Infection of legumes by rhizobia. Annu. Rev. Plant Physiol. **32,** 407—449.

Beach, K., Gresshoff, P. M., 1986: Culture and nodulation properties of legume roots transformed by *Agrobacterium rhizogenes.* Plant Science (submitted).

Bedmar, E. J., Phillips, D. A., 1983: *Pisum sativum* cultivar effects on hydrogen metabolism of *Rhizobium.* Can. J. Bot. **62,** 1682—1686.

Bedmar, E. J., Phillips, D. A., 1984: A transmissible plant shoot factor promotes uptake hydrogenase activity in *Rhizobium* symbionts. Plant Physiol. **75,** 629—633.

Bedmar, E. J., Edie, S. A., Phillips, D. A., 1983: Host plant cultivar effects on hydrogen evolution by *Rhizobium leguminosarum.* Plant Physiol. **72,** 1011—1015.

Bergersen, F. J., Goodchild, D. J., 1973: Aeration pathways in soybean root nodules. Aust. J. Biol. Sci. **26,** 729—740.

Bergersen, F. J., Turner, G. L., 1967: Nitrogen fixation by the bacteroid fraction of breis of soybean root nodules. Biochim. Biophys. Acta **141,** 507—515.

Bergersen, F. J., Turner, G. L., 1978: Activity of nitrogenase and glutamine synthetase in relation to availability of oxygen in continuous cultures of a strain of cowpea *Rhizobium* species supplied with excess ammonium. Biochim. Biophys. Acta **538,** 406—415.

Bergmann, H., Preddie, E., Verma, D. P. S., 1983: Nodulin 35: A subunit of specific uricase (uricase II) induced and localized in uninfected cells of nodules. EMBO J. **2,** 2333—2339.

Billing, E., 1982: Entry and establishment of pathogenic bacteria in plant tissues. In: Rhodes-Roberts, M. E., Skinner, F. A. (eds.), Bacteria and Plants, pp. 51—70. London: Academic Press.

Bisseling, T., Been, C., Klugkist, J., van Kammen, A., Nadler, K., 1983: Nodule specific host proteins in effective and ineffective root nodules of *Pisum sativum.* EMBO J. **2,** 961—966.

Bisseling, T., Govers, F., Gloudemans, T., Moerman, M., van Kammen, A., 1985: Expression of pea nodulins in effective and ineffective symbiosis. In: Analysis of the Plant Genes Involved in the Legume-*Rhizobium* Symbiosis, pp. 104—111. Paris: OECD Publ.

Bowen, G. D., Kennedy, M. M., 1961: Heritable variation in nodulation of *Centrosema pubescens* Beuth. Qld. J. agri. Sci. **18,** 161—170.

Braaksma, F., 1982: Genetic control of nitrate reduction in *Arabidopsis thaliana.* PhD dissertation Univ. of Groningen, Haren, The Netherlands.

Brill, W. J., 1980: Biochemical genetics of nitrogen fixation. Microbiol. Rev. **44,** 449—467.

Caldwell, B. E., 1966: Inheritance of a strain specific ineffective nodulation in soybeans. Crop Sci. **6,** 427—428.

Calvert, H. E., Pence, M., Pierce, M., Malik, N. S. A., Bauer, W. D., 1984: Anatomical analysis of the development and distribution of *Rhizobium* infections in soybean roots. Can. J. Bot. **62,** 2375—2384.

Campbell, W. H., 1976: Separation of soybean leaf nitrate reductases by affinity chromatography. Plant Sci. Lett. **7,** 239—247.

Carroll, B. J., 1985: The plant contribution to the soybean-*Rhizobium* symbiosis. PhD dissertation, Australian National University, Canberra, Australia.

Carroll, B. J., Gresshoff, P. M., 1983: Nitrate inhibition of nodulation and nitrogen fixation in white clover. Z. Pflanzenphysiol. **110,** 77—88.

Carroll, B. J., Gresshoff, P. M., 1986: Isolation after induced mutagenesis and initial characterisation of constitutive nitrate reductase mutants NR328 and NR345 of soybean. Plant Physiol., in press.

Carroll, B. J., McNeil, D. L., Gresshoff, P. M., 1984: Breeding soybeans for increased nodulation in the presence of external nitrate. In: Ghai, B. S. (ed.), Symbiotic Nitrogen Fixation, Vol. I, pp 43—50. Ludhiana, India: USG Publ. Company.

Carroll, B. J., McNeil, D. L., Gresshoff, P. M., 1985a: Isolation and properties of soybean *(Glycine max)* mutants that nodulate in the presence of high nitrate concentrations. Proc. Nat. Acad. Sci. U.S.A. **82,** 4162—4166.

Carroll, B. J., McNeil, D. L., Gresshoff, P. M., 1985b: A supernodulation and nitrate tolerant symbiotic *(nts)* soybean mutant. Plant Physiol. **78,** 34—40.

Carroll, B. J., McNeil, D. L. Gresshoff, P. M., 1986: Isolation of non-nodulation mutants of soybean by induced mutagenesis. Plant Science (submitted).

Chen, C. M., Ertl, J. R., Leisner, S. M., Chang, C. C., 1985: Localization of cytokinin biosynthetic sites in pea plants and carrot roots. Plant Physiol. **78,** 510—513.

Clark, F. F., 1957: Nodulation responses in two near isogenic lines of the soybean. Can. J. Microbiol. **3,** 113—123.

Collins, J., 1983: Anatomical investigations of nodule initiation in white clover. Honours Degree Dissertation, Botany Department, Australian National University, Canberra.

Cove, D. J., 1976: Chlorate toxicity in *Aspergillus nidulans:* studies of mutants altered in nitrate assimilation. Mol. Gen. Genet. **146,** 147—159.

Cross, J. W., 1985: Auxin action: the search for the receptor. Plant, Cell Environ. **8,** 351—359.

Cullimore, J. V., Lara, M., Lea, P. J., Miflin, B. J., 1983: Purification and properties of two forms of glutamine synthase from the plant fraction of *Phaseolus* root nodules. Planta **157,** 245—253.

Daly, J. M., 1984: The role of recognition in plant disease. Annu. Rev. Phytopathol. **22,** 273—307.

Davies, T. M., 1985: Host genes affecting nodule formation and function in chickpea *(Cicer arietinum).* In: Evans, H. J., Bottomley, P. J., Newton, W. E. (eds.), Nitrogen Fixation Research Progress, p40. Dordrecht: Nijhoff Publ.

Davies, T. M., Foster, K. W., Phillips, D. A., 1985: Non-nodulation mutants of chickpea. Crop Sci. **25,** 345—348.

Day, P. R., 1974: Genetics of host-parasite interaction. San Francisco: W. H. Freeman.

Dazzo, F. B., Gardiol, A. E., 1984: Host specificity in the *Rhizobium*-legume interaction. In: Verma, D. P. S., Hohn, Th. (eds.), Genes Involved in the Microbe-

Plant Interaction. Plant Gene Research. Vol. 1, pp. 3—31. Wien - New York: Springer

Degenhardt, T. L., LaRue, T. A., Paul, E. A., 1976: Investigation of a non-nodulating cultivar of *Pisum sativum*. Can. J. Bot. **54,** 1633—1636.

Delannay, X., Rodgers, D. M., Palmer, R. G., 1983: Relative genetic contributions among ancestral lines to North American soybean cultivars. Crop Sci. **23,** 944—949.

Delves, A. C., Day, D. A., Price, G. D., Carroll, B. J., Gresshoff, P. M., 1985: Regulation of nodulation and nitrogen fixation in nitrate tolerant, supernodulating soybeans. In: Evans, H. J., Bottomley, P. J., Newton, W. E. (eds.), Nitrogen Fixation Research Progress, p41. Dordrecht: Nijhoff Publ.

Delves, A. C., Mathews, A., Day, D. A., Carter, A. C., Carroll, B. J., Gresshoff, P. M., 1986: Regulation of the *Rhizobium*-legume symbiosis by root and shoot factors (submitted).

Devine, T. E., 1985: Nodulation of soybean plant introduction lines with fast-growing rhizobial strain USDA 205. Crop Sci. **25,** 354—356.

Devine, T. E., Weber, D. F., 1977: Genetic specificity of nodulation. Euphytica **26,** 527—535.

Devine, T. E., Palmer, R. G., Buzzell, R. I., 1983: Analysis of genetic linkage in the soybean. J. Heredity **74,** 457—460.

Ditta, G., Corbin, D., Leong, S., Barran, L., Helinski, D. R., 1983: Symbiotic nitrogen fixation genes of *Rhizobium meliloti*. In: Pühler, A. (ed.), The molecular genetics of the bacteria-plant interaction, pp. 88—97. New York - Berlin - Heidelberg: Springer.

Djordjevic, M. A., Zurkowski, W., Shine, J., Rolfe, B. G., 1983: Sym plasmid transfer to various symbiotic mutants of *Rhizobium trifolii, Rhizobium leguminosarum* and *Rhizobium meliloti*. J. Bacteriol. **156,** 1035—1045.

Dommergues, Y. R., 1978: Impact on soil management and growth. In: Dommergues, Y. R., Krupa, S. V. (eds.), Interactions Between Non-pathogenic Soil Microorganisms and Plants, pp. 443—458. Amsterdam: Elsevier.

Duke, S. H., Henson, C. A., 1985: Legume nodule carbon utilization in the synthesis of organic acids for the production of transport amides and amino acids: In: Ludden, P. W., Burris, J. E. (eds.), Nitrogen Fixation and CO_2 Metabolism, pp. 293—302.Amsterdam: Elsevier.

Elkan, G. H., 1961: A nodulation inhibiting root excretion from a non-nodulating soybean strain. Can. J. Microbiol. **7,** 851—856.

Elmerich, C., 1984: Molecular biology and ecology of diazotrophs associating with non-leguminous plants. Biotechnology **2,** 967—978.

Eskew, D. L., Schrader, L. E., 1977: Effect of rj_1rj_1 (non-nodulating) soybeans on nodulation of near isogenic Rj_1Rj_1 plants in nutrient culture. Can. J. Microbiol. **23,** 988—993.

Evans, L. S., Van't Hof, J., 1973: Cell arrest in G2 in root meristems: a control factor from the cotyledons. Exp. Cell Res. **82,** 471—473.

Feenstra, W. J., Jacobsen, E., 1985: Pea mutants with an altered response to *Rhizobium leguminosarum*. In: Analysis of Plant Genes Involved in the Legume-*Rhizobium* Symbiosis, pp. 50—51. Paris: OECD Publ.

Fischer, H. M., Hennecke, H., 1984: Linkage map of *Rhizobium japonicum nifH* and *nifDK* operons encoding the polypeptides of the nitrogenase enzyme complex. Mol. Gen. Genet. **196,** 537—540.

Flor, H. H., 1955: Host-parasite interaction in flax rust — its genetic and other implications. Phytopathol. **45,** 680—685.

Flor, H. H., 1971: Current status of the gene-for-gene concept. Annu. Rev. Phytopathol. **9,** 275—296.

Fraser, R. S. S., 1982: Are "pathogenesis-related" proteins involved in acquired systemic resistance of tobacco plants to tobacco mosaic virus? J. Gen. Virol. **58,** 305—313.

Fuller, F., Verma, D. P. S., 1984: Appearance and accumulation of nodulin mRNAs and their relationship to the effectiveness of the root nodule. Plant Mol. Biol. **3,** 21—28.

Fuller, F., Künster, P. N., Nguyen, T., Verma, D. P. S., 1983: Soybean nodulin genes: analysis of cDNA clones reveals several major tissue specific sequences in nitrogen fixing root nodules. Proc. Nat. Acad. Sci. U.S.A. **80,** 2594—2598.

Gadal, P., 1983: Phosphoenolpyruvate carboxylase and nitrogen fixation. Physiol. veg. **21,** 1069—1074.

Gelin, O., Blixt, S., 1964: Root nodulation in peas. Agric. Hort. Genet. **22,** 149—159.

Gibson, A. H., Harper, J. E., 1985: Nitrate effects on nodulation of soybean by *Rhizobium japonicum.* Crop Sci. **25,** 497—501.

Gibson, A. H., Pagan, J., 1977: Nitrate effects on the nodulation of legumes inoculated with nitrate reductase deficient mutants of *Rhizobium.* Planta **134,** 17—32.

Gorbet, D. W., Burton, J. C., 1979: A non-nodulating peanut. Crop Sci. **19,** 727—728.

Govers, F., Gloudemans, T., Moerman, M., van Kammen, A., Bisseling, T., 1985: Expression of plant genes during the development of pea root nodules. EMBO J. **4,** 861—867.

Gresshoff, P. M., Mohapatra, S. S., 1981: Legume cell and tissue culture. In: Rao, A. N. (ed.), Tissue Culture of Economically Important Crop Plants, pp. 11—24. Singapore: Univ. Singapore Press.

Gresshoff, P. M. and 14 other authors, 1984: The *Parasponia-Rhizobium* nitrogen fixing symbiosis: Genetics, biochemistry, and molecular biology of a plant and bacterium. In: Swaminathan, M. (ed.), Genetics: New Frontiers, pp. 217—226. New Delhi: Oxford and IBH Publ. Comp.

Gresshoff, P. M., McNeil, D. L., Carroll, B. J., 1985a: Symbiotic mutants of soybean capable of supernodulation in the absence and presence of nitrate. In: Szalay, A. A., Legocki, R. P. (eds.), Advances in the Molecular Genetics of the Bacteria-plant Interaction, pp. 49—51. Ithaca, N. Y.: Cornell Univ. Publ.

Gresshoff, P. M., Day, D. A., Delves, A. C., Mathews, A., Olsson, J. A., Price, G. D., Schuller, K. A., Carroll, B. J., 1985b: Plant host genetics of symbiotic nitrogen fixation and nodulation in pea and soybean. In: Evans, H. J., Bottamley, P. J., Newton, W. E. (eds.), Nitrogen Fixation Research Progress, pp. 19—25. Dordrecht: Nijhoff Publ.

Guerinot, M. L., Chelm, B., 1985: Bacterial cAMP and heme in the *Rhizobium*-legume symbiosis. In: Evans, H. J., Bottomley, P. J., Newton, W. E., (eds.), Nitrogen Fixation Research Progress, p. 220. Dordrecht: Nijhoff Publ.

Halverson, L. J., Stacey, G., 1984: Host recognition in the *Rhizobium*-soybean symbiosis: Detection of a protein factor in soybean root exudate which is involved in the nodulation process. Plant Physiol. **74,** 84—89.

Halverson, L. J., Stacey, G., 1985: Host recognition in the *Rhizobium*-soybean symbiosis: Evidence for the involvement of lectin in nodulation. Plant Physiol. **77,** 621—625.

Halverson, L. J., Stacey, G., 1986: Plant-microbial interaction, Microbiol. Rev., in press.

Hanks, J. F., Schubert, K., Tolber, N. E., 1983: Isolation and characterization of infected and uninfected cells from soybean nodules. Role of uninfected cells in ureide synthesis. Plant Physiol. **71,** 869—873.

Harper, J. E., Gibson, A. H., 1984: Differential nodulation tolerance to nitrate among legume species. Crop Sci. **24,** 797—801.

Heath, M. C., 1980: Reactions of nonsuscepts to fungal pathogens. Annu. Rev. Phytopathol. **18,** 211—236.

Heath, M. C., 1981: A generalized concept of host-parasite specificity. Phytopathol. **71,** 1121—1123.

Herridge, D. F., 1982: Use of the ureide technique to describe the nitrogen economy of field-grown soybean. Plant Physiol. **70,** 7—11.

Herridge, D. F., Betts, J., 1985: Nitrate tolerance in soybean. Variation between genotypes. In: Evans, H. J., Bottomley, P. J., Newton, W. E. (eds.), Nitrogen Fixation Research Progress, p. 32. Dordrecht: Nijhoff Publ.

Herridge, D. F., Roughley, R. J., Brockwell, J., 1984: Effect of rhizobia and soil nitrate on the establishment and functioning of the soybean symbiosis in the field. Aust. J. Agric. Res. **35,** 149—161.

Hinson, K., 1975: Nodulation response from nitrate applied to soybean half-root systems. Agron. J. **67,** 799—804.

Hodgson, A. C. M., Stacey, G., 1986: Potential for *Rhizobium* improvement. Crit. Rev. in Biotechnol., in press.

Holl, F. B., 1975: Host plant control of the inheritance of dinitrogen fixation in the *Pisum sativum-Rhizobium* symbiosis. Euphytica **24,** 767—770.

Holl, F. B., 1983: Plant genetics: manipulation of the host. Can. J. Microbiol. **29,** 945—953.

Holl, F. B., LaRue, T. A., 1976: Host genetics and nitrogen fixation. In: Hill, L. D. (ed.), Proc. World Soybean Conf. Urbana, Ill., pp. 156—163. Danville: The Interstate Printers and Publishers.

Holl, F. B., Milliron, M. L., Delafield, S. J., 1983: Quantitative variation in root nodule leghemoglobins: Interspecific variation in component problems. Plant Sci. Lett. **32,** 321—326.

Jacobsen, E., 1984: Modification of symbiotic interaction of pea *(Pisum sativum)* and *Rhizobium leguminosarum* by induced mutation. Plant and Soil **82,** 427—438.

Jacobsen, E., Feenstra, W. J., 1984: A new pea mutant with efficient nodulation in the presence of nitrate. Plant Sci. Lett. **33,** 337—344.

Jacobsen, E., Postma, J. G., Nijdam, H., 1985: Genetical and grafting experiments with pea mutants in studies of symbiosis. In: Evans, H. J., Bottomley, P. J., Newton, W. E. (eds.), Nitrogen Fixation Research Progress, p. 43. Dordrecht: Nijhoff Publ.

Johansen, E., Finan, T. M., Gefter, M. L., Signer, E. R., 1984: Monoclonal antibodies to *Rhizobium meliloti* and surface mutants insensitive to them. J. Bacteriol. **160,** 454—457.

Jones, G. D., 1962: Variation in nodule characters in S. 100 Nomark white clover. J. Sci. Food. Agric. **13,** 598—603.

Jordan, D. C., 1982: Transfer of *Rhizobium japonicum* Buchanan 1980 to *Bradyrhizobium japonicum gen. nov.,* a genus of slow-growing, root nodule producing bacteria from leguminous plants. Int. J. Syst. Bacteriol. **32,** 136—139.

Kahn, M., Krans, J., Sommerville, J. E. 1985. A model of nutrient exchange in the *Rhizobium*-legume symbiosis. 1985: In: Evans, H. J., Bottomley, P. J., Newton, W. E. (eds.), Nitrogen Fixation Research Progress, pp. 193—199. Dordrecht: Nijhoff Publ.

Kaninakis, P., Verma, D. P. S., 1985: Nodulin-24 gene of soybean encodes for a peptide of the peribacteroid membrane and was generated by tandem duplication of a sequence resembling an insertion element. Proc. Nat. Acad. Sci. U.S.A. **82,** 4157—4161.

Keen, N. T., Kennedy, B. W., 1974: Hydroxyphaseolin and related isoflavonoids in the hypersensitive resistant reaction of soybean against *Pseudomonas glycinea.* Physiol. Plant Pathol. **4,** 173—185.

Keen, N. T., Williams, P. H., 1971: Chemical and biological properties of a lipomucopolysaccharide from *Pseudomonas lachrymans.* Physiol. Plant Pathol. **1,** 247—264.

Keithley, J. H., Nadler, K. D., 1983: Protoporphyrin formation in *Rhizobium japonicum.* J. Bacteriol. **154,** 838—845.

Klement, Z., 1982: Hypersensitivity. In: Mount, M. S., Lacey, G. H. (eds.), Phytopathogenic Prokaryotes. Vol. 2., pp. 150—177. New York: Academic Press.

Kneen, B. E., LaRue, T. A., 1984a: Nodulation resistant mutant of *Pisum sativum* (L). J. Hered. **75,** 238—240.

Kneen, B. E., LaRue, T. A., 1984b: Peas *(Pisum sativum* L.) with strain specificity for *Rhizobium leguminosarum.* Heredity **52,** 383—389.

Kondorosi, A., and 11 other authors, 1985: Identification and organisation of *Rhizobium meliloti* genes relevant to the initiation and development of nodules. In: Evans, H. J., Bottomley, P. J., Newton, W. E. (eds.), Nitrogen Fixation Research Progress, pp. 73—78. Dordrecht: Nijhoff Publ.

Kondorosi, E., Baufalvi, Z., Kondorosi, A., 1984: Physical and genetical analysis of a symbiotic region of *Rhizobium meliloti.* Identification of nodulation genes. Mol. Gen. Genet. **193,** 445—452.

Kortt, A. A., Burns, J. E., Trinick, M. J., Appleby, C. A., 1985: The amino-acid sequence of hemoglobin I from *Parasponia andersonii,* a non-leguminous plant. FEBS Lett. **180,** 55—60.

Kosslak, R. M., Bohlool, B. B., 1984: Suppression of nodule development of one side of a split root system of soybeans caused by prior inoculation of the other side. Plant Physiol. **75,** 125—130.

LaFavre, J. S., Eaglesham, A. R. J., 1984: Increased nodulation of "non-nodulating" *(rj₁rj₁)* soybeans by high dose inoculation. Plant and Soil **80,** 297—300.

Lalonde, M., 1979: Immunological and ultrastructural demonstration of nodulation of the European *Alnus glutinosa* (L) Gaerthn. host plants by actinomycetal isolates from the North American *Compteria peregrina* (L.) Coult root nodule. Bot. Gaz. (suppl.) **140,** 535—543.

Lang-Unnasch, N., Ausubel, F. M., 1985: Nodule-specific polypeptides from effective alfalfa root nodules and from ineffective nodules lacking nitrogenase. Plant Physiol. **77,** 833—839.

LaRue, T. A., Kneen, B. E., Gartside, E., 1985: Plant mutants defective in symbiotic nitrogen fixation. In: Analysis of the Plant Genes Involved in the Legume-*Rhizobium* Symbiosis, pp. 39—48. Paris: OECD Publ.

Lawn, R. J., Brun, W. A., 1974: Symbiotic nitrogen fixation in soybean III: Effect of supplement nitrogen and intervarietal grafting. Crop Sci. **14,** 22—25.

Lawn, R. J., Bushby, H. V. A., 1982: Effect of root, shoot and *Rhizobium* strain on nitrogen fixation in four aseatic *Vigna* species. New Phytol. **92,** 425—434.

Lee, J. S., Brown, G. G., Verma, D. P. S., 1983: Chromosomal arrangement of leghemoglobin genes in soybean. Nuc. Acids Res. **16,** 5541—5553.

Legocki, R. P., Verma, D. P. S., 1979: A nodule specific plant protein (nodulin 35) from soybean. Science **205,** 190—193.

Legocki, R. P., Eaglesham, A. R. J., Szalay, A. A., 1983: Stem nodulation in *Aeschynomene:* A model system for bacteria-plant interaction In: Pühler, A. (ed.), The molecular genetics of the bacteria-plant interaction, pp. 210—219. New York - Berlin - Heidelberg: Springer.

Leong, S. A., Ditta, G. S., Helinski, D. R., 1982: Heme biosynthesis in *Rhizobium:* identification of a cloned gene coding for gamma-aminolevulinic acid synthase in *Rhizobium meliloti.* J. Biol. Chem. **257,** 8724—8730.

Lie, T. A., 1971: Symbiotic nitrogen fixation under stress conditions. Plant and Soil, special volume, 117—127.

Masterson, R. V., Prakash, R. K., Atherly A. G., 1985: Conservation of symbiotic nitrogen fixation gene sequences in *Rhizobium japonicum* and *Bradyrhizobium japonicum.* J. Bacteriol. **163,** 21—26.

McNeil, D. L., 1982: Variation in the ability of *Rhizobium* strains to nodulate soybeans and maintain fixation in the presence of nitrate. Appl. Environ. Microbiol. **44,** 647—652.

McNeil, D. L., LaRue, T. A., 1984: Effect of nitrate source on ureides in soybean. Plant Physiol. **74,** 227—232.

McNeil, D. L., Carroll, B. J., Gresshoff, P. M., 1984: The nitrogen fixation capacity of bacteroids extracted from soybean nodules inhibited by nitrate, ammonia and dark treatments. In: Ghai, B. S. (ed.), Symbiotic Nitrogen Fixation, Vol. I, pp. 79—88. Ludhiana, India: USG Publ.

Messager, A., 1985: Mutants for nodulation and nitrogen fixation. In: Analysis of the Plant Genes Involved in the Legume-*Rhizobium* Symbiosis, pp. 52—60. Paris: OECD Publ.

Miflin, B. J., Cullimore, J. V., 1984: Nitrogen assimilation in the legume-*Rhizobium* symbiosis. A joint endeavour. In: Verma, D. P. S., Hohn, Th. (eds.), Genes Involved in the Microbe-Plant Interaction, Plant Gene Research, Vol. 1, pp. 129—174. Wien — New York: Springer.

Minchin, F. R., Witty, J. F., Sheehy, J. E., Muller, M., 1983: A major error in the acetylene reduction assay. Decreases in nodular nitrogenase activity under assay conditions. J. Exp. Bot. **34,** 641—649.

Misaghi, I. J., 1982: Physiology and biochemistry of plant-pathogen interaction. New York: Plenum Press.

Mohapatra, S. S., Gresshoff, P. M., 1984: Sensitivity to oxygen of nitrogenase activity in *Rhizobium* strain ANU289 of the non-legume *Parasponia (Ulmaceae).* Aust. J. Biol. Sci. **37,** 31—36.

Mohapatra, S. S., Pühler, A., 1986: Detection of nodule specific polypeptides from effective and ineffective root nodules of *Medicago sativa.* J. Plant Physiol., submitted.

Mulligan, J. T., Long, S. R., 1985: Induction of *Rhizobium meliloti nodC* by plant exudate requires *nodD.* Proc. Nat. Acad. Sci. U.S.A. **82,** 6609—6613.

Mulvaney, C. S., Hageman, R. H., 1984: Acetaldehyde oxime, a product formed during the *in vivo* nitrate reductase assay of soybean leaves. Plant Physiol. **76,** 118—124.

Müller, P., Niehaus, K., Pühler, A., 1985: Isolation and characterisation of mutants of *Rhizobium meliloti* obtained by transposon Tn5 mutagenesis. In: Evans, H. J., Bottomley, P. J., Newton, W. E. (eds.), Nitrogen Fixation Research Progress, pp. 137. Dordrecht: Nijhoff Publ.

Nadler, K. D., Avissar, Y. J., 1977: Heme biosynthesis in soybean root nodules: The role of bacterial gamma-aminolevulinic acid synthase and aminolevulinic acid dehydrase in the synthesis of the heme of leghemoglobin. Plant Physiol. **60,** 433—436.

Nambiar, P. T. C., Nigam, S. N., Dart, P. J., Gibbons, R. W., 1983: Absence of root hairs in non-nodulating groundnut, *Arachis hypogaea.* (L). J. Exp. Bot. **34,** 484—488.

Nelson, R. L., Bernard, R. L., 1984: Production and performance of hybrid soybeans. Crop Sci. **24,** 549—553.

Nelson, R. S., Ryan, S. A., Harper, J. E., 1983: Soybean mutants lacking constitutive nitrate reductase I. Selection and initial plant characterization. Plant Physiol. **72,** 503—509.

Nutman, P. S., 1946: Genetical factors concerned in the symbiosis of clover and nodule bacteria. Nature **157,** 463—465.

Nutman, P. S., 1949: Physiological studies on nodule function II. The influence of delayed nodulation on the rate of nodulation in red clover. Ann. Bot. **13,** 261—283.

Nutman, P. S., 1953: Symbiotic effectiveness in nodulated red clover I. Variation in host and bacteria. Heredity **8,** 35—46.

Nutman, P. S., 1969: Genetics of symbiosis and nitrogen fixation in legumes. Proc. R. Soc. London, Ser. B **172,** 417—437.

Ohlendorf, H., 1983a: Selektion auf Resistenz von *Pisum sativum* gegen *Rhizobium*-Stamm 311d. Z. Pflanzenzücht. **90,** 204—221.

Ohlendorf, H., 1983b: Untersuchungen zur Vererbung der Resistenz von *Pisum sativum* gegen *Rhizobium-leguminosarum*-Stamm 311d. Z. Pflanzenzücht. **91,** 13—24.

Olson, E. R., Sadowsky, M. J., Verma, D. P. S., 1985: Identification of genes involved in the *Rhizobium*-legume symbiosis by D1-Mu (kan:lac)-generated transcript lesions. Biotechnology **3,** 143—149.

Pacovsky, R. S., Bayne, H. G., Bethlenfalvay, G. J., 1984: Symbiotic interactions between strain of *Rhizobium phaseoli* and cultivars of *Phaseolus vulgaris.* Crop Sci. **24,** 101—105.

Pankhurst, C. E., Sprent, J. I., 1975a: Surface features of soybean root nodules. Protoplasma **85,** 85—98.

Pankhurst, C. E., Sprent, J. I., 1975b: Effects of water stress on the respiration and nitrogen fixation ability of soybean root nodules. J. Exp. Bot. **26,** 287—304.

Peterson, M. A., Barnes, D. K., 1981: Inheritance of ineffective nodulation and non-nodulation traits in alfalfa. Crop Sci. **21,** 611—616.

Phillips, D. A., Torrey, J. G., 1970: Cytokinin production by *Rhizobium japonicum.* Physiol. Plant **23,** 1057—1063.

Phillips, D. A., Bedmar, E. J., Qualset, C. O., Teuben, L. R., 1985: Host legume control of *Rhizobium* function. In: Ludden, P. W., Burris, J. E. (eds.), Nitrogen Fixation and CO_2 Metabolism, pp. 203—212. Amsterdam: Elsevier.

Pierce, M., Bauer, W. D., 1983: A rapid regulatory response governing nodulation in soybean. Physiol. Plant **73,** 286—290.

Price, G. D., Mohapatra, S. S., Gresshoff, P. M., 1984: Structure of nodules formed by *Rhizobium* strain ANU289 in the non-legume *Parasponia* and the legume siratro *(Macroptilium atropurpureum).* Bot. Gaz. **145,** 444—451.

Ralston, E. J., Imsande, J., 1982: Entry of oxygen and nitrogen into intact nodules of soybean. J. Exp. Bot. **33,** 208—214.

Rennie, R. J., Dubetz, S., 1984: Multistrain versus single strain *Rhizobium japonicum* inoculants for early maturing (00 and 000) soybean cultivars: N_2 fixation quantified by ^{15}N isotope dilution. Agron. J. **76,** 498—502.

Rennie, R. J., Kemp, G. A., 1984: ^{15}N determined time course for nitrogen fixation in two cultivars of field bean. Agron. J. **76,** 146—154.

Robertson, J. G., Wells, B., Bisseling, T., Farnden, K. J. F., Johnston, A. W. B., 1984: Immunogold localisation of leghaemoglobin in the plant cytoplasm in nitrogen fixing root nodules of pea. Nature **311,** 254—256.

Rolfe, B. G., Shine, J., 1984: *Rhizobium-Leguminosae* symbiosis: The bacterial point of view. In: Verma, D. P. S., Hohn, Th. (eds.), Genes Involved in Microbe-plant Interaction, Plant Gene Research, Vol. 1, pp. 95—128. Wien - New York: Springer.

Ronson, C. W., Astwood, P. M., Downie, J. A., 1984: Molecular cloning and genetic organisation of C_4-dicarboxylate transport genes from *Rhizobium leguminosarum.* J. Bacteriol. **160,** 903—909.

Ronson, C. W., Lyttleton, P., Robertson, J. G., 1981: C_4-dicarboxylate transport mutants of *Rhizobium trifolii* from ineffective nodules of *Trifolium repens.* Proc. Nat. Acad. Sci. U.S.A. **78,** 4284—4288.

Ryan, S. A., Nelson, R. S., Harper, J. E., 1983: Soybean mutants lacking constitutive nitrate reductase activity II. Nitrate assimilation, chlorate resistance and inheritance. Plant Physiol. **72,** 510—514.

Sandeman, R. A., Gresshoff, P. M., 1985: Nitrogenase activity and inactivation in isolated bacteroids from the legume siratro and the non-legume *Parasponia rigida.* Plant Sci. Lett. **37,** 199—204.

Schmidt, J., John, M., Kondorosi, E., Kondorosi, A., Wieneke, U., Schröder, G., Schröder, J., Schell, J., 1984: Mapping of the protein encoding region of the *Rhizobium meliloti* common nodulation genes. EMBO J. **3,** 1705—1711.

Schofield, P. R., Djordjevic, M. A., Rolfe, B. G., Shine, J., Watson, J., 1983: A molecular linkage map of the nitrogenase and nodulation genes of *Rhizobium trifolii.* Mol. Gen. Genet. **192,** 456—466.

Schofield, P. R., Ridge, R. W., Rolfe, B. G., Shine, J., Watson, J., 1984: Host specific nodulation is encoded on a 14 Kb DNA fragment in *Rhizobium trifolii.* Plant Mol. Biol. **3,** 3—15.

Schubert, K., 1981: Enzymes of purine biosynthesis and catabolism in *Glycine max.* Plant Physiol. **68,** 1115—1122.

Schubert, K., Evans, H. J., 1976: Hydrogen evolution: a major factor affecting the efficiency of nitrogen fixation in nodulated soybeans. Proc. Nat. Acad. Sci. U.S.A. **73,** 1207—1211.

Schuller, K. A., Day, D. A., Gibson, A. H., Gresshoff, P. M., 1986: Effect of nitrate on nitrogen fixation and ammonia assimilation in soybean nodules. Plant Physiol. **80,** in press.

Scott, K. F., Rolfe, B. G., Shine, J., 1983: Biological nitrogen fixation: primary structure of the *Rhizobium trifolii* iron-protein gene. DNA **2,** 149—158.

Sen, D., Weaver, R. W., 1984: A basis for different rates of nitrogen fixation by the same strains of *Rhizobium* in peanut and cowpea nodules. Plant Sci. Lett. **34,** 239—246.

Sequira, L., 1984: Recognition systems in plant-pathogen interactions. Biol. Cell **51,** 281—286.

Sheehy, J. E., Fishbeck, K. A., DeJong, T. M., Williams, L. E. Phillips, D. A., 1980: Carbon exchange rates of shoots required to utilize available acetylene reduction capacity of soybean and alfalfa root nodules. Plant Physiol. **66,** 101—104.

Sheehy, J. E., Minchin, F. R., Witty, J. F., 1983: Biological control of the resistance to oxygen flux in nodules. Ann. Bot. **52,** 565—572.

Sinclair, T. R., Goudriaan, J., 1981: Physical and morphological constraints on transport in nodules. Plant Physiol. **63,** 143—145.

Singleton, P. W., Stockinger, K. R., 1983: Compensation against ineffective nodulation in soybean. Crop Sci. **23,** 69—72.

Skot, L., 1983: Cultivar and *Rhizobium* strain effects on the symbiotic performance of pea *(Pisum sativum).* Plant Physiol. **59,** 585—589.

Smartt, J., 1984: Gene pools of grain legume. Econ. Bot. **38,** 24—35.

Smith, G. R., Knight, W. E., 1984: Inheritance of ineffective nodulation in crimson clover. Crop Sci. **24,** 601—604.

Spreit, L., Nelson, R. S., Harper, J. E., 1985: Nitrate reductases from wild type and nr_1-soybean (*Glycine max* (L) Merr.) leaves I. Purification, kinetics and physical properties. Plant Physiol. **78,** 80—85.

Streeter, J. G., 1977: Asparaginase and asparagine transaminase in soybean leaves and root nodules. Plant Physiol. **60,** 235—239.

Summons, R. E., Letham, D. S., Gollnow, B. I., Parker, C. W., Entsch, B., Johnston, L. P., MacLeod, J. K., Rolfe, B. G., 1981: Cytokinin translocation and metabolism in species of Leguminosae: studies in relation to shoot and nodule development. In: Guern, J., Peaud-Lenoël, C. (eds.), Metabolism and Molecular Activity of Cytokinins, pp. 69—79. New York - Berlin - Heidelberg: Springer

Sutton, W. D., Paterson, A. D., 1983: Further evidence for plant host effect on *Rhizobium* bacteroid viability. Plant Sci. Lett. **30,** 33—41.

Tanner, J. W., Anderson, I. C., 1963: Investigation on non-nodulating soybean strains. Can. J. Plant Sci. **43,** 542—546.

Thimann, K. V., 1936: On the physiology of the formation of nodules on legume roots. Proc. Nat. Acad. Sci. U.S.A. **22,** 511—514.

Thomas, R. J., Jokinen, K., Schrader, L. E., 1983: Effect of *Rhizobium japonicum* mutants with enhanced nitrogen fixation activity on nitrogen transport and photosynthesis of soybean during vegetative growth. Crop Sci. **23,** 453—456.

Tjepkema, J. D., Yocum, C. S., 1974: Measurement of oxygen partial pressure within soybean nodules by microelectrode. Planta **119,** 351—360.

Trinick, M. J. 1973: Symbiosis between *Rhizobium* and the non-legume *Trema aspera.* Nature (London) **244,** 459—460.

Tran, T. V. K., Toabart, P., Cousson, A., Darvil, A. G., Gollin, D. J., Chelf, P., Albersheim, P., 1985: Manipulation of the morphogenetic pathways of tobacco explants by oligosaccharides. Nature **314,** 615—617.

Tsein, H. C., Dreyfus, B. L., Schmidt, E. L., 1983: Initial stages in the morphogenesis of nitrogen fixing stem nodules of *Sesbania rostrata.* J. Bacteriol. **156,** 888—897.

Tudge, C., 1984: Whatever happens to nitrogen? New Scientist **9,** 13—15.

Van den Bos, R., Schots, A., Hentelez, J., van Kammen, A., 1983: Constitutive nitrogenase synthesis from de novo transcribed mRNA in isolated *Rhizobium leguminosarum* bacteroids. Bioch. Biophys. Acta **740,** 313—322.

Vance, C. P., Johnson, L. E. B., 1983: Plant induced ineffective nodules in alfalfa (*Medicago sativa* L): structural and biochemical comparisons. Can. J. Bot. **61,** 93—106.

Verma, D. P. S., Nadler, K. D., 1984: Legume-*Rhizobium* symbiosis: The host's point of view. In: Verma, D. P. S., Hohn, Th. (eds.) Genes Involved in Microbe-Plant Interaction, Plant Gene Research, Vol. 1, pp. 58—84. Wien - New York: Springer.

Verma, D. P. S., Lee, J. S., Katinakis, P., Sutton, B., 1985: Nodule specific genes of soybean. In: Analysis of Plant Genes Involved in the Legume-*Rhizobium* Symbiosis, pp. 74—84. Paris: OECD Publ.

Vest, G., 1970: *Rj*3 — a gene controlling ineffective nodulation in soybean. Crop Sci. **10,** 34—35.

Vest, G., Caldwell, B. E., 1972: *Rj*4 — a gene controlling ineffective nodulation in soybean. Crop Sci. **12,** 692.

Viands, D. R., Vance, C. P., Heichel, G. H., Barnes, D. K., 1979: An ineffective nitrogen fixation trait in alfalfa. Crop Sci. **19,** 905—908.

Weber, C. R., 1966: Nodulating and non-nodulating soybean isolines II. Response to applied nitrogen and modified soil conditions. Agron. J. **58,** 46—49.

Weinman, J. J., Fellows, F. F., Gresshoff, P. M., Shine, J., Scott, K. F., 1984: Structural analysis of the genes encoding the molybdenum-iron protein of nitrogenase in the *Parasponia Rhizobium* strain ANU289. Nuc. Acids Res. **12,** 8329—8344.

Werner, D., Mörschel, E., Kort, R., Mellor, R. B., Bassarab, S., 1984: Lysis of bacteroids in the vicinity of the host cell nucleus in an ineffective *(fix$^-$)* root nodule of soybean *(Glycine max).* Planta **162,** 8—16.

Werner, D., Christensen, T., Mellor, R. B., Mörschel, E., 1985: Glycosyltransferases and the peribacteroid membrane. In: Analysis of the Plant Genes Involved in the Legume-*Rhizobium* Symbiosis, pp. 61—71. Paris: OECD Publ.

Whitmore-Smith, D., 1985: Studies on the constitutive nitrate reductase activity of different soybean mutants. Honours dissertation. Botany Department, Australian National University, Canberra, Australia.

Williams, L. F., Lynch, D. L., 1954: Inheritance of a non-nodulation character in soybean. Agron. J. **46,** 28—29.

Williams, L. E., DeJong, T. M., Phillips, D. A., 1982: Effect of changes in shoot carbon exchange rate on soybean nodule activity. Plant Physiol. **69,** 432—436.

Winarno, R., Lie, T. A., 1979: Competition between *Rhizobium* strains in nodule formation. Interaction between nodulating and non-nodulating strains. Plant and Soil **51,** 135—142.

Witty, J. F., Minchin, F. R., Sheehy, J. E., Minguez, M. I., 1984: Acetylene-induced changes in the oxygen diffusion resistance and nitrogenase activity of legume root nodules. Ann. Bot. **53,** 13—20.

Witty, J. F., Skot, L., Revsbech, N. P., 1985: Direct evidence for a variable barrier to oxygen diffusion into legume nodules. In: Evans, H. J., Bottomley, P. J., Newton, W. E. (eds.), Nitrogen Fixation Research Progress, p. 355. Dordrecht: Nijhoff Publ.

Chapter 7

Endosperm Proteins

Peter I. Payne

Plant Breeding Institute, Maris Lane, Trumpington, Cambridge CB2 2LQ, U.K.

With 7 Figures

Contents

I. Introduction

Nearly all the major crop plants of the world are cereals, comprising in decreasing order of production, wheat, maize, rice, barley, sorghum, oats, millet and rye (Harlan and Starks, 1980). The major organ by volume of the cereal grain is the endosperm which serves virtually exclusively as a store of food reserves for the germinating seedling.

The major macromolecule in the endosperm of all cereals is starch and it amounts to some 80—90 % of the total dry weight. The next most abundant macromolecule is protein. Its concentration varies according to the cereal and to the method of cultivation but is generally highest in wheat and oats, with a usual range of 10—17 %, and lowest in maize and rice, when it can fall to 6 % (Altschul, 1965). The concentration of protein in the cereal endosperm is appreciably lower than in leguminous seeds, where the storage organ is not the endosperm but the cotyledon and the range of protein content is usually from 20—40 % of the dry weight (Altschul, 1965). Nevertheless, because of the relative production of cereals and legumes, the decreasing order of protein production amongst these crops is wheat, maize, soybean, rice, barley, oats, sorghum, peanut, millet, rye, peas and beans (Harlan and Starks, 1980). The majority of the peoples of the world, therefore, obtain most of their protein by eating cereals and it is not surprising that protein amount and protein quality feature strongly in many cereal breeding programmes and that their biochemistry and genetics are being studied actively.

The object of this chapter is to show how mutations involving endosperm protein structural genes or genes which control their output are being exploited, not only by the plant breeder but also by the plant biochemist and the molecular biologist as a means to understanding the structure, synthesis and regulation of these proteins.

II. Origin and Development of the Endosperm

The endosperm originates at the time of fertilisation in the embryo sac of the ovum by a two-step process. First, the two haploid, polar nuclei of the embryo sac coalesce to form the diploid, fusion nucleus. This in turn fuses with one of the two sperm nuclei derived from a single pollen grain to form the triploid endosperm nucleus. The remaining sperm nucleus fuses with the egg-cell nucleus to form a diploid zygote which gives rise to the embryo. The endosperm's triploidy has important consequences in biochemical and genetic analyses.

With the exception of a few species such as orchids, the endosperm tissue grows rapidly during the early stages of seed development. It becomes senescent in most plant species well before seed maturity, is rapidly digested and is used as a food source for the developing embryo. But in other species, and particularly in all the cereals, the endosperm continues to develop and becomes the primary food store of the mature seed. The formation of the endosperm and its development during seed maturation have been described in detail for wheat by Simmonds and O'Brien (1981). In the remainder of this section, I will restrict myself to a very brief description of the synthesis and accumulation of protein in cereal endosperm, with the emphasis on wheat.

Storage protein, to be discussed and defined in detail in the next section, accumulates in the cell expansion phase of endosperm development. For wheat, this is from about 12 to 40 days after anthesis on a time scale of 50 days from anthesis to grain maturity (Parker, 1980). The molecular events in the synthesis of storage protein have been elucidated and they resemble the synthesis of protein in animal cells which are destined for export. The messenger RNAs encoding storage proteins pass from nucleus to cytoplasm and combine with ribosomes to form polysomes. Protein synthesis commences with the formation of a signal sequence at the N-terminus. This enables these polysomes to specifically attach to the endoplasmic reticulum. As synthesis of the protein chain continues, the leader sequence and the rest of the protein pass into the lumen of the endoplasmic reticulum, where the signal sequence is removed (barley: Cameron-Mills *et al.*, 1978; maize: Burr and Burr, 1981).

From the endoplasmic reticulum the proteins pass into storage organelles, the protein bodies, but the mechanism of the movement is disputed in some cereals. In maize the mechanism is clear: storage protein accumulates in localised regions of the endoplasmic reticulum which swell and eventually bud off as membrane-bound protein bodies (Larkins and Hurkman, 1978). The majority of work in barley and wheat (e. g. Cameron-Mills and von Wettstein, 1980; Parker, 1982, respectively) has implicated the involvement of Golgi bodies. For these species it is suggested that storage proteins pass to the Golgi bodies, with which the endoplasmic reticulum is directly connected. The protein is then packaged into vesicles and transferred to vacuoles which develop into protein bodies. This mechanism is very similar to that accepted for storage protein accumulation in legumes. However, Shewry and Miflin (1985) believe that storage protein accumulates in barley and wheat as it does in maize without the involvement of Golgi. Very recently, Parker (unpublished observations) has presented evidence that both mechanisms exist, with endoplasmic-reticulum budding being prevalent in those cells in the centre of the endosperm which accumulate least protein.

The major function of the storage proteins of the endosperm is to serve the embryo as a supply of elements, particularly nitrogen, in the early stages of germination prior to the seedling becoming established. Since a specific amino-acid sequence is not essential to fulfil the role of these proteins, change through mutation will be tolerated more than changes in the sequence of enzymic proteins or structural proteins in multi-molecular complexes such as ribosomes. Nevertheless, some constraints on storage-protein amino-acid sequences must exist. Other than the maintenance of specific sequences in the structural genes which are essential for the production of all proteins, the leader sequence must be preserved and all parts of protein structure involved in the transport and packaging of the proteins. For maximum efficiency, storage proteins contain large percentages of nitrogen-rich amino-acid residues (asparagine or glutamine) and, to aid packing, high percentages of hydrophobic amino-acid residues such as proline, leucine and isoleucine and low percentages of hydrophylic amino -acid residues such as lysine, glutamic and aspartic acids. Major changes in the balance of these amino acids are unlikely to be tolerated. The rate of mutation of different storage-protein groups will be discussed in a later section of this chapter.

III. Classification of the Major Endosperm Proteins

The classification of the endosperm proteins of cereals has been dominated by Osborne (1907) and his contemporaries and is based on the proteins' relative solubilities in different solvents:

1. albumin: soluble in water;
2. globulin: soluble in salt solutions but insoluble in water;
3. prolamin: soluble in aqueous alcohols;
4. glutelin: soluble in dilute acids and alkalis.

For the prolamin group of proteins, specific names have been given for each cereal. Thus the prolamin of maize is zein, that of barley is hordein, that of wheat is gliadin and that of oats avenin. In wheat the glutelin fraction is usually called glutenin. From the above definitions, wheat and sorghum were considered to contain about equal amounts of prolamin and glutelin, maize and barley rather more prolamin than glutelin, whereas rice was thought to be dominanted by glutelin and oats by globulin.

In 1982, Payne and Rhodes stated: "These terms for the different groups of proteins are still used today although modern workers use chemicals and extraction procedures which are different from those used at the beginning of the century. This resulted in changed meanings for these protein groups and unfortunately there are no simple, modern definitions that are acceptable to all. The major problems lie with prolamin and glutenin."

Fortunately the last two or three years have seen rapid advances in our understanding of these proteins from all the major cereals. This has been due partly to N-terminal amino-acid sequencing of isolated proteins but mainly to DNA sequencing of storage-protein genes cloned in bacterial vectors. Our present understanding of these storage proteins has been covered very recently by an extensive review (Shewry and Miflin, 1985).

In wheat, the sequencing of storage-protein genes is well advanced. Extensive sequences for α-gliadins, γ-gliadins and low-molecular-weight (LMW) subunits of glutenin have been published (Bartels and Thompson, 1983; Kasarda *et al.*, 1984; Rafalski *et al.*, 1984). From this work it is likely that the genes of all the protein groups above originated from the duplication of a common ancestral gene. Thus, with the possible exception of high-molecular-weight (HMW) glutenin subunits whose sequence appears to be basically unrelated (Thompson *et al.*, 1983) there is only one major class of storage protein in wheat endosperm and this would most appropriately be called prolamin. Within the prolamin group are proteins with prolamin-type and glutenin-type solubility in spite of the similarity of their primary sequences. This is because glutenin subunits occur as disulphide bonded aggregates which greatly reduces their solubility, whereas gliadins occur as simple, monomeric molecules and contain only intramolecular disulphide bonds. The distinction between gliadin and glutenin is certain to continue to be used because the two groups have different functionalities in food processing, particularly in bread making, but it should be remembered that both lie within the prolamin subgroup of proteins.

A further complication in the old classification of proteins by solubility is that some proteins in the glutenin fraction are not true storage proteins but are probably structural proteins of macromolecular complexes such as ribosomes or membranes. These therefore are not part of the modern prolamin fraction and they have not been studied extensively.

Sequencing has demonstrated a close evolutionary relationship between the prolamins of barley, rye and wheat (Shewry *et al.*, 1980; Miflin *et al.*, 1984) which are all members of the subfamily, Triticeae. The other cereals which contain mainly prolamins are maize, sorghum and millet, three cereals which are closely related to each other but distant from barley, rye and wheat. This evolutionary distance is reflected in the very different prolamin amino-acid sequences of maize (Hu *et al.*, 1982) and the three cereals of the Triticeae. The two types of prolamins nevertheless have similar amino-acid compositions being rich in glutamine and hydrophobic amino acids and deficient in acidic and basic amino-acid residues such as lysine. Prolamin molecules are characteristic in containing repeated sequences each several amino-acid residues long.

Oat is unique amongst cereals in containing protein which is mainly soluble in salt solutions. Peterson (1978) demonstrated that the protein occurs as a large protein, 12 S with a molecular weight of 322,000. He gave evidence that it is built up from six subunits of molecular weight 21,700 and six subunits of 31,700. This and other evidence shows that the protein resembles the 12 S, legumin group of globulin storage proteins found in leguminous seeds.

A particularly exciting development recently has been the finding that at least some of the major components of rice glutenin have a subunit structure which resembles that of oat and legume 12 S globulin (Yamagata *et al.*, 1982). Furthermore, amino-acid sequencing of one component has demonstrated a 25 % homology with a subunit of pea legumin (Zhae *et al.*, 1983). Presumably this rice globulin must have undergone mutational change during evolution, resulting in a greater proportion of hydrophobic amino-acid residues in the protein and causing it no longer to be soluble in salt solutions.

For the purposes of this chapter, we can conclude in the light of recent research that cereals can contain two major storage protein groups, globulins and prolamins, the former being related to counterparts in leguminous seeds.

IV. Biochemical Complexity and Genetic Variation of Endosperm Proteins

During the course of evolution the prolamin genes in particular have duplicated and undergone mutational change to form families of different but related genes. These families have, during the course of time, become partially split up by chromosomal rearrangements. The outcome has resulted (1) in individual cereal varieties containing complex mixtures of endosperm proteins and (2) significant variation occuring between varieties for protein type.

In Fig. 1 the endosperm proteins of a variety of bread wheat have been fractionated by two-dimensional electrophoresis. In total there are about 60 major components and about double that number of minor components.

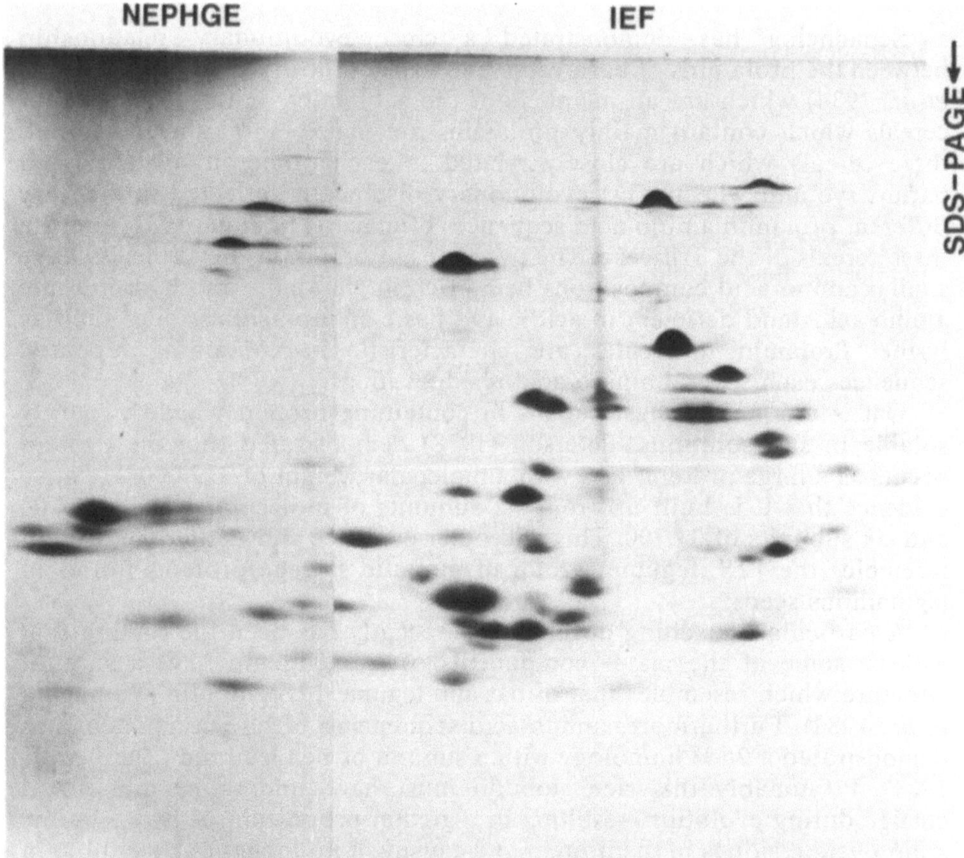

Fig. 1. Two-dimensional fractionation of the endosperm proteins of the wheat variety Sicco. There are two first dimensions, IEF (isoelectric focussing) to fractionate the acidic proteins and NEPHGE (non-equilibrium, pH gradient electrophoresis) to resolve the more basic proteins. The two separations were fractionated side-by-side on the same second-dimension gel of SDS-PAGE. Further details of the method of grouping the proteins into glutenin subunits, gliadins and non-storage groups are given in Payne *et al.* (1984b)

The complexity in wheat has been exaggerated as it possesses three different genomes, the storage proteins from each genome being distinguishable. For comparison, a separation of a variety of diploid barley is also shown (Fig. 2). The complexity is less, though at least 40 major components have resolved. A similar order of complexity to that in barley has been shown to occur in maize (Burr and Burr, 1981).

In Figs. 3 and 4 the proteins of different varieties of wheat and barley, respectively, have been fractionated by one-dimensional electrophoresis to display the great variation, presumably due to the presence of different alleles at each storage-protein gene locus. Interestingly, the variation revealed by the same technique is very much less for millet, sorghum (Fig. 5)

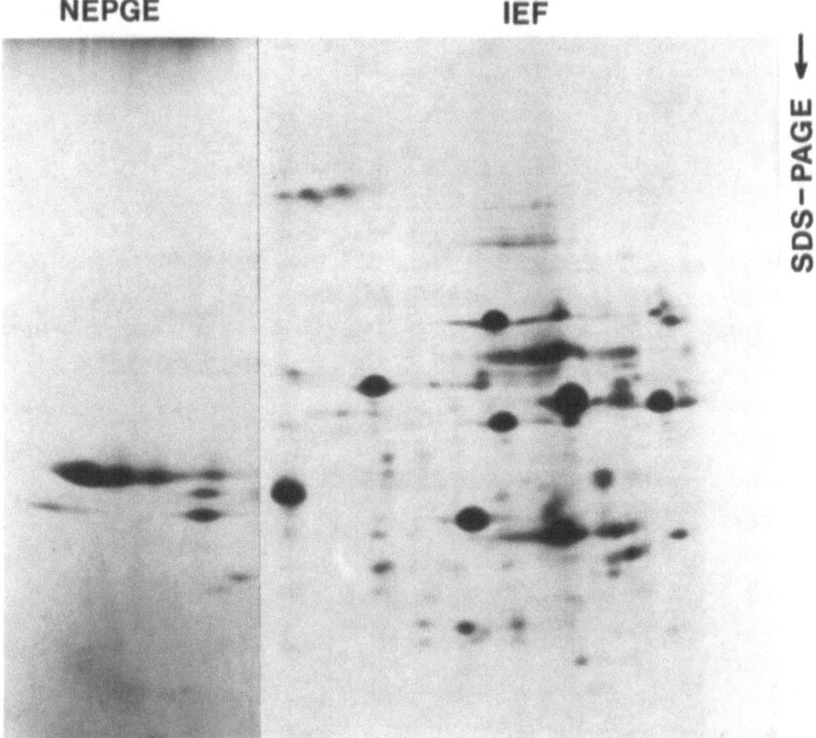

Fig. 2. Two-dimensional fractionation of the endosperm proteins of barley variety, Sundance. The electrophoretic procedure is described in the legend to Fig. 1

and maize. The technique, SDS-PAGE, fractionates mainly according to molecular weight. Variation in the prolamins of the above three cereals is much more by charge than molecular weight and such techniques as isoelectric focussing do show extensive allelic variation (Rhighetti *et al.*, 1977).

The globulin storage proteins of cereals show much less variation amongst varieties than do prolamins. Thus, amongst nine varieties of oats, there was very little variation in the globulin fraction when assessed by SDS-PAGE and only slightly more when fractionated by IEF (Robert *et al.*, 1983 a). When the prolamins (avenins) from the same set of varieties were analysed by SDS-PAGE and IEF, variation was much more prominent (Robert *et al.*, 1983 b). Similarly there is very little electrophoretic variation amongst globulins (glutelins) from varieties of rice (Park and Stegemann, 1979).

The likely interpretation of these contrasting results with globulins and prolamins is that the latter are undergoing more rapid evolutionary change that the former. This is probably because the globulins have a far more elaborate packaging system in the protein bodies. Globulin polypeptides are arranged into an ordered structure of an approximate molecular weight of 300,000 (Peterson, 1978) imparting sequence constraints to conserve

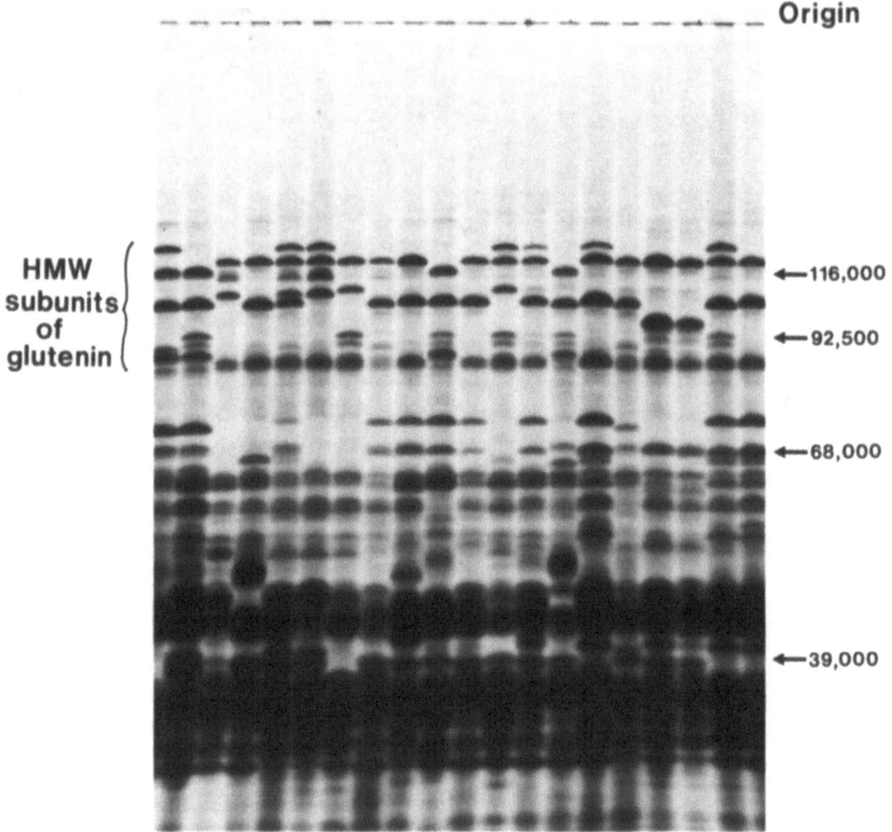

Fig. 3. Fractionation of the total endosperm proteins of 20 varieties of wheat by SDS-PAGE (sodium dodecyl sulphate, polyacrylamide-gel electrophoresis). The arrows indicate molecular weights obtained from the fractionation of standard proteins

molecular shape. For prolamins, which in the main have a much more hydrophobic nature, deposition in protein bodies is probably only by precipitation.

V. Gene Mutations

The variation in cereal endosperm proteins amongst varieties, described in Section IV, has occured over the millenia as a result of the mutation of structural genes and probably also of genes regulating endosperm protein synthesis. Further variation in the regulatory genes has been induced either with chemicals or by irradiation. Mutations in single structural genes have in general very little effect in changing the composition of grain storage proteins because the genes occur in nearly all cases in large families. However, with chromosome mutations (see Section IV) effects can be marked.

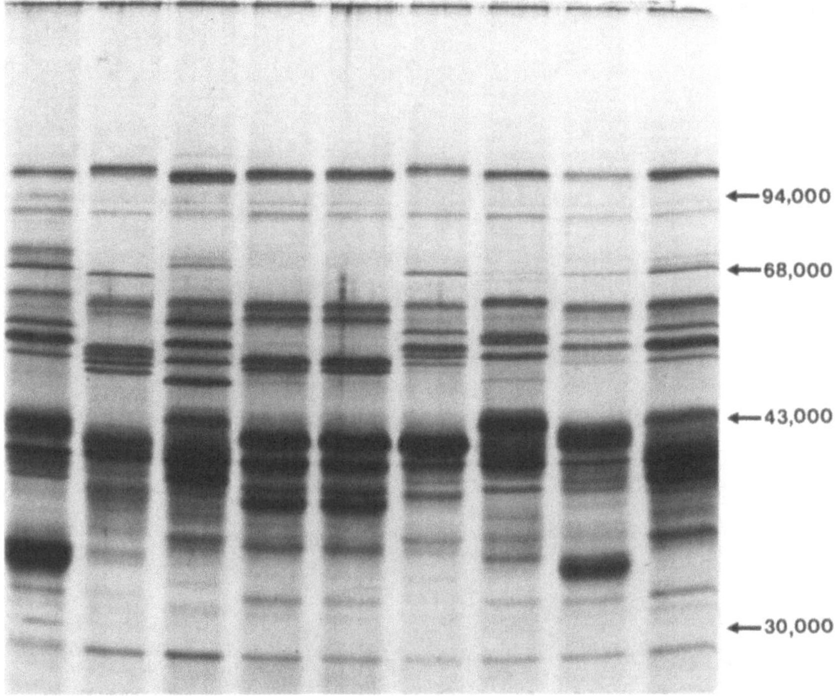

Fig. 4. Fractionation of endosperm proteins of nine varieties of barley by SDS-PAGE

A. Role in Plant Breeding

The cereals contain a small percentage of protein compared with legume seeds and those which store predominantly prolamins have in particular a poor balance of amino acids for human nutrition. The first limiting amino acid is invariably lysine (Altschul, 1965). Much effort has therefore gone into screening for higher protein and high lysine content in germplasm collections and in lines treated with mutagens. Success was achieved, rather easily for protein content, and this brought a flurry of activity throughout the world by cereal breeders to bring these so-called "high-protein" and "high-lysine" genes into élite, commercial varieties. The results have so far been disappointing, in that few if any novel varieties with higher-than-normal protein or lysine have been developed which have outclassed and replaced conventional varieties in agriculture. Some of the problems have included low or erratic grain yield, reduced protein yield, small grain size and increased susceptibility to certain diseases (IAEA, 1984). Nevertheless in several major breeding institutes, continued and detailed work is steadily reaping rewards. At CIMMYT for instance the opaque-2 genes of maize, which confer the high lysine trait, have been transferred from undesirable floury endosperm types to a vitreous endosperm type which is hard

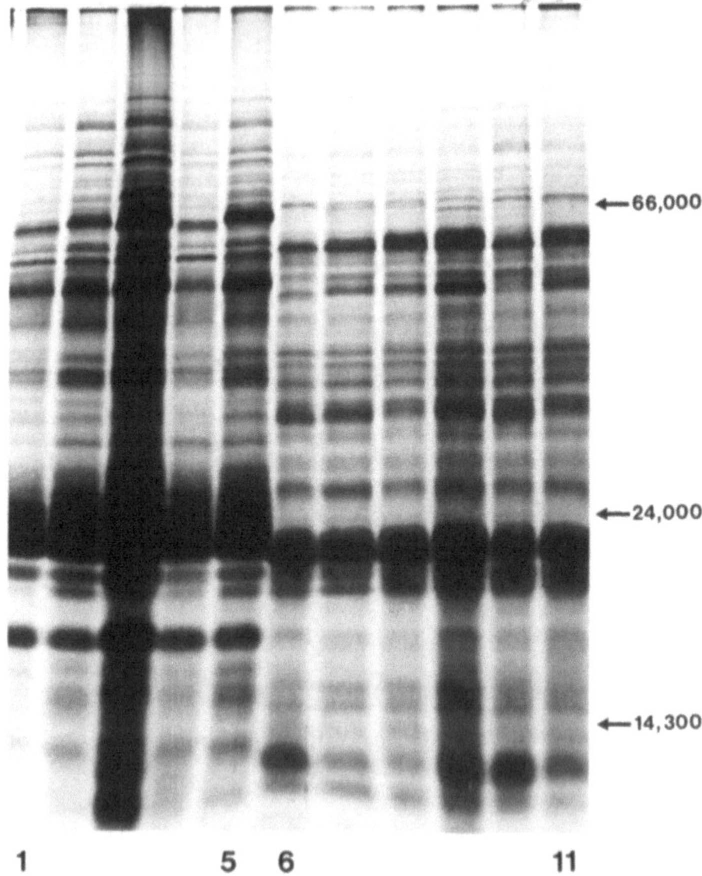

Fig. 5. Fractionation of grain protein of five sorghum varieties (slots 1—5) and six millet varieties (slots 6—11) by SDS-PAGE

milling and new lines are beginning to compare favourably with commercial varieties for yield and protein content (Vasal *et al.*, 1984). Transfer of high protein genes from emmer *(Triticum dicoccoides)* to bread-wheat also looks promising (e. g. Grama *et al.*, 1984). Indeed in the most recent proceedings of a conference on cereal grain improvement (IAEA, 1984), various groups working on maize, barley and wheat are still optimistic about transferring high-protein and high-lysine genes into commercially acceptable varieties.

The great majority of high-lysine genes have so far been detected in barley and maize as described below in Section V.C, but none have been found in wheat. Vogel *et al.* (1973) screened some 20,000 entries in the world wheat collection of the United States Department of Agriculture and found only 0.5 % of genetic variation in total grain protein content was due to lysine. A likely reason for this failure is probably that wheat is hexaploid: a mutant gene conferring high lysine would only occur in one of the three genomes and so might not be detected.

Nevertheless, the endosperm proteins of wheat are becoming more important in the breeding of new varieties, not with respect to improved nutritional quality, but to improved bread-making quality. In wheat-growing areas of the world where water is plentiful and the growing season is long, as in Western Europe, grain yield tends to be high but protein content low (11 % of dry weight or lower) even with the application of nitrogen fertilizers to attain optimal yield. Low protein content has an adverse affect on bread-making quality and if varieties are to be grown in these areas for bread production they must have a good protein quality. Variation in protein quality is primarily caused by the great variation in protein components which occurs between genotypes (see Fig. 3). Gliadins are monomeric proteins which give extensibility to a dough whereas glutenins consist of approximately 16 different subunits (Payne and Corfield, 1979) disulphide-bonded into large, elongated fibrils which impart strength and elasticity (Wall, 1979; Payne *et al.*, 1984b). It is generally agreed that it is the elastic component which is often limiting for bread-making (Orth and Bushuk, 1972). Payne *et al.* (1981) were the first to demonstrate convincingly, by the analysis of segregating progeny, that the different subunits which make up glutenin have different associations with good bread-making quality. By the production of more accurately defined genotypes these associations are being studied in more detail (Payne *et al.*, 1984b).

Glutenin subunit genes occur at six different major loci on three of the 21 chromosomes of bread-wheat. The genes at each locus display extensive allelic variation and, for several loci, different alleles are associated with different degrees of elasticity in native glutenin (Payne *et al.*, 1984b). Currently, wheat breeders at the Plant Breeding Institute, Cambridge, are selecting parents in their crossing programmes which have complementary, good-quality glutenin subunits (i. e. their genes occur at different loci) so that they can be combined in a few progeny which will have better quality than either parent. Gel electrophoresis is being used in the breeding programme at fairly early generations as a secondary screen to select the few progeny which have the ideal protein composition for a particular cross (Payne *et al.*, 1984b).

B. Role in Genetics

As will be described in Section VI, chromosome mutations have been exploited by geneticists to locate the genes which code for storage proteins to chromosomes. Gene mutations have been used to study the location and distribution of the genes on individual chromosomes. In most cereals, the prolamin genes occur as a series of tightly linked gene families at complex loci. Genetic analysis of crosses has shown that the variation in protein pattern which occurs between varieties is due to the presence of numerous complex alleles at each locus.

In wheat, there are nine major complex loci and a few minor loci coding for storage proteins, all of which occur on the homoeologous group one and group six chromosomes. The relative positions of the loci on the

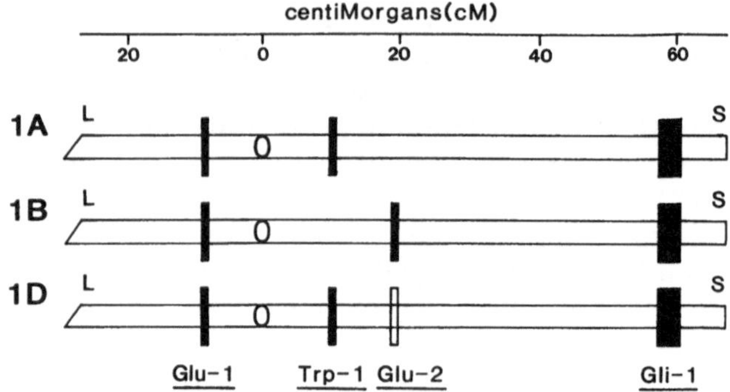

Fig. 6. Chromosomal location and position of the storage protein loci on the group 1 chromosomes of bread wheat. The two major groups of loci are *Glu-1* and *Gli-1*. The former code for HMW glutenin subunits and the latter for families of LMW glutenin subunits, ω-gliadins and γ-gliadins. The *Trp-1* loci probably code for minor globulin-type proteins (Singh and Shepherd, 1985) and the *Glu-2* loci minor LMW subunits of glutenin. The position of *Glu-2* on chromosome 1 D has not been determined. The other main loci, *Gli-2*, occur on the short arms of chromosomes 6 A, 6 B and 6 D and code for α- and β-gliadins (L = long, S = short)

chromosomes was determined by making crosses between lines with contrasting storage-protein alleles and determining the frequency of recombination in the progeny. The chromosomal position of *Glu-1*, coding for HMW subunits of glutenin (Fig. 6), appears to have been remarkably stable during the course of evolution and is one of several stable markers which are being used to understand chromosome homologies in the *Triticeae*. In rye the equivalent locus, *Glu-R 1*, occurs on chromosome 1 R (formerly E) and in barley the locus, called *Hor 3*, is on chromosome 5 (Shewry *et al.*, 1984 a; see Fig. 2). Both loci lie close to the centromeres on the long arm of their respective chromosomes and also contain one or more prolamin loci on the short arms of the same chromosomes which are at least partially equivalent to *Gli-1* of wheat (Shewry *et al.*, 1984 a). It is therefore concluded that chromosomes 1 A, 1 B and 1 D of wheat are homologous to chromosomes 1 R of rye and 5 of barley. Simply from the chromosomal location of other loci homologous to *Glu-1*, other homologous chromosomes are likely to be 1 U of *Aegilops umbellulata*, I of *Ae. elongatum* (Lawrence and Shepherd, 1981) and G of *Hordeum chilense* (Payne, Holt and Miller, unpublished results).

C. Role in Biochemistry and Molecular Biology

The so-called high-protein genes discussed in Section V.A probably vary widely in their functions, for instance by causing improved nitrogen uptake by roots or nitrogen translocation from senescing leaves and stems, or by changing plant height. In contrast the high-lysine genes directly affect storage-protein accumulation and they are being actively exploited by biochemists and molecular biologists to understand the synthesis and regulation of prolamin and globulin in the developing endosperm.

In barley, several lines containing higher-than-normal lysine contents have been described (reviewed by Miflin and Shewry, 1979) and three of them are well characterised. These are Risφ 56, obtained from treating variety Carlsberg II with γ-rays, Risφ 1508, derived from ethyleneimine-treated grain of Bomi, and Hiproly, an Ethiopian landrace naturally high in lysine. All contain reduced levels of hordein but this is achieved by different mechanisms and to different degrees (Table 1).

Table 1. Properties and chromosome location of the commonly studied "high-lysine" genes of cereals

Cereal	Line	Mutation	Locus	Chromosome	Expression	Prolamin Synthesis inhibition %	Prolamin Synthesis groups suppressed
Barley	Hiproly	Spontaneous	*lys*	7	Recessive	—	All
Barley	Risφ 56	Induced	*Hor2ca*	5	Co-dominant	25	B-hordein
Barley	Risφ 1508	Induced	*lys3a*	7	Recessive	69	All
Maize	Opaque-2	Spontaneous	O_2	7	Recessive	47	Mainly 20,000
Maize	Opaque-6	Spontaneous	O_6	—	Recessive	89	All
Maize	Opaque-7	Spontaneous	O_7	10	Recessive	78	Mainly 20,000
Maize	Floury-2	Spontaneous	Fl_2	4	Semi-dominant	35	All

Data taken from Doll (1984) and Soave and Salamini (1984).

The high lysine content of Risφ 56 was shown by Doll (1980) to be due to the presence of a non-functional allele, *Hor2ca*, at the *Hor2* locus which codes for B hordeins, one of two major prolamin groups in barley. By SDS-PAGE, it was shown that the synthesis of the major B hordeins was completely suppressed. Several minor bands were still present but it was not determined whether their genes actually occur at *Hor2*. Recently, by methods in molecular biology, Kries *et al.* (1983) showed that in Risφ 56, the *Hor2* locus was partially or completely deleted. It is very likely, therefore, that Risφ 56 has arisen from a chromosome mutation which resulted in the loss of a very small, interstitial segment of chromosome. In this mutant there is a doubling in the production of C hordein (Koie and Doll, 1979), the other major prolamin, compared to the parent variety Carlsberg II, and a great increase in four salt-soluble proteins; protein z, β-amylase and chymotrypsin inhibitors CI-1 and CI-2. The salt-soluble proteins are responsible for raising the lysine content of the grains for they contain between 5 and 11 % lysine.

Unlike the *Hor2ca* locus of Risφ 56, the *lys3a* locus of Risφ 1508 and the *lys* locus of Hiproly are not alleles of prolamin genes and they also occur on different chromosomes (Table 1). The Risφ mutant 1508 shows a drastic reduction in hordein accumulation and has a correspondingly high lysine content (Doll, 1984). Hiproly itself also contains a much reduced hordein content but when crossed and backcrossed to commercial varieties, selecting for the *lys* locus at each generation, the hordein content is only slightly lower than that of the recurrent parent. This led Tallberg (1984) to conclude that the *lys* gene has little direct effect on hordein accumulation. The two genes, *lys* and *lys3a,* also have different effects on the rate of synthesis of the lysine-rich proteins: in Hiproly, protein z, β-amylase and chymotrypsin inhibitors CI-1 and CI-2 are over-produced as in Risφ 56 (Doll, 1984). In Risφ 1508 however, CI-1 is preferentially increased together with free lysine but β-amylase is strongly inhibited (Doll, 1983). Clearly a comparative analysis of the *lys* and *lys3a* genes at the molecular level may give insight into the derepression and expression of the hordein genes. Research in this area has barely started but Kries *et al.* (1984) have indicated that the effect of *lys3a* on hordein synthesis is either at the level of transcription or on the early processing of messenger RNA.

In maize, like barley, several high-lysine lines have been developed and described and those most widely studied are included in Table 1. Work at the biochemical and molecular level is probably more advanced than in barley. Approximately 80% of the zein polypeptides seperate into two groups by SDS-PAGE, one having a molecular weight of 22,000 and the other, 20,000 (Soave and Salamini, 1984). Sequence analysis of cDNA clones specifying these two groups indicate that they are distantly related to each other and that their structural genes probably originated from the duplication and subsequent divergence of a single, ancestral gene (Marks and Larkins, 1982).

The maize mutants lised in Table 1 show a wide range of zein inhibition. Furthermore, specificity of inhibition can differ. Thus, opaque-7 primarily suppresses the 22,000 zein group, opaque-2 the 20,000 group, whereas opaque-6 and floury-2 appear to suppress each protein group to the same extent (Soave and Salamini, 1984). The reduced production of zein in all the mutants shown in Table 1 is due to a reduced level of zein messenger RNA in the endosperm cells (Langridge *et al.,* 1982).

Soave *et al.* (1981) speculated that the loci regulating zein synthesis might operate by producing factors which interfere with zein production and, being diffusable, interact with all the zein genes dispersed in the maize genome. A search was made for regulatory proteins using specially constructed lines which had various regulatory genes inserted in the same genetic background as the selected low-lysine (control) line. It was demonstrated that the control contained a major salt-soluble protein which was virtually absent in the mutants. The authors provided evidence that the protein was coded for at the opaque-6 locus and that it probably also interacts in some unknown way with the opaque-2 locus.

In related experiments, Galante *et al.* (1983) showed that an albumin-type protein of molecular weight 70,000, present in small amounts in controls, is overproduced in the floury-2 mutants. The protein occurs mainly in protein bodies with some evidence that it is located on the protein-body membrane. The authors suggested that the protein may interfere with the transport and accumulation of zein in the protein bodies. These results are of great interest but more work is required to elucidate the precise mechanisms of action of these regulatory genes.

In wheat in particular, manipulation of both gene and chromosome mutants will help in the understanding of the structure of glutenin and the molecular basis for its elasticity. The HMW subunits of glutenin are probably the main determinants of the elastic properties of dough (Payne *et al.*, 1981; Tatham *et al.*, 1985). These subunits, as well as showing extensive allelic variation (Fig. 3), are also occasionally deleted. Thus the line shown in Fig. 7, slot 3, lacks a protein presumably because of a null allele at *Glu-1* on chromosome 1 D and that in slot 9 a protein at the equivalent locus on chromosome 1 B. When these two lines are crossed, a quarter of the progeny will have double nulls. In the author's laboratory, a line is

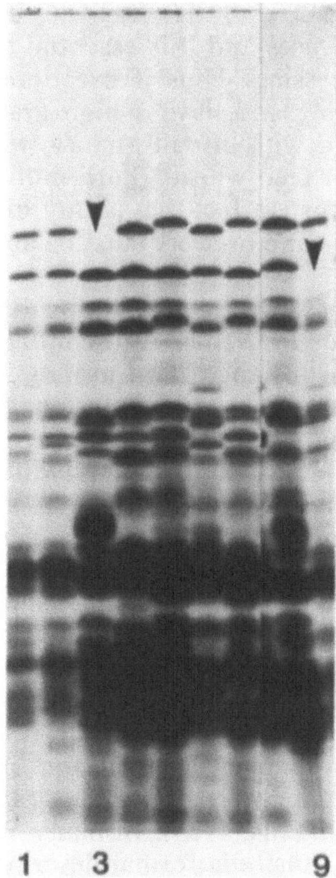

Fig. 7. SDS-PAGE of wheat landraces from Afghanistan. Deleted HMW subunits of glutenin occur in slot 3 (chromosome 1 D) and slot 9 (chromosome 1 B). The expected positions are arrowed

being developed by repetitive backcrossing which will have the same glutenin subunit composition as the backcrossed parent except for the deletion of four of the five HMW subunits. Similarly another line is being produced, exploiting chromosome mutants with deleted *Gli-1* loci (Fig. 6) and thus deficient in low-molecular-weight (LMW) glutenin subunits. Such lines, with drastically altered proportions of glutenin subunit types, should help in the elucidation of glutenin structure. Very recently, Graveland *et al.* (1985) described a new model for glutenin structure in which subunit 10, a chromosome 1D-encoded HMW subunit, plays the key role in the complex. Currently in the author's laboratory, a line is being developed which lacks subunit 10, or any allelic variant of it, which otherwise has the same subunit composition as a control line containing subunit 10. Analysis of the two lines should enable the hypothesis of Graveland *et al.* (1985) to be tested.

D. Role in the Food Industry

In the U.K. and many other countries, varieties of wheat and varieties of barley have different end uses and it is essential that they do not become intermixed prior to processing. Thus if bread-quality wheats which are hard milling, strong mixing and have a high grain protein content were significantly contaminated with biscuit quality wheats (soft milling, weak mixing, low in protein) or feed wheats (hard milling, weak mixing, low in protein) they would seriously affect the quality of loaves produced at the bakery. Similarly maltsters and brewers must be able to distinguish malting barleys from feed barleys.

For wheat there is not sufficient variation in the morphological characters of grain to distinguish varieties. At the mill, rapid chemical tests can be used to distinguish, for instance, hard-milling wheats from soft-milling wheats and wheats rich and poor in protein, but for unequivocal identification of varieties, protein electrophoresis is used. The standard, international method used is electrophoresis of gliadins at pH 3.1—3.2 using either starch or polyacrylamide gels (Ellis, 1984). Gliadins are coded by genes at six, unlinked complex loci in the wheat genome and through spontaneous mutation there are many alleles at each locus (Payne *et al.,* 1984b). The great majority of varieties can be distinguished by this technique and those which cannot may be identified by an additional electrophoretic test, SDS-PAGE (Shewry *et al.,* 1978), which reveals allelic variation at three more loci coding for HMW subunits of glutenin (Payne *et al.,* 1984b). Thus, individual analyses on 100 grains can determine the varietal composition of an incoming shipment destined for bread-making. An unacceptably high contamination by feed or biscuit-quality wheat would result in its rejection at the mill (Ellis, 1984). In many Western European countries, North America and Australia formal descriptions of newly accepted varieties include gliadin banding patterns, with varietal identification in mind.

Unfortunately, identification of barley varieties by electrophoresis is much less effective. For instance, Cooke *et al.* (1983) were only able to divide 68 barley cultivars into 19 groups on the basis of hordein composition. There are two main reasons for this: first, as barley is a diploid it would be expected to contain only about one third the variation of that found in wheat. Second, the two hordein gene loci are closely linked (Shewry *et al.*, 1984a), thus making it likely that the final, élite line selected from a varietal cross would have both B- and C-hordeins from one of the parents. For these reasons, electrophoresis has not become nearly as important to the maltster as it has to the miller.

VI. Chromosome Mutations

A chromosome mutation is usually defined as a structural change in the genome resulting in the gain, loss or translocation of a chromosome segment. Many types of such mutations, notably chromosome loss, can only be tolerated in polyploid organisms so it is not surprising that the most widely studied cereal is hexaploid wheat.

In wheat, deletions are a common form of chromosome mutation. Riley and Kimber (1961) examined the mitotic chromosomes of root-tip nuclei from 2027 seedlings in four wheat varieties. Twenty-two were atypical: 14 were monosomics, three were trisomics and one was telocentric. Similarly in the other polyploid cereal, oats, a low level of abnormal karyotypes occurs. These whole chromosome effects are most likely due to a failure of chromosome pairing at meiosis.

Using the wheat variety Chinese Spring, Sears (1954) obtained a complete series of lines monosomic for the 21 different chromosomes ($2n = 41$). These were either obtained as a result of spontaneous mutations or from Chinese Spring lines nullisomic for chromosome 3B, which induces pairing failure in other chromosomes at meiosis.

Monosomic series have been used to determine the chromosomal location of storage-protein genes in wheat (Payne *et al.*, 1980) but not extensively because separate electrophoretic analyses of several grains are required and conclusions are based partially on the relative amount of protein under study which is produced in the developing grain. The reason for this is that selfed grains of monosomics (those used for analysis) have different chromosome numbers. On average, 73% of the progeny are monosomics, 24% are disomics and 3% nullisomics (Law and Worland, 1973). Thus, a protein not controlled by the monosomic chromosome will be present at equal intensity in all the progeny analysed but proteins which are controlled by genes on this chromosome will occur at 1, 3 and 0 dose levels in the respective progeny (Payne *et al.*, 1980). Monosomic series are now available for over 60 different varieties (Law *et al.*, 1983) so enabling the chromosomal location of many protein variants to be determined.

Having established the complete series of Chinese Spring monosomics, Sears (1954) went on to obtain, by selfing and selection, the complete series of nullisomics. A complete series of trisomics ($2n = 43$) was also obtained by the same methods as those used to obtain monosomics. By selfing the trisomics, tetrasomics (2n = 44) were obtained and collected in the segregating progeny.

None of these stocks have been used to any extent to study the genetics of the storage proteins in wheat. Although nullisomic grains would be ideal for locating genes to chromosomes they nearly all produce feeble plants with few grains and several are sterile. Fortunately, Sears (1966) developed a series of compensating nullisomic-tetrasomics (2n = 42). In these lines a chromosome pair (e.g. 1A) is missing but the normal, euploid chromosome number of 42 is maintained by the presence of an extra pair of homoeologous chromosomes (either 1B or 1D). These plants have, in the main, near-normal fertility and have been used extensively in locating storage protein genes to chromosomes (for example, Wrigley and Shepherd, 1973; Bietz *et al.*, 1975; Brown *et al.*, 1979).

As well as whole chromosome deletions or additions, Sears (1954) developed various types of line for Chinese Spring with chromosome arms deleted. The ditelocentrics, lacking either the short or long arms of a chromosome pair, have been studied extensively to locate storage-protein genes to chromosome arms.

Compensated nullisomic-tetrasomics and ditelocentrics can also be used to advantage in recombination and linkage studies on storage protein genes (see also Section V.B) but they have only been exploited rarely. In one example, Payne *et al.* (1982) estimated recombination between storage-protein loci *Glu-A1* and *Gli-A1* on the long and short arms respectively of chromosome 1A by analysing progeny of the cross:

Chinese Spring x Chinese Spring (Hope 1A) (Primary cross)

↓

F_1 ♀ x Chinese Spring ditelo 1AL ♂ (Secondary cross)

In the primary cross, recombination occurs between the 1A chromosomes of Chinese Spring and Hope. The F_1 plants were then crossed to a ditelocentric line lacking the short arm of chromosome 1A and therefore the *Gli-A1* locus. Segregation of the chromosome 1A encoded gliadins can therefore be determined unambiguously by gel electrophoresis without interference from the equivalent proteins from the male parent of the secondary cross, or from other proteins coded by genes on other chromosomes.

When ditelocentrics are used in the primary cross, the distance between the centromere and any gene on the remaining arm can be estimated in terms of recombination units, provided there is allelic variation. This type of cross was used by Payne *et al.* (1982) to measure the distances from *Glu-A1*, *Glu-B1* and *Glu-D1* to the centromere of the group 1 chromosomes (see Fig. 6) and by Rybalka and Sozinov (1979) to map *Gli-B1* to the centromere of chromosome 1B (see Fig. 6).

Chromosome deficiencies resulting from breakage during meiosis have been described for chromosomes 6BL (Giorgi, 1981) and 5AL (Miller and

Reader, 1982). In the author's laboratory, chromosome deficiencies of the homoeologous group 1 have been detected in about 10 out of several thousand grains and two of them have been well characterised (Payne *et al.*, 1984a; Ainsworth *et al.*, 1984). Chromosome mutants involving the loss of storage-protein genes are easily detected by gel electrophoresis of progeny from crosses. An example in which a segment of the short arm of chromosome 1B containing *Gli-B1* has been deleted is shown in Fig. 1 of Payne *et al.* (1984a). Lines with chromosome deficiencies have been very useful in deletion mapping of various genes and in relating gene recombination maps with physical maps of chromosomes (Payne *et al.*, 1984a; Snape *et al.*, 1985).

An inter-chromosomal translocation has probably contributed to the distribution of storage-protein genes in the wheat genome. Recent sequencing of cloned genes for α-gliadins has shown a low but significant sequence homology to γ-gliadin genes (Rafalski *et al.*, 1984; Kasarda *et al.*, 1984) and it is now generally recognised that they are derived from a common ancestral gene. By comparison with other cereals (Shewry *et al.*, 1984a) it is clear that a chromosome homoeologous to the group 1 chromosomes of wheat carried the ancestral gliadin genes. It has been speculated that the α-/β-gliadin genes were transferred from chromosome 1 to chromosome 6 in wild diploid species by an interchromosomal translocation well before the origin of hexaploid wheat (Shepherd and Jennings, 1970; Kasarda, 1982). Interchromosomal translocations were probably also responsible for the presence of secalin genes on chromosome 2R of rye, *Secale cereale* (Shewry *et al.*, 1984b) and 5R of *S. montanum* (Shewry *et al.*, 1985).

A chromosome inversion could account for the separation of hordein genes on chromosome 5 of barley. They occur at two loci, about eight recombination units apart (Shewry *et al.*, 1984a), *Hor1* coding for C-hordein and *Hor2* for B-hordein. Recent cDNA sequencing studies of these two hordein groups (Forde *et al.*, 1985) strongly suggest a common evolutionary origin and so a common gene locus.

Oats are also hexaploid, containing three different genomes each with seven pairs of chromosomes. Progress in assembling chromosome mutants has been slow. By 1974, all 21 monosomics had been produced but in six different genotypes (Rajhathy and Thomas, 1974). A disadvantage with oats in the development of further genetic stocks is that selfing monosomics produces a variable number of nullisomics and for certain chromosomes none are produced. So far, no related work on the genetics of oat globulin and prolamin has been fully reported although some work is now in progress.

Some types of chromosomal mutants occur in diploid cereals although, of course, lines with whole or partial chromosome deletions are not available. For barley ($2n = 14$) several complete sets of trisomics are available for varieties Betzes (Eslick and Ramage, 1969) and Shin Ebisu No. 16 (Tsuchiya, 1967). Monotelotrisomics are stable in several cereals although complete sets have not yet been produced (Singh and Tsuchiya, 1977).

For maize, trisomics and interchromosomal translocation were analysed by electrophoresis amd it was shown that the zein genes are located on chromosomes 4, 7 and 10 (Valentini *et al.*, 1979). For barley, equivalent stocks were used indirectly to determine the chromosome location of the hordein genes. By recombination mapping, the hordein genes were shown to be fairly closely linked to *Ml-a* genes giving resistance to powdery mildew (Oram *et al.*, 1975). Previously *Ml-a* had been shown using chromosome mutants to be located on chromosome 5 (Burnham and Hagberg, 1956) which must also be the chromosome carrying the hordein genes.

VII. Conclusions

The storage-protein genes of cereals are tissue-specific in their output and are developmentally regulated. That is they are derepressed only in the endosperm (or for some genes additionally the embryo) and at only a specific period in the growth and development of this tissue. For these and other reasons the genes have attracted the attention of molecular biologists and significant advances have already been made in terms of the structure and evolutionary relationships. The formidable tasks ahead in understanding the tissue specificity and regulation of these genes will very likely be aided by the geneticist who, by using natural and induced variation in the structural and regulatory genes, will be able to construct genotypes to rigorously test future hypotheses on the molecular mechanisms of plant development.

In wheat, a greater understanding of the relationship between breadmaking quality and allelic variation of storage-protein genes is needed and this will be achieved from the analysis of specifically constructed genetic lines. Molecular biologists are already close to understanding at the sequence level why certain HMW subunits of glutenin impart greater elasticity than their allelic counterparts (Flavell *et al.*, 1984). In future, this knowledge will be exploited in screening for good-quality glutenin subunits from diverse sources including landraces of ancient agriculture and wild diploid species related to wheat. The genes coding for them would then be transferred to commercial varieties by recurrent backcrossing.

Acknowledgements

I am extremely grateful to several of my colleagues at the Plant Breeding Institute: Mrs. L. M. Holt for Fig. 1, Dr. E. A. Jackson for Fig. 2, Mr. D. B. Smith for Fig. 3 and Dr. R. B. Austin for critically reading the text. Thanks are also due to Dr. M. L. Parker (Food Research Institute, Norwich) for permission to quote some of her unpublished work.

VIII. References

Ainsworth, C. C., Johnson, H. M., Jackson, E. A., Miller, T. E., Gale, M. D., 1984: The chromosomal locations of leaf peroxidase genes in hexaploid wheat, rye and barley. Theor. Appl. Genet. **69,** 205—210.

Altschul, A. M., 1965: Proteins: their chemistry and politics. London: Chapman and Hall.

Bartels, D., Thompson, R. D., 1983: The characterisation of cDNA clones coding for wheat storage proteins. Nuc. Acids Res. **11,** 2961—2977.

Bietz, J. A., Shepherd, K. W., Wall, J. S., 1975: Single-kernel analysis of glutenin: used in wheat genetics and breeding. Cereal Chem. **52,** 513—532.

Brown, J. W. S., Kemble, R. J., Law, C. N., Flavell, R. B., 1979: Control of endosperm proteins in *Triticum aestivum* (var. Chinese Spring) and *Aegilops umbellulata* by homoeologous group 1 chromosomes. Genetics **93,** 189—200.

Burnham, C. R., Hagberg, A., 1956: Cytogenetic notes on chromosome interchanges in barley. Hereditas **42,** 467—482.

Burr, F. A., Burr, B., 1981: *In vitro* uptake and processing of prezein and other maize preproteins by maize membranes. J. Cell Biol. **90,** 427—434.

Cameron-Mills, V., Ingversen, J., Brandt, A., 1978: Transfer of *in vitro* synthesised barley endosperm proteins into the lumen of the endoplasmic reticulum. Carlsberg Res. Commun. **43,** 91—102.

Cameron-Mills, V., von Wettstein, D., 1980: Protein body formation in the developing barley endosperm. Carlsberg Res. Commun. **45,** 577—594.

Cooke, R. J., Cliff, E. M., Draper, S. R., 1983: Barley cultivar characterisation by electrophoresis II. Classification of hordein electrophoregrams. J. Nat. Inst. Agric. Bot. (G. B.) **16,** 197—206.

Doll, H., 1980: A nearly non-functional mutant allele of the storage protein locus *Hor2* in barley. Hereditas **93,** 217—222.

Doll, H., 1983: Barley seed proteins and possibilities for their improvement. In: Gottschalk, W., Muller, H. P. (eds.), Seed Proteins, pp. 207—223. The Hague: Martinus Nijhoff.

Doll, H., 1984: Nutritional aspects of cereal proteins and approaches to overcome their deficiencies. Philos. Trans. R. Soc. London, Ser.B **304,** 373—380.

Ellis, J. W. S., 1984: The cereal grain trade in the United Kingdom: the problem of cereal variety. Philos. Trans. R. Soc. London, Ser.B **304,** 395—407.

Eslick, R. F., Ramage, R. T., 1969: Primary trisomics in the variety Betzes. Barley Newsl. **12,** 17.

Flavell, R. B., Payne, P. I., Thompson, R. D., Law, C. N., 1984: Strategies for the improvement of wheat-grain quality using molecular genetics. Biotechnol. Genet. Eng. Rev. **2,** 157—173.

Forde, B. G., Kries, M., Williamson, M. S., Fry, R. P., Pywell, J., Shewry, P. R., Bunce, N., Miflin, B. J., 1985: Short tandem repeats shared by B- and C-hordein cDNAs suggest a common evolutionary origin for two groups of cereal storage protein genes. EMBO J. **4,** 9—15.

Galante, E., Vitale, A., Manzocchi, L., Soave, C., Salamini, F., 1983: Genetic control of a membrane component and zein deposition in maize endosperm. Mol. Gen. Genet. **192,** 316—321.

Giorgi, B., 1981: A line with a deletion on the long arm of chromosome 6B isolated in *Triticum aestivum* cv. Chinese Spring. Wheat Inf. Serv. **50,** 22—23.

Grama, A., Gerechter-Amitai, Z. K., Blum, A., 1984: Breeding bread wheat cultivars for high protein content by transfer of protein genes from *Triticum dicoccoides*. In: Cereal Grain Protein Improvement, pp. 145—153. Vienna: International Atomic Energy Agency.

Graveland, A., Bosveld, P., Lichtendonk, W. J., Marseille, J. P., Moonen, J. H. E., Scheepsta, A., 1985: A model for the molecular structure of the glutenins from wheat flour. J. Cereal Sci. **3,** 1—16.

Harlan, J. R., Starks, K. J., 1980: Germplasm resources and needs. In: Maxwell, F. G., Jennings, R. (eds.): Breeding Plants Resistant to Insects, pp. 254—273. New York: John Wiley.

Hu, N. T., Peifer, M. A., Heidecker, G., Messing, J., Rubenstein, I., 1982: Primary structure of a genomic zein sequence of maize. EMBO J. **11,** 1337—1342.

IAEA, 1984: Cereal Grain Protein Improvement, 385 pp. Vienna: International Atomic Energy Agency.

Kasarda, D. D., 1982: Toxic proteins and peptides in celiac disease: relations to cereal genetics. In: Walcher, D., Kretchmer, N. (eds.), Food, Nutrition and Evolution. New York: Masson.

Kasarda, D. D., Okita, T. W., Bernadin, J. E., Baecker, P. A., Nimmo, C. C., Lew, E. J.-L., Dietler, M. D., Greene, F. C., 1984: Nucleic acid (cDNA) and amino acid sequences of α-type gliadins from wheat *(Triticum aestivum)*. Proc. Nat. Acad. Sci. U.S.A. **81,** 4712—4716.

Koie, B., Doll, H., 1979: Protein and carbohydrate components in the Risϕ high-lysine barley mutants. In: Seed Improvement in Cereal and Grain Legumes, I, pp. 205—215. Vienna: International Atomic Energy Agency.

Kries, M., Shewry, P. R., Forde, B. G., Rahman, S., Miflin, B. J., 1983: Molecular analysis of a mutation conferring the high lysine phenotype on the grain of barley *(Hordeum vulgare)*. Cell **34,** 161—167.

Kries, M., Shewry, P. R., Forde, B. G., Rahman, S., Bahramian, M. B., Miflin, B. J., 1984: Molecular analysis of the effect of the *lys 3a* gene on the expression of *Hor* loci in developing endosperms of barley *(Hordeum vulgare)*. Biochem. Genet. **22,** 231—255.

Langridge, P., Pintor-Toro, J. A., Feix, G., 1982: Transcriptional effects of the opaque-2 mutation of *Zea mays* L. Planta **156,** 166—170.

Larkins, B. A., Hurkman, W. G., 1978: Synthesis and deposition of zein proteins in maize endosperm. Plant Physiol. **62,** 256—263.

Law, C. N., Worland, A. J., 1973: Aneuploidy in wheat and its use in genetic analysis. In: Annual Report, 1972, pp. 25—65. Cambridge: Plant Breeding Institute.

Law, C. N., Snape, J. W., Worland, A. J., 1983: Quantitative genetic studies in wheat. In: Proc. 6th International Wheat Genetics Symposium, Kyoto, Japan 1983, pp. 539—547. Kyoto: Germplasm Institute.

Lawrence, G. J., Shepherd, K. W., 1981: Chromosomal location of genes controlling seed proteins in species related to wheat. Theor. Appl. Genet. **59,** 25—31.

Marks, M. D., Larkins, B. A., 1982: Analysis of sequence microheterogeneity among zein messenger RNAs. J. Biol. Chem. **257,** 9976—9983.

Miflin, B. J., Shewry, P. R., 1979: The synthesis of proteins in normal and high lysine barley seeds. In: Laidman, D. L., Wyn Jones, R. G. (eds.), Recent Advances in the Biochemistry of Cereals, pp. 239—273. London: Academic Press.

Miflin, B. J., Forde, B. G., Kreis, M., Rahman, S., Forde, J., Shewry, P. R., 1984: Molecular biology of the grain storage proteins of the Triticeae. Philos. Trans. R. Soc. London, Ser.B **304,** 333—339.

Miller, T. E., Reader, S. M., 1982: A major deletion of part of chromosome 5A of *Triticum aestivum*. Wheat Inf. Serv. **55,** 10—12.

Oram, R. N., Doll, H., Koie, B., 1975: Genetics of two storage protein variants in barley. Hereditas **80,** 53—58.

Orth, R. A., Bushuk, W., 1972: A comparative study of the protein of wheats of diverse baking qualities. Cereal Chem. **49,** 268—275.

Osborne, T. B., 1907: The proteins of the wheat kernel. Carnegie Inst. Washington Publ. 84. Washington: Judd and Dutweiler.

Park, W. M., Stegemann, H., 1979: Rice protein patterns. Comparison by various PAGE — techniques in slabs. Z. Acker Pflanzenbau **148,** 446—454.

Parker, M. L., 1980: Protein body inclusions in developing wheat endosperm. Ann. Bot. **46,** 29—36.

Parker, M. L., 1982: Protein accumulation in developing endosperm of a high-protein line of *Triticum dicoccoides*. Plant Cell Environ. **5,** 37—43.

Payne, P. I., Corfield, K. G., 1979: Subunit composition of wheat glutenin proteins, isolated by gel filtration in a dissociating medium. Planta **145,** 83—88.

Payne, P. I., Law, C. N., Mudd, E. E., 1980: Control by homoeologous group 1 chromosomes of the high-molecular-weight subunits of glutenin, a major protein of wheat endosperm. Theor. Appl. Genet. **58,** 113—120.

Payne, P. I., Corfield, K. G., Holt, L. M., Blackman, J. A., 1981: Correlations between the inheritance of certain high-molecular-weight subunits of glutenin and bread-making quality in progenies of six crosses of bread wheat. J. Sci. Food. Agric. **32,** 51—60.

Payne, P. I., Rhodes, A. P., 1982: Cereal storage proteins: structure and role in agriculture and food technology. Encyl. Plant. Physiol. **14A,** 346—369.

Payne, P. I., Holt, L. M., Worland, A. J., Law, C. N., 1982: Structural and genetical studies on the high-molecular-weight subunits of wheat glutenin. Part 3. Telocentric mapping of the subunit genes on the long arms of the homoeologous group 1 chromosomes. Theor. Appl. Genet. **63,** 129—138.

Payne, P. I., Holt, L. M., Hutchinson, J., Bennett, M. D., 1984a: Development and characterisation of a line of bread wheat, *Triticum aestivum,* which lacks the short-arm satellite of chromosome 1B and the *Gli-B1* locus. Theor. Appl. Genet. **68,** 327—334.

Payne, P. I., Holt, L. M., Jackson, E. A., Law, C. N., 1984b: Wheat storage proteins: their genetics and their potential for manipulation by plant breeding. Philos. Trans. R. Soc. London, Ser.B **304,** 359—371.

Peterson, D. M., 1978: Subunit structure and composition of oat seed globulin. Plant Physiol. **62,** 506—509.

Rafalski, J. A., Scheets, K., Metzler, M., Peterson, D. M., Hedgcoth, C., Soll, D. G., 1984: Developmentally regulated plant genes: the nucleotide sequence of a wheat gliadin clone. EMBO J. **3,** 1409—1415.

Rajhathy, T., Thomas, H., 1974: Cytogenetics of oats (*Avena* L). Ottawa: The Genetics Society of Canada.

Rhighetti, P. G., Gianazza, E., Viotti, A., Soave, C., 1977: Heterogeneity of storage proteins in maize. Planta **136,** 115—123.

Riley, R., Kimber, G., 1961: Aneuploids and the cytogenetic structure of wheat varietal populations. Heredity **16,** 275—290.

Robert, L. S., Matlashewski, G. J., Adeli, K., Nozzolillo, C., Altosaar, I., 1983a: Electrophoretic and developmental characterisation of oat (*Avena sativa* L) globulins in cultivars of different protein content. Cereal Chem. **60,** 231—234.

Robert, L. S., Nozzolillo, D., Altosaar, I., 1983b: Molecular weight and charge heterogeneity of prolamins (avenins) from nine oat (*Avena sativa* L) cultivars of different protein content and from developing seeds. Cereal Chem. **60,** 438—442.

Rybalka, A. I., Sozinov, A. A., 1979: Mapping the locus of Gli 1B, which controls the biosynthesis of reserve proteins in soft wheat. Tsitol. Genet. **13,** 276—282.

Sears, E. R., 1954: The aneuploids of common wheat. Research Bulletin Agricultural Experimental Station, University of Missouri, Columbia, Missouri, **572.**

Sears, E. R., 1966: Nullisomic-tetrasomic combinations in hexaploid wheat. In: Riley, R., Lewis, K. R. (eds.), Chromosome Manipulation and Plant Genetics, pp. 29—45. Edinburgh: Oliver and Boyd.

Shepherd, K. W., Jennings, A. C., 1970: Genetic control of rye endosperm proteins. Experientia **27,** 98—99.

Shewry, P. R., Faulks, A. J., Pratt, H. M., Miflin, B. J., 1978: The varietal identification of single seeds of wheat by sodium dodecyl sulphate polyacrylamide gel electrophoresis of gliadin. J. Sci. Food. Agric. **29,** 847—849.

Shewry, P. R., Autran, J.-C., Nimmo, C. C., Ellen, J.-L., Kasarda, D. D., 1980: N-terminal amino acid sequence homology of protein components from barley and a diploid wheat. Nature (London) **286,** 520—522.

Shewry, P. R., Miflin, B. J., Kasarda, D. D., 1984a: The structural and evolutionary relationships of the prolamin storage proteins of barley, rye and wheat. Philos. Trans. R. Soc. London, Ser.B **304,** 297—308.

Shewry, P. R., Bradberry, D., Franklin, J., White, R. P., 1984b: The chromosomal locations and linkage relationships of the structural genes for the prolamin storage protein (secalins) of rye. Theor. Appl. Genet. **69,** 63—69.

Shewry, P. R., Miflin, B. J., 1985: Seed storage proteins of economically important cereals. Adv. Cereal. Sci. Technol. **7,** 1—83.

Shewry, P. R., Parmar, S., Miller, T. E., 1985: Chromosomal location of the structural genes for the M_r 75,000 γ-secalins in *Secale montanum* Guss: evidence for a translocation involving chromosomes 2R and 6R of cultivated rye *(Secale cereale).* Heredity **54,** 381—383.

Simmonds, D. H., O'Brien, T. P., 1981: Morphological and biochemical development of the wheat endosperm. Adv. Cereal Sci. Technol. **4,** 5—70.

Singh, N. K., Shepherd, K. W., 1985: The structure and genetic control of a new class of disulphide-linked proteins in wheat endosperm. Theor. Appl. Genet., **71,** 79—92.

Singh, R. J., Tsuchiya, T., 1977: Morphology, fertility and transmission in seven monotelotrisomics of barley. Z. Pflanzenzücht. **78,** 327—340.

Snape, J. W., Flavell, R. B., O'Dell, M., Hughes, W. G., Payne, P. I., 1985: Intrachromosomal mapping of the nucleolar organiser region relative to three marker loci on chromosome 1B of wheat *(Triticum aestivum).* Theor. Appl. Genet. **69,** 263—270.

Soave, C., Tardani, L., Di Fonzo, N., Salamini, F., 1981: Zein level in maize endosperm depends on a protein under control of the opaque-2 and opaque-6 loci. Cell **27,** 403—410.

Soave, C., Salamini, F., 1984: Organisation and regulation of zein genes in maize endosperm. Philos. Trans. R. Soc. London, Ser.B **304,** 341—347.

Tallberg, A., 1984: Biochemical and genetic characterisation of lysine genes and their utilization in breeding barley for improved grain protein. In: Cereal Grain Protein Improvement, pp. 205—214. Vienna: International Atomic Energy Agency.

Tatham, A. S., Miflin, B. J., Shewry, P. R., 1985: The β-turn conformation in wheat gluten proteins: relationship to gluten elasticity. Cereal Chem., **62,** 405—411.

Thompson, R. D., Bartels, D., Harberd, N. P., Flavell, R. B., 1983: Characterisation of the multigene family coding for HMW glutenin subunits in wheat using cDNA clones. Theor. Appl. Genet. **67,** 87—96.

Tsuchiya, T., 1967: The establishment of a trisomic series in a two-rowed cultivated variety of barley. Can. J. Genet. Cytol. **9,** 667—682.

Valentini, G., Soave, C., Ottaviano, E., 1979: Chromosomal location of zein genes in *Zea mays*. Heredity **42,** 33—40.

Vasal, S. K., Villegas, E., Tang, C. Y., 1984: Recent advances in the development of quality protein maize germplasm at CIMMYT. In: Cereal Grain Protein Improvement, pp. 167—189. Vienna: International Atomic Energy Agency.

Vogel, K. P., Johnson, V. A., Mattern, P. J., 1983: Results of systematic analyses for protein and lysine composition of common wheats (*Triticum aestivum* L) in the USDA world collection. Nebr. Res. Bull. **258,** 27.

Wall, J. S., 1979: The role of wheat proteins in determining baking quality. In: Laidman, D. L., Wyn Jones, R. G. (eds.), Recent advances in the Biochemistry of Cereals, pp. 275—311. London: Academic Press.

Wrigley, C. W., Shepherd, K. W., 1973: Electrofocussing of grain proteins from wheat genotypes. Ann. N. Y. Acad. Sci. **209,** 154—162.

Yamagata, H., Tanaka, K., Kasai, Z., 1982: Evidence for a precursor form of rice glutelin subunits. Agric. Biol. Chem. **46,** 321—322.

Zhae, W. M., Gatehouse, J. A., Boulter, D., 1983: The purification and partial amino acid sequence of a polypeptide from the glutelin fraction of rice grains: homology to pea legumins. FEBS Lett. **162,** 96—102.

Chapter 8

Molecular Approaches to Plant and Pathogen Genes

Richard I. S. Brettell and Anthony J. Pryor

CSIRO, Division of Plant Industry, P. O. Box 1600, Canberra City, ACT 2601, Australia

Contents

I. Introduction

Host-pathogen combinations among higher plants cover a continuous wide range of diversity. At one end is the extreme specialisation of obligate biotrophs such as the rust and mildew fungi, whose existence depends on an association with living tissue of the host. At the other end are the facultative necrotrophs which are able to grow in the absence of host tissue but under certain circumstances will infect a plant and feed off the dead and dying tissues.

These types of interaction have been well studied in descriptive terms, in genetics, and to a lesser extent in biochemical terms, but have received little attention from the recent development and application of molecular biological techniques. In this article we will discuss some areas of plant-pathogen interaction that are currently undergoing analysis by molecular techniques and attempt to point to other promising approaches.

For the obligate biotrophs, many examples of a gene-for-gene relationship have been established since the studies by Flor (1947, 1956) on flax and the rust *Melampsora lini.* Resistance in the host is conditioned by a single, usually dominant, gene which matches a gene for avirulence in the pathogen. There is a wealth of genetic information about major gene resistances to a number of obligate biotrophs, yet an almost complete ignorance of the nature of the products of these genes and how they interact with the genes of the pathogens. It is frequently noted that gene-for-gene resistance is not the only level of plant resistance that is important. This is undoubtedly true and there is perhaps no better illustration of this than the demonstration by Wynn (1976) of the role of leaf epidermal topography on the ability of a rust germ tube to locate a stomate and infect the host plant. There must be a variety of levels, of varying importance, which all contribute to effective resistance. However, molecular analysis is limited at present to phenotypes with single gene inheritance, and traits such as leaf surface topography are unlikely to fall into this class.

Moving from biotrophic to necrotrophic diseases, it is apparent, in one sense, that the necrotrophic pathogens are less specialised than the obligately parasitic organisms, in not needing to maintain the host cells in a metabolically active condition. However, given the limited ranges of most necrotrophs, virulence to a particular host plant must still be considered the exception rather than the rule and, therefore, represents a specialised level of adaptation. This in turn is consistent with genes in the host that condition resistance and genes in the pathogen that condition virulence being critical factors in disease development, even though the result is scarcely a balanced relationship. In some cases virulence may be ascribed to toxic telepathogenic substances produced by the pathogen, and such examples may provide useful models for the study of host-pathogen gene interactions.

At the molecular level, perhaps the best characterised plant and pathogen relationship involves the bacterium *Agrobacterium tumefaciens*. This is the causal agent of crown gall disease in a number of dicotyledonous plants, which results from infection by the bacterium at a wound site and transfer of a small piece of DNA, the T-DNA, from a plasmid within the bacterium to the host cell (Gheysen *et al.*, 1985). This T-DNA is integrated into the host cell genome and signals the proliferation of host tissue through the production of phytohormones. Recent work has established that the T-DNA carries genes that code for enzymes which are involved in auxin and cytokinin biosynthesis and which represent new synthetic pathways which cannot be regulated by the host (Schröder *et al.*, 1984; Inzé *et al.*, 1984; Buchmann *et al.*, 1985). Other genes on the T-DNA alter the host cell's metabolism by coding for enzymes involved in the synthesis and catabolism of opines, a class of arginine derivatives which can be used as an energy source by the bacterium. This may be a unique example of parasitism at the DNA level, as there is no evidence that transfer of genetic material from pathogen to host is a widespread phenomenon. Nevertheless, it serves to illustrate that the relationship between host and pathogen depends essentially on an interaction between host DNA and pathogen DNA; the intermediates in this interaction may be RNA, proteins or substances produced as a result of the activity of specific proteins, yet the balance between the two organisms rests at the DNA level.

II. A Molecular Approach to Gene-for-Gene Resistance

The gene-for-gene hypothesis, first elaborated in the analysis of the inheritance of resistance and virulence respectively in flax and its rust *Melampsora lini* (Flor, 1956), is the model system most readily accessible to presently available molecular techniques. Perhaps the most glaring anomaly in our understanding of plant disease is the discrepancy between our knowledge of the genetics and the almost complete absence of information of the underlying biochemical mechanisms. Most major gene resistance to plant pathogens has its expression in the hypersensitive response, yet the mechanisms that initiate this response are unknown (Hooker, 1967; Ingram, 1978). Several attempts (Torp and Andersen, 1982; Manners *et al.*, 1985) to analyse the early events in the resistance response all suffer from the difficulty involved in distinguishing cause from effect. Molecular analysis of plant genes has focussed mainly on genes with abundant or easily recognisable gene products. This approach has not been possible for gene-for-gene resistance and methods must be found which do not depend on a prior knowledge of the gene products involved.

As stated by Flor (1956), the gene-for-gene hypothesis concluded that "for each gene conditioning rust reaction in the host, there is a specific gene conditioning pathogenicity in the parasite". With few exceptions, both the host resistance gene and the pathogen avirulence gene specifically recognised in the incompatible or resistant reaction are dominant. Some of the apparent exceptions are due to genetic complexities superimposed on the basic gene-for-gene model. For example, dominant virulence in flax rust is due to an independent locus which is a dominant inhibitor of the avirulence gene function (Lawrence *et al.*, 1981). However, for the vast majority of gene-for-gene resistances the resistant reaction seems to occur when the product of an avirulence gene is recognised by a product of an allele of a host resistance gene. This view leads to several expectations crucial in the development of methods of molecular analysis.

1. Transformation with a dominant gene should result in the acquisition of a new function. Thus by transformation it should be possible to change a susceptible plant to a resistant one and a virulent pathogen into an avirulent race, but the reciprocal conversions are not expected.

2. Mutation of a dominant gene should result in a loss of function. Thus a resistant plant will become susceptible and an avirulent pathogen will become virulent.

At present there are two general methods that exploit these expectations. The first, called the "shotgun method" because of its inherently random nature, involves random cloning of genomic DNA followed by transformation and selection of the newly acquired character. The second method is called "transposon mutagenesis" in micro-organisms or "gene tagging" in higher plants. The success of both methods depends on rare events and on the absolute size of the genome. It is not surprising then that these approaches have been successfully employed in the molecular analysis of avirulence genes of pathogens, which have small genomes, but not in the cloning of a resistance gene from a host plant with its much larger genome size.

A. The Shotgun Method

The general approach is to produce a genomic library of the pathogen or the host plant DNA, to use this library of DNA sequences to transform cells of the host or pathogen and then to screen for the acquisition of resistance or avirulence respectively. At present this can be done in a cosmid vector in *E. coli.* Such vectors are based on the ability of bacteriophage Lambda to package 37—53 kb of DNA, with about 10—25 kb of DNA comprising the vector specific sequences and the remaining 20—40 kb being available for library sequences. Table 1 compares the number of clones of transformants required to be 95 % certain of recovering a gene for resistance from several plant species, together with yeast and *E. coli.* Clearly from a consideration of genome size alone, the ability to isolate avirulence genes from plant pathogens such as fungi or bacteria is going to be a much simpler job than isolating resistance genes from plant hosts with their much larger genomes.

Recently, Staskawicz *et al.* (1984) have used this approach to isolate a bacterial avirulence gene which determines incompatibility on resistant soybean leaves. The bacterium, *Pseudomonas syringae* pv. *glycinea,* is the causal agent for bacterial blight of soybean, and host resistance is thought to be of the gene-for-gene type already outlined. A cosmid library, with an average insert size of about 25 kb, was constructed from restriction fragments of the DNA isolated from an avirulent race of the bacterium. Staskawicz *et al.* first demonstrated that the particular cosmid vector could be mobilized to various races of *P.s.* pv. *glycinea,* permitting the cosmid library to be transduced into a virulent bacterial race and DNA sequences

Table 1. Number of transformants needed to contain a resistance gene

Species[a]	Haploid Genome Size x 10^6 bp	[b]Number of Transformants (N) 40 kb	25 kb
Wheat	17,300	1,296,000	2,073,000
Rye	8,300	622,000	995,000
Barley	5,500	412,000	659,000
Pea	4,800	360,000	552,000
Maize	3,900	292,000	467,000
Flax	700	52,000	84,000
Yeast	30	2,200	3,600
E. coli	7.5	560	890

[a] Sources are from Bennett and Smith (1976) and Bennett *et al.* (1982), and on the approximation that 1 pg of DNA is equivalent to about 1 × 10^9 bp.

[b] The number of transformants (N) needed to give a 95 % chance of recovering a resistance gene is calculated according to Clark and Carbon (1976) for cosmid inserts of two sizes, 25 and 40 kb.

screened for the avirulence phenotype. Bacterial colonies, each containing a different hybrid cosmid derived from the virulent race, were inoculated onto soybean leaves from a plant susceptible to the virulent race but resistant to the avirulent race whose DNA was used for the construction of the cosmid library. Provided the avirulence gene was expressed in the recipient bacterial cells, the expectation was that the rare transduced colonies would cause a resistant reaction on the leaf. This would be readily visible as a hypersensitive reaction of plant cells in the region of the inoculation. From a consideration of the above table it is evident that Staskawicz *et al.* needed to screen around 500—1000 cosmid clones to have a chance of recovering the avirulence gene. In fact a single clone was recovered from the 680 independently transduced bacterial colonies that were tested. This 27.2 kb clone was also mobilized into other bacterial races and conferred on these transconjugants the same avirulence phenotype that was present in the parental or donor race. Thus the shotgun method works for small genomes for which methods of genetic manipulation such as transformation are available. Unfortunately for most of the obligate pathogens such as rusts and smuts these techniques are not available.

When it comes to using the shotgun method for cloning a plant resistance gene the problems are even greater, not only because of the large number of cosmid clones that need to be screened, but also because of the difficulty in producing a sufficient number of plant transformants. Transformation of dicotyledons is now feasible particularly with methods based on the Ti vector system (Gheysen *et al.,* 1985) and improvements are being added continually. The initial fear that expression might be an insurmountable problem has not been realised since the available evidence suggests that genes present in transformed DNA will be expressed if they are provided with appropriate 5′ promoter and 3′ terminal sequences (Herrera-Estrella *et al.,* 1984). Transformation of monocotyledons has not yet reached this stage of development but the difficulties of transformation with gene expression for angiosperms in general are likely to be overcome in the near future. However, this still leaves the problem of the enormous number of transformants that have to be screened to recover a unique gene sequence. Either the size of the DNA in the (cosmid) library must be substantially increased or methods must be developed to fractionate or to decrease the size of the genome. Working with small genomes such as that of flax will help and this could be enhanced by using donor DNA from a plant carrying more than one functional resistance gene. For example, there are five loci that confer resistance to flax rust.

In plants with larger genomes it might be possible to fractionate whole chromosomes if methods for isolating populations of metaphase chromosomes were developed. In maize the tip of the short arm of the smallest chromosome (chromosome 10) carries a gene complex of four to five loci that confer rust resistance (Hooker, 1967). Furthermore, the tip of this chromosome, consisting of a small region of euchromatin containing the genes for rust resistance, can be translocated to a B chromosome (Beckett, 1978). If this euchromatin could be separated from the heterochromatin in a preparation of purified B chromosomes and used to create the cosmid library, the number of transformants needed to recover a resistance gene would be reduced 100 fold.

B. Transposon Mutagenesis or Gene Tagging

Transposon mutagenesis in bacteria results in inactivation of a gene by the insertion of a transposable element into the gene. Generally the transposon carries a selectable drug resistance marker. Thus if drug resistant mutations of avirulence to virulence are selected in a pathogen, they are probably the result of insertional inactivation of the avirulence gene. This mutation of the avirulence gene is now accessible to molecular analysis without any prior knowledge of the function of the gene. A genomic clone of the mutation can be isolated on the basis of homology to the transposon base sequences. Flanking sequences, those sequences adjacent to but not homologous to the transposon sequences, should be sequences normally involved in the expression of the avirulence phenotype and can be used to isolate the intact gene. Although transposon mutagenesis can be considered part of the normal repertoire for the genetic and molecular analysis of microorganisms, it has yet to see widespread application to the analysis of plant pathogens. The technique has been used to isolate sequences essential for pathogenesis in strains of *Pseudomonas syringae* (Anderson and Mills, 1985; Niepold *et al.,* 1985) but these may have little relationship to avirulence genes involved in pathogen specificity such as occurs in the gene-to-gene system. Many gene products must be required for pathogen growth. Bacteria of the genus *Rhizobium* might not be considered pathogenic, however the use of transposon mutagenesis to isolate genes required for nodulation is perhaps one of the best examples of the potential of this approach (see de Bruijn and Lupski [1984] for a general discussion of the use of transposon Tn 5 mutagenesis). The major difficulty in the analysis of avirulence will be in transferring the genetic technology to pathogenic organisms. As we have already seen from the preceeding discussion, this has been possible with a number of non-obligate pathogens, particularly bacteria, but it is difficult to see how this technology will be developed in an obligate biotrophic pathogen such as a plant rust.

In plants the equivalent approach to transposon mutagenesis is called "gene tagging" and is at present only feasible in one plant species, *Zea mays.* The first transposable genetic elements described by McClintock were those in maize plants (see Federoff, 1983 for a recent review) and, while similar elements certainly occur in other plants (Fincham and Sastry, 1974), only in maize has their analysis reached a stage where they can be used as gene tags. The feasibility of this approach was demonstrated by the isolation of the *bronze 1* locus from maize (Federoff *et al.,* 1984 a) using an inactive mutation of the gene which experimentally had been determined to be a result of insertion of the transposable element *Ac.*

Taking this example as a model, it should be possible to isolate a resistance gene from maize if certain further conditions are met. The resistance must be specified by a single dominant gene and the system must be amenable to mutagenesis by a transposable element system. The *Rp*1 gene complex specifying resistance against *Puccinia sorghi*, the causal agent of maize rust (Hooker, 1967), is a good example, and so too are resistances to the HC-toxin of *Cochliobolus carbonum* (Nelson and Ullstrup, 1964). Next, it is necessary to recover a susceptible mutation of the resistance gene and show that it is due to insertional mutagenesis by a given transposable element. Unlike the bacterial transposons, the maize elements do not carry a selectable drug-resistance marker and recovery of a mutation requires the screening of a large number of progeny. Such a screen will extract all susceptible mutations caused by a variety of mechanisms, including those due to insertion of a transposable element. These may be difficult or impossible to distinguish except that insertional mutations are expected to be unstable (Freeling, 1984), reverting to the resistance phenotype because of the propensity of the element to undergo secondary transposition events. A number of independent transposable element systems have been identified in maize (Federoff, 1983) and many may be useful as gene tags, although, at present cloned probes are only available for a limited few (Federoff *et al.*, 1984b; Freeling, 1984; Pereira *et al.*, 1985). At present it is not clear whether all target loci in the plant are equally accessible to insertional mutagenesis or even at what frequency such mutations might occur. Unlike the internal portion of the *Ac* element, many transposable elements are present in high copy number (> 50) (Freeling, 1984; Sutton *et al.*, 1984) and this can superimpose an additional difficulty in distinguishing these background genomic locations from genomic clones in which the element is inserted into the resistance gene. To clone the *a1* locus from maize, O'Reilly *et al.* (1985) overcame this difficulty by isolating clones from independent genomic libraries made from two insertional mutations of the *a1* locus which had been induced by the unrelated transposable elements *En* and *Mu*1. The *En* and *Mu*1 selected clones were cross hybridized and clones which had sequences in common could be used to identify those clones carrying all or part of the *a1* gene.

III. The Role of Toxins in Plant Disease

Most plants are resistant to most pathogens. Mechanisms of defence include preformed barriers such as the cuticle and existing chemical substances which antagonise the growth of potential pathogens. Alternatively, the barriers to disease development may be generated in response to the presence of the pathogen and these induced barriers may also be structural or chemical in nature. Specific chemical compounds produced by plants in response to infection have been termed phytoalexins (Müller, 1956) and have been identified in a range of plant species (Kuc, 1972). The onus is then on the pathogen to circumvent the components of defence carried by the host plant. The production of phytotoxic substances by a pathogen may be seen as part of an overall strategy to get beyond host systems of defence, particularly those formed in response to infection. Not surprisingly the production of phytotoxins is seen more often among necrotrophic organisms where the host-pathogen interaction does not depend on the plant cells remaining metabolically active.

For a small number of plant diseases, the pathogen is found to produce a host-specific or host-selective toxin which is an absolute determinant of pathogenicity. Thus, production of the toxin is required for full disease symptoms to develop on a susceptible host plant. Examples of pathogens in this category have been collated by Yoder (1980), Kono *et al.* (1981), Scheffer (1983) and Nishimura and Kohmoto (1983) and predominantly comprise species of the fungi *Alternaria* and *Drechslera (Cochliobolus).* The necessity of the toxin for disease development may be seen both for the pathogen and for the host. Mutants of the pathogen which fail to produce the host-specific toxin will either not infect the plant or at most produce minor disease symptoms. This is illustrated by the experiments of Scheffer *et al.* (1967) on *Cochliobolus victoriae* (perfect stage of *Drechslera,* or *Helminthosporium victoriae*) and *Cochliobolus carbonum* (perfect stage of *Drechslera zeicola, Helminthosporium carbonum*). *C. victoriae* produces the host-specific HV-toxin and is the causal agent of Victoria blight of oats, whereas *C. carbonum* race 1 produces HC-toxin and causes leaf spot of maize. The fungi are heterothallic and sexually compatible, and progeny of crosses were found to segregate in a manner consistent with the production of toxin being under the control of a single dominant locus in each species. In virtually all cases tested, pathogenicity of the resulting isolates was correlated with the ability to produce toxin specific to the host.

Genotypes of the host which are resistant to the effects of toxin are also resistant to the development of the particular disease. Susceptibility to Victoria blight of oats is associated with a dominant gene governing resistance to crown rust caused by *Puccinia coronata* (Welsh *et al.,* 1954). The locus for sensitivity to HV-toxin appears to be either allelic with or very tightly linked to the rust resistance gene *Pc-2* (see Day, 1974). In maize, resistance to the HC-toxin of *C. carbonum* race 1 is determined by the two genes, *Hm* on chromosome 1 and *Hm-2* on chromosome 9 (Nelson and Ullstrup, 1964), whereas sensitivity to T-toxin of *Cochliobulus heterostrophus* (perfect stage of *Drechslera,* or *Helminthosporium victoriae*) race T is associated with a genetic alteration in the mitochrondrial genome (Leaver and Gray, 1982; Laughnan and Gabay-Laughnan, 1983). In tomato, resistance to *Alternaria alternata* f. sp. *lycopersici* is controlled by a single dominant nuclear gene which also conditions resistance to AL-toxin(s) produced by the fungus (Gilchrist and Grogan, 1976).

Many questions still remain concerning the relationship between the pathogen and host genes. Among related pathogenic fungi, such as the *Cochliobolus* species discussed above, there is as yet no clear structural relationship between the host-specific toxins produced (Macko, 1983), although the apparent chemical instability of HV-toxin has until now precluded a structural determination (Kono *et al.,* 1981). It is thus not clear whether the toxin production represents an amplification of an existing pathway of secondary metabolism or a completely novel synthetic pathway. In addition, there is no simple pattern regarding the sites of action of the toxins in the respective host cells (Yoder, 1980; Daly, 1981). Further genetic investigation of both pathogen and host should be able to provide some of the answers, particularly as it becomes possible to examine genetic differences at the DNA level. For example, in the *Cochliobolus* species where the production of toxin appears to be determined by few genes, it would be informative to perform some molecular analysis of those genes and their polypeptide products. Such genes might be identified following transformation of avirulent toxin-less strains of a fungus with a related toxin-producing strain. For instance, there is good evidence that a single gene, *tox1*, determines virulence and T-toxin production in *C. heterostrophus* (Leach *et al.*, 1982). Thus it should be possible to screen for toxin production among homozygous *tox1*$^-$ lines genetically transformed with cloned DNA from a *tox1*$^+$ strain, and thereby identify at a structural level a major determinant of toxin production. If successful, this approach might provide evidence for or against the hypothesis that T-toxin production and the resulting virulence of race T of *C. heterostrophus* is due to a recent mutation in race 0 (Leonard, 1973), thus shedding light on the evolution of the capacity to produce a host-specific toxin.

Similarly, in the host plant, application of a molecular approach may make it possible to determine precisely the genetic difference between resistant and susceptible lines. Steps have been made in this direction for the southern corn leaf blight disease caused by *C. heterostrophus,* where susceptible T-toxin sensitive lines of maize invariably carry the Texas (T) source of cytoplasmic male-sterility (Hooker *et al.*, 1970; Hilty and Josephson, 1971). Mitochondria from T cytoplasm were found to be very sensitive to T-toxin of *C. heterostrophus* at concentrations which had no effect on mitochondria from N (normal) cytoplasm (Miller and Koeppe, 1971). Restriction fragment differences have been identified between DNA of T-toxin sensitive mitochondria from T cytoplasm and mitochondrial DNA from T-toxin resistant revertants derived from tissue culture, although not all of the differences correlated with the change in toxin sensitivity (Gengenbach *et al.*, 1981; Kemble *et al.*, 1982). The polypeptide translation products of isolated mitochondria have also been examined and found to have differences associated with the toxin sensitive phenotype (Forde and Leaver, 1980; Dixon *et al.*, 1982). However, identification of the site of action of T-toxin needs further information regarding the structure and function of the alterations observed at the molecular level.

The link between HV-toxin sensitivity and crown rust resistance at the *Pc-2* locus may also be amenable to molecular analysis once techniques for cloning single disease-resistance genes become generally available (see preceding section). Transformation studies using cloned fragments from and adjacent to the *Pc-2* locus should allow identification of the DNA sequences responsible for imparting HV-toxin sensitivity to Victoria blight susceptible oat plants.

In contrast to host-specific toxins, there are a number of necrotrophic plant pathogens where the role of toxins is less clear. Here toxic substances produced by the pathogen, although not absolute determinants of pathogenicity, have been implicated as having an important function in disease development (Scheffer, 1983). These include fungal pathogens such as *Septoria nodorum* (perfect stage of *Leptosphaeria nodorum*), the causal agent of glume blotch in wheat; *Rhyncosporium secalis,* the agent of barley scald disease; *Cochliobolus miyabeanus* (perfect stage of *Drechslera,* or *Helminthosporium oryzae*); and bacterial pathogens such as *Pseudomonas syringae* pv. *tabaci* and *Pseudomonas syringae* pv. *phaseolicola.*

The fungal pathogen *Septoria nodorum* can be considered as an example of how a genetical approach could help determine the contribution of toxin production to disease development. There is evidence that toxin substances produced by the fungus are to some extent determinants of pathogenicity (Bousquet and Skajennikoff, 1974; Kent and Strobel, 1976). Tolerance to the disease in wheat has been correlated with resistance of plant tissues to the effects of sterile culture filtrates of the fungus (Hann, 1978). Yet there are several difficulties associated with determination of the importance of toxins in development of the disease. First, genetic analysis is made difficult by the unavailability of wheat genotypes carrying full resistance to the pathogen (Shipton *et al.,* 1971; Bockmann *et al.,* 1975). Second, disease development is influenced by factors such as architecture and development of the plant. Tall and late-maturing genotypes tend to escape infection (Brönnimann *et al.,* 1973; Tavella, 1978; Scott *et al.,* 1982). Third, there is evidence for a degree of host specialisation among different isolates of the fungus (Rufty *et al.,* 1981). The absence of near isogenic host material differing only in disease susceptibility is a serious disadvantage. One approach to resolving this difficulty is to use toxins produced by the fungus to select for resistance in cultured tissues (Brettell and Ingram, 1979), as was done in the selection of alfalfa with resistance to *Fusarium oxysporum* (Hartman *et al.,* 1984), another pathogen for which non-specific toxins may have a role in disease development. A second possibility would be to identify resistance in wheat lines carrying addition chromosomes from related alien grass species. If such resistance was confined to a single locus, then it would be feasible to examine how this resistance interacts with pathogen genes or gene products. A truly integrated approach would also involve genetic and biochemical analysis of the pathogen which, in contrast to studies of biotrophic fungi such as the rusts and mildews, would be made easier by axenic culture of the pathogen.

IV. Conclusions

We have reviewed just a few of the many possibilities for the application of molecular techniques to the study of host-pathogen interactions. We are not recommending a wholly molecular approach but rather trying to make the obvious point that a number of relatively new molecular techniques may provide fresh insights into what has been a largely intractable problem. These techniques will help attack, but not necessarily solve, the problem.

Cloning a gene for toxin production will not tell us the function of this gene without some more detailed understanding of the biochemistry and the biology of the host-pathogen interaction. Recognizing differences in restriction fragments between DNA of T-toxin sensitive mitochondria and normal mitochondria provides a clear direction for future work to identify the site of action of T-toxin.

Shotgun cloning and gene tagging are methods which could provide isolated DNA sequences specifying resistance and avirulence gene functions without any prior knowledge of the functions of these genes. The determination of these functions will be the next step. The base sequences may contain some obvious signals, such as an open reading frame indicating that the gene product is a protein, something which is not known at the moment. This sequence may code for certain protein structural features which indicate a membrane location or a particular enzymatic function. The gene sequence or a cDNA clone of the mRNA might be placed into an expression vector producing sufficient gene product to elicit antibody production which could be used for cytochemical localization studies. But knowing that a gene product is, for example, a membrane located protein, is only a part of the understanding of how a pathogen and its host interact.

Acknowledgement

The authors wish to thank their colleagues at CSIRO Division of Plant Industry for their helpful comments during the preparation of the manuscript.

V. References

Anderson, D. M., Mills, D., 1985: The use of transposon mutagenesis in the isolation of nutritional and virulence mutants in two pathovars of *Pseudomonas syringae*. Phytopathology **75**, 104—108.

Beckett, J. B., 1978: B-A translocations in maize. I. Use in locating genes by chromosome arms. J. Hered. **69**, 27—36.

Bennett, M. D., Smith, J. B., 1976: Nuclear DNA amounts in angiosperms. Philos. Trans. R. Soc. London, Ser.B **274**, 227—274.

Bennett, M. D., Smith, J. B., Heslop-Harrison, J. S., 1982: Nuclear DNA amounts in angiosperms. Proc. R. Soc. London, Ser.B **216**, 179—199.

Bockmann, H., Mielke, H., Wachholz, G., 1975: Investigations on the susceptibility of various winter and spring wheat cultivars to *Septoria nodorum* Berk. and *Fusarium culmorum* Link. Z. Pflanzenzücht. **74**, 39—47.

Bousquet, J. F., Skajennikoff, M., 1984: Isolement et mode d'action d'une phytotoxine produite en culture par *Septoria nodorum* Berk. Phytopathol. Z. **80**, 355—360.

Brettell, R. I. S., Ingram, D. S., 1979: Tissue culture in the production of novel disease-resistant crop plants. Biol. Rev. **54**, 329—345.

Brönnimann, A., Fossati, A., Hani, F., 1983: Spreading of *Septoria nodorum* Berk. and damage to artificially induced straw-length mutants of the winter wheat Zenith (*Triticum aestivum* L.). Z. Pflanzenzücht. **70**, 230—245.

Buchmann, I., Marner, F. J., Schröder, G., Waffenschmidt, S., Schröder, J., 1985: Tumour genes in plants: T-DNA encoded cytokinin biosynthesis. EMBO J. **4**, 853—859.

Clarke, L., Carbon, J., 1976: A colony bank containing synthetic Col E1 hybrid plasmids representative of the entire *E. coli* genome. Cell **9**, 91—99.

Daly, J. M., 1981: Mechanisms of action. In: Durbin, R. D. (ed.), Toxins in Plant Disease, pp. 331—394. New York: Academic Press.

Daniels, M. J., Barber, C. E., Turner, P. C., Cleary, W. G., Sawczyc, M. K., 1984: Isolation of mutants of *Xanthomonas campestris* pv. *campestris* showing altered pathogenicity. J. Gen. Microbiol. **130**, 2447—2455.

Day, P. R., 1974: Genetics of Host-Parasite Interaction, pp. 17—19. San Francisco: W. H. Freeman and Company.

de Bruijn, F. J., Lupski, J. R., 1984: The use of transposon Tn5 mutagenesis in the rapid generation of correlated physical and genetic maps of DNA segments cloned into multicopy plasmids — a review. Gene **27**, 131—149.

Dixon, L. K., Leaver, C. J., Brettell, R. I. S., Gengenbach, B. G., 1982: Mitochondrial sensitivity to *Drechslera maydis* T-toxin and the synthesis of a variant mitochondrial polypeptide in plants derived from maize tissue cultures with Texas male-sterile cytoplasm. Theor. Appl. Genet. **63**, 75—80.

Federoff, N., 1983: Controlling elements in maize. In: Shapiro, J. (ed.), Mobile Genetic Elements, pp. 1—63. New York: Academic Press.

Federoff, N., Furtek, D., Nelson, O. E., 1984a: Cloning of the *bronze* locus in maize by a simple and generalised procedure using the transposable controlling element *Ac*. Proc. Nat. Acad. Sci., U.S.A. **81**, 3825—3829.

Federoff, N., Wessler, S., Shure, M., 1984b: Isolation of the maize controlling elements *Ac* and *Ds*. Cell **35**, 235—242.

Fincham, J. R. S., Sastry, G. R. K., 1974: Controlling elements in maize. Annu. Rev. Genet. **8**, 15—50.

Flor, H. H., 1947: Inheritance of reaction to rust in flax. J. Agric. Res. **74**, 241—262.

Flor, H. H., 1956: The complementary genic systems in flax and flax rust. Adv. Genet. **8**, 29—54.

Forde, B. G., Leaver, C. J., 1980: Nuclear and cytoplasmic genes controlling synthesis of variant mitochondrial polypeptides in male-sterile maize. Proc. Nat. Acad. Sci., U.S.A. **77**, 418—422.

Freeling, M., 1984: Plant transposable elements and insertion sequences. Annu. Rev. Plant Physiol. **35**, 277—298.

Gengenbach, B. G., Connelly, J. A., Pring, D. R., Conde, M. F., 1981: Mitochondrial DNA variation in maize plants regenerated during tissue culture selection. Theor. Appl. Genet. **59**, 161—167.

Gheysen, G., Dhaese, P., Van Montagu, M., Schell, J., 1985: DNA flux across genetic barriers: the crown gall phenomenon. In: Hohn, B., Dennis, E. S. (eds.), Plant Gene Research: Genetic Flux in Plants, pp. 11—47. Wien — New York: Springer.

Gilchrist, D. G., Grogan, R. G., 1976: Production and nature of a host-specific toxin from *Alternaria alternata* f. sp. *lycopersici*. Phytopathology **66**, 165—171.

Hann, A. C., 1978: Studies on the epidemiology of *Septoria* species on wheat. Ph. D. Thesis, Aberystwyth: University College of Wales.

Hartman, C. L., McCoy, T. J., Knous, T. R., 1984: Selection of alfalfa *(Medicago sativa)* cell lines and regeneration of plants resistant to the toxins(s) produced by *Fusarium oxysporum* f. sp. *medicaginis*. Plant Sci. Lett. **34**, 183—194.

Herrera-Estrella, L., Van den Broeck, G., Maenhaut, R., Van Montagu, M., Schell,

J., 1984: Light-inducible and chloroplast-associated expression of a chimaeric gene introduced into *Nicotiana tabacum* using a Ti-plasmid vector. Nature (London) **310,** 115—120.

Hilty, J. W., Josephson, L. M., 1971: Reaction of corn inbreds with different cytoplasms to *Helminthosporium maydis.* Plant Dis. Rep. **55,** 195—198.

Hooker, A. L., 1967: The genetics and expression of resistance in plants to rusts of the genus *Puccinia.* Annu. Rev. Phytopathol. **5,** 163—182.

Hooker, A. L., Smith, D. R., Lim, S. M., Beckett, J. B., 1970: Reaction of corn seedlings with male-sterile cytoplasm to *Helminthosporium maydis.* Plant Dis. Rep. **54,** 708—712.

Ingram, D. S., 1978: Cell death and resistance to biotrophs. Ann. Appl. Biol. **89,** 291—295.

Inzé, D., Follin, A., Van Lijsebettens, M., Simoens, C., Genetello, C., Van Montagu, M., Schell, J., 1984: Genetic analysis of the individual T-DNA genes of *Agrobacterium tumefaciens;* further evidence that two genes are involved in indole-3-acetic acid synthesis. Mol. Gen. Genet. **194,** 265—274.

Kemble, R. J., Flavell, R. B., Brettell, R. I. S., 1982: Mitochondrial DNA analyses of fertile and sterile maize plants derived from tissue culture with the Texas male sterile cytoplasm. Theor. Appl. Genet. **62,** 213—217.

Kent, S. S., Strobel, G. A., 1976: Phytotoxin from *Septoria nodorum.* Trans. Br. Mycol. Soc. **67,** 354—358.

Kono, Y., Knoche, H. W., Daly, J. M., 1981: Structure: fungal host-specific. In: Durbin, R. D. (ed.), Toxins in Plant Disease, pp. 221—257. New York: Academic Press.

Kuć, J., 1972: Phytoalexins. Annu. Rev. Phytopath. **10,** 207—232.

Laughnan, J. R., Gabay-Laughnan, S., 1983: Cytoplasmic male sterility in maize. Annu. Rev. Genet. **17,** 27—48.

Lawrence, G. J., Mayo, G. M. E., Shepherd, K. W., 1981: Interactions between genes controlling pathogenicity in the flax rust fungus. Phytopathology **71,** 12—19.

Leach, J., Tegtmeier, K. J., Daly, J. M., Yoder, O. C., 1982: Dominance at the *Tox1* locus controlling T-toxin production by *Cochliobolus heterostrophus.* Physiol. Plant Pathol. **21,** 327—333.

Leaver, C. J., Gray, M. W., 1982: Mitochondrial genome organization and expression in higher plants. Annu. Rev. Plant Physiol. **33,** 373—402.

Leonard, K. J., 1983: Association of mating type and virulence in *Helminthosporium maydis,* and observations on the origin of the race T population in the United States. Phytopathology **63,** 112—115.

Macko, V., 1983: Structural aspects of toxins. In: Daly, J. M., Deverall, B. J. (eds.), Toxins and Plant Pathogenesis, pp. 41—80. Sydney: Academic Press Australia.

Manners, J. M., Davidson, A. D., Scott, K. J., 1985: Patterns of post-infectional protein synthesis in barley carrying different genes for resistance to the powdery mildew fungus. Plant Mol. Biol. **4,** 275—283.

Miller, R. J., Koeppe, D. E., 1971: Southern corn leaf blight: Susceptible and resistant mitochondria. Science **173,** 67—69.

Müller, K. O., 1956: Einige einfache Versuche zum Nachweis von Phytoalexinen. Phytopathol. Z. **27,** 237—254.

Nelson, O. E., Ullstrup, A. J., 1964: Resistance to leaf spot in maize. J. Hered. **55,** 195—199.

Niepold, F., Anderson, D., Mills, D., 1985: Cloning determinants of pathogenesis from *Pseudomonas syringae* pathovar *syringae.* Proc. Nat. Acad. Sci., U.S.A. **82,** 406—410.

Nishimura, S., Kohmoto, K., 1983: Roles of toxins in pathogenesis. In: Daly, J. M., Deverall, B. J. (eds.), Toxins and Plant Pathogenesis, pp. 137—157. Sydney: Academic Press Australia.

O'Reilly, C., Shepherd, N. S., Pereira, A., Schwarz-Sommer, Z., Bertram, I., Robertson, D. S., Peterson, P. A., Saedler, H., 1985: Molecular cloning of the *a1* locus of *Zea mays* using the transposable elements *En* and *Mu1*. EMBO J. **4,** 877—882.

Pereira, A., Schwarz-Sommer, Zs., Gierl, A., Bertram, I., Peterson, P. A., Saedler, H., 1985: Genetic and molecular analysis of the Enhancer (En) transposable element system of *Zea mays*. EMBO J. **4,** 17—23.

Rufty, R. C., Herbert, T. T., Murphy, C. F., 1981: Variation in virulence in isolates of *Septoria nodorum*. Phytopathology **71,** 593—596.

Scheffer, R. P., 1983: Toxins as chemical determinants of plant disease. In: Daly, J. M., Deverall, B. J. (eds.), Toxins and Plant Pathogenesis, pp. 1—40. Sydney: Academic Press Australia.

Scheffer, R. P., Nelson, R. R., Ullstrup, A. J., 1967: Inheritance of toxin production and pathogenicity in *Cochliobolus carbonum* and *Cochliobolus victoriae*. Phytopathology **57,** 1288—1291.

Schröder, G., Waffenschmidt, S., Weiler, E. W., Schröder, J., 1984: The T-region of Ti plasmids codes for an enzyme synthesising indole-3-acetic acid. Eur. J. Biochem. **138,** 387—391.

Scott, P. R., Benedikz, P. W., Cox, C. J., 1982: A genetic study of the relationship between height, time of ear emergence and resistance to *Septoria nodorum* in wheat. Plant Pathol. **31,** 45—60.

Shipton, W. A., Boyd, W. R. J., Rosielle, A. A., Shearer, B. I., 1971: The common *Septoria* diseases of wheat. Bot. Rev. **37,** 231—262.

Staskawicz, B. J., Dahlbeck, D., Keen, N. T., 1984: Cloned avirulence gene of *Pseudomonas syringae* pv. *glycinea* determines race-specific incompatibility on *Glycine max* (L.) Merr. Proc. Nat. Acad. Sci., U.S.A. **81,** 6024—6028.

Sutton, W. D., Gerlach, W. L., Schwartz, D., Peacock, W. J., 1984: Molecular analysis of *Ds* controlling element mutations at the *Adh1* locus of maize. Science **223,** 1265—1268.

Tavella, C. M., 1978: Date of heading and plant height of wheat varieties as related to *Septoria* leaf blotch damage. Euphytica **27,** 577—580.

Torp, J., Andersen, B., 1982: Two-dimensional electrophoresis of proteins from cultures of *Erysiphe graminis* f. sp. *hordei*. Physiol. Plant Pathol. **21,** 151—160.

Welsh, J. N., Peturson, B., Machacek, J. E., 1954: Associated inheritance of reaction to races of crown rust, *Puccinia coronata avenae* Erikss., and to Victoria blight, *Helminthosporium victoriae* M. and M., in oats. Can. J. Bot. **32,** 55—68.

Wynn, W. K., 1976: Appressorium formation over stomates by the bean rust fungus: response to a surface contact stimulus. Phytopathology **66,** 136—146.

Yoder, O. C., 1980: Toxins in pathogenesis. Annu. Rev. Phytopathol. **18,** 103—129.

Note added in proof:

Further to the discussion on methods for identifying genes involved in pathogenicity, Daniels *et al.* (1984) have isolated non-pathogenic mutants of *Xanthomonas campestris* pv. *campestris* following treatment of the bacteria with chemical mutagen. As with the non-pathogenic strains of *Pseudomonas syringae* obtained by transposon mutagenesis (Anderson and Mills, 1985; Niepold *et al.*, 1985), it has yet to be established whether the mutations reside in the genes involved in pathogen specificity.

Chapter 9

Gametophytic Gene Expression

David L. Mulcahy

Department of Botany, University of Massachusetts, Amherst, MA 01003, U.S.A.

Contents

I. Introduction

Gametophytic gene expression is a topic of both basic and applied interest. It has been subjected to many reviews, starting with Buchholz (1922) and continuing with unabated interest to the present (Zamir, 1983; Mulcahy, 1984). Consequently, this review will focus on general characteristics and principles.

When a diploid organism, heterozygous at N loci, undergoes meiosis, it has the potential of producing 2^N different haploid genotypes. In the case of pollen, this great variety is often associated with an equally impressive overabundance of individuals. Insect-pollinated angiosperms, for example, produce on the average about six thousand pollen grains for each ovule and, in wind-pollinated species, the ratio averages one million to 1 (Cruden, 1977). The picture that these facts convey is one of a vast array of pollen genotypes competing for a relatively small number of ovules. Certainly such competition holds the potential for extremely intense selection. However, this potential will be realized only if genes are expressed in the gametophyte. Futhermore, such pollen competition will effect the sporophyte only if the genes which are expressed in the gametophyte are, at least in part, the same genes which are expressed in the sporophyte. Thus we must ask if genes are expressed in the gametophyte, and, if so, are they among the genes which are expressed in the sporophyte? Although spermatozoa in animals correspond to the sperm produced by the microgametophyte and comparisons of spermatozoa with pollen grains are misleading, post-meiotic gene expression is apparently rare in animals (Braden, 1972). Haldane (1932) speculated that it would be disfavoured by natural selection in plants. He suggested that the potential intensity of pollen selection could overwhelm selective values of the sporophyte and thus allow the spread of gametophytically favoured, but sporophytically disfavoured, alleles. Selection should thus either reduce gene expression in the pollen or limit it to genes which are not expressed in the sporophytic portion of the life cycle. These concepts were apparently affirmed by the fact that the gene *waxy,* expressed in the pollen of *Zea mays,* is not expressed in the sporophyte (Nelson and Tsai, 1964). On the other hand, Schwartz (1971) demonstrated conclusively that the Adh (alcohol dehydrogenase) loci of *Zea mays* are expressed in both phases of the life cycle.

Apparently examples can be found which do support Haldane's suggestions and others which do not. A more general test of his ideas was needed, not limited to single loci. The first such test was provided by Ter-Avanesian (1949). He varied the intensity of pollen competition in *Gossypium, Vigna* and *Triticum* by applying limited or excessive quantities of pollen to stigmas. In the first case, fertilizations will be accomplished by gametes of both fast and slow growing pollen tubes since ovules are not limiting. With excessive pollinations, however, only the fastest growing pollen tubes will reach unfertilized ovules. If pollen tube growth rate is determined by genes which are expressed in the pollen, then it should be possible to select among the segregating pollen genotypes of a single plant. Furthermore, if the genes which are expressed, and thus subject to selection in the pollen, are different from the genes which are expressed in the sporophyte, then selecting among pollen tubes should have no effect upon the resulting sporophytes. In each of the species he studied, Ter-Avanesian found that the range of morphological variation exhibited in the resulting sporophytic generation was greatest when limited amounts of pollen were used. Apparently, genes *are* expressed in the pollen and many of these same genes are expressed also in the sporophytic portion of the life cycle. A summary of these and later studies was published by Ter-Avanesian (1978) and a similar study was described by Matthews (see Lewis, 1954). Despite their significance, these studies were largely overlooked.

Starting in 1975, further evidence was presented that the quality of the next sporophytic generation is indeed influenced by the outcome of pollen competition. For example, in *Petunia hybrida,* increasing the quantity of pollen used in pollinations resulted in increased growth rates of F1 plants, as indicated by the number of leaves produced and the total above-ground fresh weight (Mulcahy *et al.,* 1975). This study, and also the studies by Ter-Avanesian, was complicated by the fact that varying the quantity of pollen used in pollinations often modifies the number of seeds produced, which in turn introduces variation in the weights of seeds. However, at least in the *Petunia* study, this added variable was apparently outgrown within 60 days of planting. Furthermore, the influence of pollen selection persisted in the plants resulting from selfing the F1 plants (Mulcahy *et al.,* 1978). Nevertheless, another study was run in order to avoid altogether the possibility of persistent maternal influences. The species used in this latter study was *Dianthus chinensis,* chosen because its elongated stigmatic surfaces allow pollination to take place at different distances from the ovary. This is useful because the greater the distance travelled by a group of competing pollen tubes, the greater will be the separation between the faster and the slower individuals. In this way, the intensity of pollen competition can be varied without modifying either the number of pollen grains applied or, more importantly, the sizes of the resulting seeds. Again, in this study it was found that pollen selection would modify the quality of the resulting sporophytic generation. Specifically, sporophytes which resulted from the gametes of the most rapidly growing pollen tubes showed significantly greater speeds of germination and growth than did other sporophytes (Mulcahy and Mulcahy, 1975). These studies cast serious doubts on Haldane's suggestions about the nature (or lack) of gene expession in the pollen. They also prompted a renewed interest in the extent to which specific genes were or were not expressed in the pollen.

II. Overlap Between Sporophytic and Gametophytic Genotypes

A series of studies has tested for the presumed overlap between haploid and diploid expressed genotypes. The first of these was by Tanksley *et al.,* (1981). Pollen is known to contain many gene products but these could as well be products of the diploid pollen source instead of the individual pollen genotypes. To distinguish between these two alternatives, Tanksley and associates utilized the fact that if an individual is heterozygous for a gene encoding a dimeric enzyme, that individual will exhibit a gene product unique to heterozygotes, the heterodimer. If the gene contents of pollen are the products of sporophytic transcription and translation, pollen of heterozygotes should contain heterodimers. Alternatively, if the pollen contents are the results of postmeiotic gene expression, the pollen should exhibit only homodimers. They examined thirty loci which code for dimeric enzymes in *Lycopersicon esculentum* and closely related species. They concluded that 60 % (18/30) of the genes which were expressed in the sporophyte were expressed, without heterodimers, also in the pollen. This was very strong evidence that a substantial fraction of the structural genes which are expressed in the sporophyte of tomato are expressed also postmeiotically. This conclusion would explain why selection in the pollen could modify the subsequent sporophytic generation.

In view of the potential significance of this conclusion, Sari-Gorla *et al.* (1983) tested the possibility that perhaps the absence of the heterodimers in pollen could be due, not to postmeiotic gene expression but rather to something in the gametophytic cellular environment which prevented the formation of heterodimers. Thus they constructed B-A translocations in *Zea mays* in order to produce some pollen grains which would carry two different alleles of selected loci. Functionally, this provided some pollen which was partially diploid and heterozygous for a few loci, and thus allowing an opportunity for the formation of heterodimers within the pollen. Using this method, they concluded that some sporophytically expressed loci, for example, glutamate-oxaloacetate-transaminase, *got-1,* did form heterodimers in the pollen. Thus the absence of heterodimers in Tanksley's study was, as they had concluded, due not to the inability to form heterodimers but to postmeiotic gene expression.

Sari-Gorla and colleagues at the University of Milan have used the B-A translocations also to test for overlap between sporophytically and gametophytically expressed loci (Frova *et al.*, in prep.). They also concluded that a clear majority of the sporophytically expressed structural genes of *Zea mays* are expressed also in the pollen (see Table 1).

Table 1. Three Estimates of Overlap Between Pollen and Sporophytically Expressed Genes.

Expressed in Pollen Only	Expressed in both Pollen and Sporophyte	Expressed in Sporophyte Only	Species	Ref.
3 %	58 % (59 %)	39 %	*Lycopersicon esculentum*	Tanksley *et al.*, 1981
<15 %	54 % (<60 %)	>31 %	*Tradescantia*	Willing *et al.*, 1984
6 %	73 % (78 %)	21 %	*Zea mays*	Frova, (pers. comm.)

Figures in parentheses indicate which percentage of the sporophytically expressed genes are expressed in the pollen.

The most recent study of overlap between sporophytic and gametophytic genomes involved a molecular survey and comparison of the mRNAs produced in the pollen and stems of *Tradescantia paludosa* (Willing and Mascarenhas, 1984). It thus differed from previous studies in the number of genes which could be monitored. mRNA extracted from pollen and stems was used to prepare cDNAs for those messages. Hybridization of homologous cDNA and RNA showed that pollen contained approximately 20,000 diverse mRNAs while shoots contained 30,000. This demonstrates quite clearly that there is substantial gene activity in the pollen. Heterologous hybridizations, i. e., pollen cDNA to shoot poly(A)RNA and shoot cDNA to pollen poly(A)RNA suggested that about 64 % of the pollen sequences are similar to those in shoots and about 60 % of the sequences in shoots are similar to those in pollen. These analyses were based on total cellular poly(A)RNA, which allows the possibility that both tissues may yield a great proportion of nuclear RNAs which may or may not be translated. However, the authors point out that pollen is dehydrated at anthesis and no RNA or protein synthesis is occurring. The use of total cell poly(A)RNA seems justified also in the case of vegetative tissue, since Galau *et al.* (1981) found in cotton that hybridization kinetics of DNA complementary to total cell poly(A)RNA were virtually identical to the kinetics when polysomal poly(A)RNA was used.

Considering the above studies, summarized in Table 1, it seems apparent that a large proportion of the structural genes expressed in the sporophyte are expressed, and thus subject to selection, also in the pollen. The sporophytic and gametophytic genotypes thus appear to overlap extensively. This overlap may be significant since selection among haploid genotypes is, in several situations, far more effective than selection among diploid genotypes (Pfahler, 1983). For example, a rare recessive is immediately exposed to selection in a haploid genotype. Furthermore, multilocus adaptations are more easily assembled in haploidy than in diploidy (Zamir, 1983). These considerations, and the fact that pollen is often produced in very large numbers, mean that pollen selection is strictly comparable to the mass selection which contributes so greatly to the adaptability of microbial organisms. With overlap, these aspects of haploid selection could impinge on substantial portions of the sporophytic genotype.

III. Gametophytic Gene Expression and the Angiosperms

Overlap between gametophytic and sporophytic genotypes brings the potential of haploid selection to the sporophytic portion of the life cycle. However, this potential will be realized only to the extent that individual gametophytes are able to complete with each other. Two unique features of the angiosperms, evolved in response to other selective pressures (Crepet, 1984, for a review), serve to greatly accentuate pollen tube competition (Mulcahy, 1979), namely insect pollination and the closed carpel. Insect pollination increases the number of pollen grains reaching stigmatic surfaces, which should intensify pollen tube competition. Insect pollination also means that pollen grains will tend to arrive in clumps, rather than as individuals carried by the wind. This simultaneity of pollination reduces the influence of random starting times and thus allows a more consistent separation of faster from slower pollen tubes. The closed carpel increases the distance through which pollen tubes must grow and, as discussed earlier, this will improve the separation of faster and slower pollen tubes. As a result, the consequences of gene expression in the gametophyte and of genetic overlap may be greatly enhanced in the angiosperms.

IV. The Influence of Haploid Genotype on Pollen Size

Johnson *et al.* (1976) found that pollen diameter in *Zea mays* was influenced by both the sporophytic and the gametophytic genotypes. The sporophytic influence was shown by a steady decline ($Y = 98 + (-0.107) X$) in the average pollen grain diameter from the F1 ($x = 100.85\ \mu$) to the F7 ($x = 94.03\ \mu$). Apparently, the inbred plants produce smaller pollen grains than do the vigorous hybrids. Superimposed on this sporophytically determined decline in average diameter is a gametophytic influence. This reflects the fact that inbreeding reduces heterozygosity and the pollen genotypes thus become more homogeneous as inbreeding progresses. As a result, the coefficient of variation in pollen diameter also declines with inbreeding ($r = -0.5530$, $P < 0.01$). Variance in pollen diameters reflects not only the influence of the genes carried within the individual pollen grains but also competitive gametophytic interactions. A pollen lethal allele on the fourth chromosome of *Zea mays* greatly reduces the diameter of pollen grains carrying it (Singleton and Mangelsdorf, 1940). Furthermore, the maximum diameter of the wild-type segregants is clearly greater than that of plants not segregating for the pollen lethal, indicating that in the segregating plants the wild-type gametophytes are able to pre-empt resources relinquished by those carrying the pollen lethal factor. It is not known if these greater resources allow the functional grains any competitive advantages either through greater longevity or increased pollen tube growth rates.

V. Time of Gene Expression in Pollen

Stinson and Mascarenhas (1985) found that no ADH activity was detectable when tetrads of *Zea mays* were first formed. Activity was first detected soon after the tetrads began to break apart but before separation was complete. In *Medicago sativa,* however, glutamate dehydrogenase activity first appeared in the mitochondria of pollen of the binucleate stage, that is, close to pollen maturity (Vienne *et al.,* 1985). Apparently, different enzymes become active at different stages of pollen development. A systematic investigation of this subject could suggest which enzymes are essential for pollen development and which function during pollen tube growth. This information would ultimately provide useful information in programmes designed to select particular qualities by means of pollen selection.

VI. Methods of Haploid Selection

Considering the microgametophytic portion of the life cycle, two subsets of selection phenomena should be considered: microspore selection and pollen selection. The first of these includes all selection occurring between the completion of meiosis and that of pollen maturity. Pollen selection, in contrast, extends from pollen maturity until zygote formation. In all cases, only phenomena which are controlled by genes transcribed and translated after meiosis should be considered.

Zamir and Vallejos (1983) contrasted the relative effectiveness of the two subsets of microgametophytic selection in *Lycopersicon spp.* In one case, ramets of an Fl between *Lycopersicon esculentum* and a cold tolerant accession of *L. hirsutum* were maintained at controlled environments of either 24/19° C and 12/6° and then backcrossed to *Lycopersicon esculentum* grown in a greenhouse (approximately 24/18° C). In another case, pollen from a greenhouse maintained Fl clone was backcrossed to *Lycopersicon esculentum* in the controlled environments. In the first case, any differential effects of the two controlled environments would be applied only during microspore development. In the second case, differential was applied during pollen germination, pollen tube growth, and fertilization.

In each treatment, 8 isozyme loci were analyzed in the backcross progeny to determine if either *Lycopersicon esculentum* or *L. hirsutum* loci deviated from the expected 1:1 ratio. When controlled temperatures were applied to the microspores, one locus, *Aps-2,* deviated significantly with the higher temperature, *Got-3* and *Est-4* deviated with the lower temperature, and two loci, *Pgm-2* and *Pgi-1,* deviated with both temperature regimes. In each of these five cases, the significant deviations were in favour of the *L. esculentum* isozyme. When the different controlled temperatures were applied to the pollen, significant deviations in favour of the *L. esculentum* isozymes (*Aps-2, Pgi-1* and *Est-4*) occurred at the higher temperature. At the lower temperature, significant deviations occurred in favour of *L. hirsutum* isozymes (*Adh-2, Pgi-1* and *Est-4*). From these data, it is clear that selection for isozymes associated with cold tolerance was much more effective with pollen than with microspores.

In contrast to the clear superiority of pollen selection observed by Zamir and Vallejos (1983), Searcy and Mulcahy (1985) found that microspore selection was the more effective means of selecting for tolerance to heavy metals. Working with metal-tolerant and metal-nontolerant ecotypes of *Silene dioica* and *Mimulus guttatus,* they produced hybrids between tolerant and nontolerant parents segregating for metal tolerance. When the Fl plants were grown in the presence of copper, in the case of *Mimulus,* and zinc, for *Silene,* the proportion of potentially functional pollen was reduced to 54.0 % and 61.4 % of the control values, respectively. This by itself suggests that the hybrids are each producing two subpopulations of pollen, one tolerant and one nontolerant to the heavy metal. When backcrossed to nontolerant pistillate parents, pollen of segregating plants of *Mimulus* and *Silene,* normally produced 32.9 % and 54.2 % tolerant progeny, respectively. However, if the segregating pollen sources are treated with heavy metal during pollen development, the proportion of tolerant progeny resulting from backcrosses increases to 59.1 % in the case of *Mimulus* and to 79.2 % in the case of *Silene.* This amounts to a 79.6 % increase in the proportion of tolerant progeny in the case of *Mimulus* and a 46.1 % increase with *Silene.*

The efficacy of pollen selection, as opposed to microspore selection, was tested by growing pistillate plants which were tolerant to heavy metals either with or without heavy metals. Growing *Mimulus* in nutrient solutions containing 0, 1 or 2 ppm copper resulted in 6, 23, and 48 ppm copper in the flowers, respectively. For *Silene,* growth on 0 and 5 ppm zinc resulted in 34—40 and 150—200 ppm zinc within the styles, respectively. Pollen for backcrosses to these plants was obtained from tolerant and nontolerant plants grown without heavy metals. The test of pollen selection was to measure the growth rates of pollen tubes in the treated and untreated styles. In *Silene,* neither the tolerance of the pollen source nor the concentration of heavy metal within the style influenced pollen tube growth rates. With *Mimulus,* pollen from the nontolerant pollen source grew significantly faster than did that from the tolerant source. However, here too, the concentration of heavy metal in the style failed to have a significant effect upon pollen tube growth rate.

For whatever reason, it appears that microspore selection is far more effective in the case of heavy metals and that pollen selection is more effective for isozymes which are associated with cold tolerance.

An alternative method of selecting among pollen genotypes is to treat the mature pollen with *in vitro* techniques. Schwartz and Osterman (1976) took advantage of the fact that alcohol dehydrogenase (ADH) converts allyl alcohol to acrylaldehyde, a highly toxic compound, to select for ADH mutants. Pollen from *Zea mays* which lacks alcohol dehydrogenase survives exposure to allyl alcohol whereas wild-type pollen is killed.

An additional if undirected form of selecting pollen grains is to subject them to storage. Pfahler (1967) found that pollen from separate sporophytes of *Zea mays* showed significant differences in the ability to tolerate storage. This could reflect sporophytic as well as gametophytic genotypes. Later investigations, however, indicated that specific alleles from a segregating plant showed significant effects on storage tolerance (Pfahler *et al.*, 1981). Furthermore, in *Petunia hybrida,* the pollen grains which are best able to survive storage give rise to sporophytes which grow significantly more rapidly than do sporophytes resulting from unselected pollen genotypes (Mulcahy *et al.*, 1982). This last result may seem rather unexpected until considered in light of the Singleton and Mangelsdorf (1940) study, which showed that more vigorous gametophytes are able to pre-empt resources from others within the same anther locule. Perhaps such preemption is associated with a greater ability to tolerate storage and also contributes to increased growth rates in the resultant sporophytic generation.

VII. Gametophytic Gene Expression and the Style

In all of the above studies, the style is either ignored or considered to be a passive arena within which microgametophytic gene expression is manifested. This convenient assumption neglects extensive evidence that the style has a very significant influence on pollen and pollen gene expression.

Self-incompatibility is the most widely recognized expression of pollen-style interactions but there are others. In *Zea mays,* the stylar genotype has a significant influence on pollen tube growth rate. Furthermore, there are significant interactions between pollen and stylar genotypes, meaning that not all pollen types are affected in the same way (Pfahler, 1967; Sari-Gorla and Bellintani, 1976). Within the style there are also interactions between different pollen tubes (Sari-Gorla and Rovida, 1980).

VIII. Gene Expression in the Megagametophyte

Owing to the facts that ovules are both less numerous and less accessible than pollen grains, the megagametophyte has received relatively little attention. However, Schwemmle (1968) and colleagues have demonstrated there too that genes are expressed. Using the well known Renner complexes in *Oenothera* spp., they showed for several cases that the excess of heterozygotes usually obtained was not due to abortion of pollen grains or embryo sacs which carry one complex or the other. Rather the excess results from selective fertilization; the tendencies for pollen tubes and ovules to unite is determined by the Renner complex carried in each. It is particularly significant that selective attraction between different ovule and pollen types could be demonstrated also *in vitro.* Grant (1973) quotes studies by Lamprecht indicating that selective fertilization occurs also in *Pisum sativum.* It would seem that with the ready availability of isozyme markers, selective fertilization and megagametophytic gene expression are now accessible for study.

IX. Conclusions

Studies of several angiosperm species have indicated that many structural genes are expressed in the pollen. By one estimate, the number of such genes is approximately two thirds the number of structural genes expressed in the sporophyte. Genes expressed in the microgametophyte influence competitive interactions between developing microspores, pollen diameter, pollen germination, pollen tube growth rate, interactions between pollen tubes, and nonrandom fertilization. A large fraction (64—95 %) of microgametophytically expressed genes are expressed also in the sporophyte. Thus, up to 60 % of the sporophytic structural genes are exposed to selection in the pollen. This extensive overlap between gametophytic and sporophytic genomes explains why pollen selection influences sporophytic qualities such as seed germination, plant growth rate, competitive ability, yield, and tolerance to heavy metals, as well as other forms of stress. The efficacy of pollen selection stems largely from the fact that pollen is haploid and available in large populations. It thus provides a functional equivalent of the mass selection which contributes so greatly to the adaptability of microorganisms. The angiosperms possess characteristics which accentuate pollen selection and this may have contributed to the adaptability and success of that group. Megagametophytic gene expression has been less studied but data available suggest it too may be an arena of extensive gene expression and selection.

Acknowledgements

Sincere thanks are due to my collaborators, Gabriella Bergamini Mulcahy and Douglas MacMillan, for many helpful discussions in the preparation of this paper. Our efforts are supported, in part, by USDA Biotechnology Grant No. 8502969.

Dedication

This manuscript is dedicated to Prof. H. F. Linskens. The breadth and significance of his many contributions to pollen biology stand as an admirable model for us all. I look forward to the continuation of his productivity for many years to come.

X. References

Braden, A. W. H., 1972: T-Locus in mice: segregation distortion and sterility in the male. In: Beatty, R. A., Glueckshon-Waelsch, S. (eds.), Proc. Int. Symp. The Genetics of the Spermatozoon, pp. 289—305. Edinburgh: University of Edinburgh.

Buchholz, J. T., 1922: Developmental selection in vascular plants. Bot. Gaz. (Chicago) **73,** 249—286.

Crepet, W. L., 1984: Advanced (constant) insect-pollination mechanisms: pattern of evolution and implications vis-a-vis angiosperm diversity. Ann. Mo. Bot. Gerd. **71,** 607—630.

Cruden, R. W., 1977: Pollen-ovule ratios: a conservative indicator of breeding systems in flowering plants. Evolution **31,** 32—46.

Galau, G. A., Legocki, S. C., Greenway, S. C., Dure, L. S., 1981: Cotton messenger RNA sequences exist in both polyadenalated and non-polyadenalated forms. J. Biol. Chem. **256,** 2552—2560.

Grant, V., 1973: Genetics of flowering plants, p. 239. New York: Columbia University Press.

Haldane, J. B. S., 1932: The Causes of Evolution. London: Longmans, Green and Co.

Johnson, C. M., Mulcahy, D. L., Galinat, W. C., 1976: Male gametophyte in maize: influence of the gametophytic genotype. Theor. Appl. Genet. **48,** 299—303.

Lewis, D., 1954: Annual report of the department of genetics. Annu. Rep. John Innes Hort. Inst. **45,** 12—17.

McKenna, M., Mulcahy, D. L., 1983: Ecological aspects of gametophytic competition in *Dianthus chinensis.* In: Mulcahy, D. L., Ottaviano, E. (eds.), Pollen: Biology and Implications for Plant Breeding, pp. 419—424. New York: Elsevier.

Mulcahy, D. L., Robinson, R. W., Ihara, M., Kesseli, R., 1981: Gametophytic transcription for acid phosphatases in pollen of *Cucurbita* species hybrids. J. Hered. **72,** 353—354.

Mulcahy, D. L., 1979: The rise of the angiosperms: a genecological factor. Science **206,** 20—23.

Mulcahy, D. L., 1984: Manipulation of gametophytic populations. In: Lange, W. *et al.* (eds.), Efficiency in Plant Breeding, pp. 167—175. Wageningen: Pudoc.

Mulcahy, D. L., Mulcahy, G. B., 1975: The influence of gametophytic competition on sporophytic quality in *Dianthus chinensis.* Theor. Appl. Genet. **46,** 277—280.

Mulcahy, D. L., Mulcahy, G. B., Ottaviano, E., 1975: Sporophytic expression of gametophytic competition in *Petunia hybrida.* In: Mulcahy, D. L., Ottaviano, E., (eds.), Pollen: Biology and Implications for Plant Breeding, pp. 227—232. New York: Elsevier.

Mulcahy, D. L., Mulcahy, G. B., Ottaviano, E., 1978: Further evidence that gametophytic selection modifies the genetic quality of the sporophyte. Soc. bot. Fr., Actualites botaniques **1978,** 57—60.

Mulcahy, G. B., Mulcahy, D. L., Pfahler, P. L., 1982: The effect of delayed pollination in *Petunia hybrida.* Acta Bot. Neerl. **31,** 97—103.

Nelson, O. E., Tsai, C. Y., 1964: Glucose transfer from adeninosine diphosphate-glucose in preparations of waxy seeds. Science **145,** 1194—1195.

Ottaviano, E., Sari-Gorla, M., Mulcahy, D. L., 1980: Pollen tube growth rates in *Zea mays:* implications for genetic improvement of crops. Science **210,** 437—438.

Pfahler, P. L., 1967: Fertilization ability of maize pollen grains II. Pollen genotype, female sporophyte, and pollen storage interactions. Genetics **57,** 513—521.

Pfahler, P. L., 1983: Comparative effectiveness of pollen genotype selection in higher plants. In: Mulcahy, D. L., Ottaviano, E. (eds.), Pollen: Biology and Implications for Plant Breeding, pp. 361—366. New York: Elsevier.

Pfahler, P. L., Linskens, H. F., Mulcahy, D. L., 1981: Effect of pollen ultraviolet radiation on the abortion frquency and segregation patterns at various endosperm mutant loci in maize (*Zea mays* L.). Environ. Exp. Biol. **21,** 5—13.

Sari-Gorla, M., Bellintani, R., 1976: Variation in pollen fertilizing ability in relation to the genotype of the stylar tissue. Maize Genet. News **50,** 77—79.

Sari-Gorla, M., Rovida, E., 1980: Competitive ability of maize pollen: intergametophytic effects. Theor. Appl. Genet. **57,** 37—42.

Sari-Gorla, M., Frova, C. V., Ottaviano, E., Soave, C., 1983: Gene expression at the gametophytic phase in maize. In: Mulcahy, D. L., Ottaviano, E. (eds), Pollen: Biology and Implications for Plant Breeding, pp. 323—328. New York: Elsevier.

Schwartz, D., 1971: Genetic control of alcohol dehydrogenase — a competition model for regulation of gene action. Genetics **67,** 411—425.

Schwartz, D., Osterman, J., 1976: A pollen selection system for alcohol dehydrogenase-negative mutants in plants. Genetics **83,** 63—65.

Schwemmle, J., 1968: Selective fertilization in *Oenothera.* Adv. Genet. **14,** 225—324.

Searcy, K., Mulcahy, D. L., 1985: Selection for heavy metal tolerance during pollen development among pollen grains from a single individual. Amer. J. Bot. **72,** 1700—1706.

Singleton, W. R., Mangelsdorf, P. C., 1940: Gametic lethals on the fourth chromosome of maize. Genetics **25,** 366—390.

Stinson, J., Mascarenhas, J. P., 1985: Onset of alcohol dehydrogenase synthesis during microsporogenesis in maize. Plant Physiol. **77,** 222—224.

Tanksley, S., Zamir, D., Rick, C. M., 1981: Evidence for extensive overlap of sporophytic and gametophytic gene expression in *Lycopersicon esculentum.* Science **213,** 453—455.

Ter-Avanesian, D., 1949: The influence of the number of pollen grains used in pollination. Bull. Appl. Genet. Plant Breeding Leningrad. **28,** 119—133.

Ter-Avanesian, D., 1978: The effect of varying the number of pollen grains used in fertilization. Theor. Appl. Genet. **52,** 77—79.

Vienne, D., Savina, F., Daussant, J., 1985: La glutamate deshydrogenase au cours de la gametogenese mâle chez *Medicago sativa.* Physiol. Plant. **63,** 208—214.

Willing, R. P., Mascarenhas, J. P., 1984: Analysis of the complexity and diversity of mRNAs from pollen and shoots of *Tradescantia.* Plant Physiol. **75,** 865—868.

Zamir, D., Vallejos, E. C., 1983: Temperature effects on haploid selection of tomato microspores and pollen grains. In: Mulcahy, D. L., Ottaviano, E. (eds.), Pollen: Biology and Implications for Plant Breeding, pp. 335—342. New York: Elsevier.

Zamir, D., 1983: Pollen gene expression and selection: applications in plant breeding. In: Tanksley, S. D., Orton, T. J. (eds.), Isozymes in Plant Genetics and Breeding. Part A., pp. 313—330. Amsterdam: Elsevier.

Chapter 10

Auxotroph Isolation *In Vitro*

Anne D. Blonstein

Friedrich Miescher-Institut, P. O. Box 2543, CH-4002 Basel, Switzerland

Contents

I. Introduction

A. Some Preliminary Comments

Following the first report of a nutritional mutant in higher plants (Langridge & Brock, 1961), just ten years elapsed before auxotrophic mutants were declared to have been produced from plant somatic cell culture (Carlson, 1970), and the future looked bright for the application of *in vitro* techniques in the induction and isolation of "novel" plant variation. Nevertheless, in 1980 Robert Shields, reviewing a book on plant cell and tissue culture, commented that there is a "paucity of useful selectable mutations in defined genes in plant cells" even though with the availability of haploid material "recessive mutations should be obtainable at reasonable frequency". He went on to say, "one can only assume that not enough people have tried hard enough."

This review, concentrating on auxotrophic mutants (which can be regarded as selectable in culture in that they can be selected against), will examine the accuracy of this statement. It will show that much time and effort has been put into *in vitro* mutant isolation and when problems and setbacks have been encountered some have been solved, others not. There has been both success and failure.

However, it will also highlight some of the approaches that seem to have been neglected or under-exploited in this field. One of the penalties of scientific specialization has been marked lack of co-operation and collaboration between tissue-culture technologists on the one hand and biochemists, physiologists, and developmental biologists on the other. A better awareness by the former of "need" and by the latter of available techniques could result in some significant advances in knowledge and understanding in key areas of plant biology.

B. Why Auxotrophs?

There is an extensive literature on the role auxotrophic mutants have played in the elucidation of biochemical processes in micro-organisms. Enzyme lesions may lead to the accumulation of precursors which can be identified directly or by the feeding of radiolabelled compounds. Altered enzymes themselves can be extracted and analysed. Genetic complementation studies between phenotypically similar mutants will establish the complexity of the system under study and can ultimately lead to an idea of the physical organization and distribution of the related genes within the genome. In addition to biosynthetic lesions, nutritional mutants may show alterations in uptake, transport, degradation etc. Auxotrophs are also a powerful tool for gene isolation via transformation.

Although nutritional mutants were found with comparative ease in bacteria and fungi, they proved somewhat more difficult to isolate in simple lower plants, but success has been reported in *Chlamydomonas* (requirements for carbon sources, nicotinamide, para-aminobenzoic acid (PABA), thiamine and purines — Gowans, 1960), *Physcomitrella patens* (thiamine, PABA, niacin and yeast extract auxotrophs — Engel, 1968), and the liverwort *Sphaerocarpos donnelli* (nicotinic acid, choline, glucose and arginine requiring — Schieder, 1976).

Until the introduction of *in vitro* techniques, higher plants proved even more recalcitrant, and auxotrophs were available for only a very restricted range of compounds, and were frequently leaky: tomato (thiamine — Langridge and Brock, 1961); *Arabidopsis* (thiamine: over 200 independent mutations at at least 4 loci — Feenstra, 1964; Rédei, 1965, 1975; Li and Rédei, 1969), *Plantago insularis* (thiamine — Murr and Spurr, 1973; Murr and Stebbins, 1971), *Zea mays* (proline — Gavazzi *et al.*, 1975; Racchi *et al.*, 1978).

The various reasons advocated for this apparent inability to isolate plant auxotrophs by conventional seed mutagenesis are worth reiterating because they both justify the use of *in vitro* technology and highlight complications that even tissue culture cannot resolve.

Feenstra (1964) predicted that repairable recessive mutants would only be found for compounds non-essential for male gamete development and seed formation or, if essential, for ones which can be supplied by the maternal tissue or receiving pistil and diffuse to the pollen or seed. Mutant plants themselves must be able to take up the nutrient from the substrate (or through their leaves, perhaps, if they are sprayed) and transport it to the locations where it is needed. Nelson and Burr (1973) again emphasized the failure of most auxotrophic mutants to survive embryonic development when borne on heterozygous, phenotypically normal plants.

In addition, Li *et al.* (1967) speculated that auxotrophs may be rare in green plants because a) their genetic material may be extensively duplicated, b) alternative metabolic pathways co-exist and c) basic biochemical reactions may be coupled to photosynthesis or other photochemical processes.

C. The Characteristics, Advantages and Disadvantages of In vitro *Systems for Mutant Isolation*

Although they have been noted many times before, it is important to review and comment on the factors which have encouraged the use of cell-culture systems for the induction of biochemical mutants.

First and foremost, a very large (i. e. several million) number of genomes can be subjected to mutagenesis and screened for a desired trait. In the case of recessive mutations (and most auxotrophs can be expected to be recessive) as long as a chimera is not produced, selection can be applied in culture and regenerated plants will be homozygous recessive. There will be no need to wait until an F 2 generation to see segregation of the mutant. Manipulations to "rescue" an auxotroph might be quite easy in the cell culture environment, but feeding problems may still arise at and after plant regeneration: the trait may not be sexually transmitted due to gametic lethality. It may have to be accepted that some traits can *only* be studied in culture (but see IV. B). In such cases it may be very difficult to distinguish epigenetic from mutational events.

Freshly isolated leaf protoplasts are the preferred source of material for mutagenesis and subsequent culture, selection and regeneration. Established cell cultures have a variety of disadvantages: their regeneration capacity is usually very low, possibly due to an accumulation of chromosomal alterations, mutations and biochemical (especially hormonal) "conditioning"; cells aggregate; and they are difficult to synchronize. In contrast, leaf-derived protoplasts are usually highly synchronous and "frozen" in G1 at the time of isolation (Magnien *et al.*, 1980; Magnien and Devreux, 1980; Galbraith *et al.*, 1983).

The availability of haploid plants as a source of protoplasts was further impetus in the search for auxotrophic mutants because only haploid material allows the immediate expression of recessive mutations induced early in culture. Most haploid cultures diploidize quite rapidly. Although stable haploid cultures have been reported (e. g. Furner *et al.*, 1978), it is questionable whether they possess any advantages over haploid leaf-derived protoplasts, and are subject to all the drawbacks of continuous culture that have already been mentioned.

D. Species Suitable for Auxotroph Isolation

About eight species, most of them Solanaceae, are currently used, or probably will be in the very near future, as *in vitro* systems for biochemical mutant isolation. These are described briefly below.

i) *Nicotiana tabacum*

The first plant species for which *in vitro* culture from protoplast to plant was reported (Nagata and Takebe, 1971). Its advantages as an experimental system include well-established genetics and cytogenetics, and very high regeneration capacity. However, as an allotetraploid possibly containing the duplication of many genes it is less than ideal as starting material when screening for recessive mutations, although it should be added that these have been found.

ii) *Nicotiana sylvestris*

One of the two true diploid *Nicotiana* species that have been used as alternatives to tobacco. Although protocols have been reported for the culture from protoplasts to plants (Nagy and Maliga, 1976; Durand, 1979; Bourgin *et al.*, 1979) there continue to be various reports of manipulation problems with this species (e. g. Magnien and Devreux, 1980).

iii) *Nicotiana plumbaginifolia*

Currently one of the most popular "model" species and a true diploid, for which sterile haploid shoot cultures are easily maintained. The passage from protoplast to flowering plant takes about sixteen weeks (Bourgin *et al.*, 1979; Negrutiu, 1981), but in contrast to tobacco, *N. plumbaginifolia* appears to lose its regeneration capacity quite rapidly in culture and regenerants are frequently infertile or self-sterile. It may be that true diploids, spending a period of time in culture as haploids, may be more susceptible and sensitive than polyploid species to mutagenesis and factors generating somaclonal variation.

iv) *Petunia hybrida*

Haploid plants are available in this genetically well characterized species, and plant regeneration from haploid and diploid protoplasts has been reported (Durand *et al.*, 1973; Binding, 1974).

v) *Datura innoxia*

In use in some laboratories (Schieder, 1975).

vi) *Hyoscyamus muticus*

A diploid species for which a protocol for culture of haploid protoplasts to plants was reported (Lörz *et al.,* 1979). Since then it has very successfully been used for auxotroph isolation *in vitro* but fertile plant regeneration of both mutants and wild type has been very difficult to obtain to date.

vii) *Solanum tuberosum*

Although di- and mono-haploids have been available in potato for some years (Lamm, 1938; Dunwell and Sunderland, 1973; van Breukelen *et al.,* 1975; Sopory *et al.,* 1978) and techniques for protoplast culture and plant regeneration are well established (Behnke, 1976; Binding *et al.,* 1978; Thomas, 1981), many cultivars are often not sexually fertile, a drawback to their application as a model genetic system (Shepard, 1980).

viii) *Lycopersicon esculentum*

Tomato would be an ideal species for *in vitro* work having a very well-defined genetic system. In addition, biochemical markers already exist at the whole plant level which may express in culture. However, as yet, haploid material is not generally available, and the techniques for the regeneration of fertile plants from protoplasts of 14 diverse cultivars reported recently (Shahin, 1985) remain to be shown to be universally applicable. Nevertheless, tomato is set to become a very useful experimental system in the near future.

II. Mutagenesis

The mutagenesis of plant cells — its theory and practical methodology — has recently been reviewed in detail (Blonstein and King, 1985) and this section will provide just a brief overview of the subject.

Mutagens should only be applied to plant cells if it seems absolutely necessary. When a mutagen is applied, the generation of undesirable secondary mutations, and the (probably associated) frequently observed reduction in plant regeneration capacity (Negrutiu *et al.,* 1983) may often outweigh the benefits of increased mutation frequency. However, if the spontaneous frequency of a desired mutation is low, in the absence of an effective selection scheme (see below) mutagenesis will probably be opted for to reduce the effort involved in screening.

Cell irradiation (by UV, γ- or X-rays) is preferable to the use of chemical mutagens if only because less culture manipulation (centrifugation, washing etc.) is required. However, both irradiation and chemicals are believed to have been successful in inducing mutants (Blonstein and King, 1985).

Low killing rates are desirable to preserve regeneration capacity, and have been shown to be associated with an adequately high mutant frequency. Indeed, Colijn *et al.* (1979) only found drug-resistant calli at nitrosoguanidine concentrations that had no observable effect on the survival of mutagenized cultures.

One further factor of great importance is the timing of mutagen application. There are two aspects to this problem: First, the stage(s), in the case of synchronized cultures, in the cell cycle which are most mutagen susceptible, and second, the necessity to generate pure clones i. e. to ensure that the DNA strands which segregate into daughter cells both carry the mutation. Very little is known about the precise biochemical modes of action of mutagens in plant cells and information must, perforce, be extrapolated from the whole plant and microorganism literature.

Some mutagens (e. g. MNNG) seem to be most effective on replicating DNA (Gebhardt *et al.*, 1981) and, in the case of mesophyll-derived protoplasts, maximal DNA replication usually occurs 1 to 2 days after isolation and is paralleled by an increase in irradiation sensitivity (Galun and Raveh, 1975; Magnien *et al.*, 1980). However, there are few data for plant cells establishing that mutation frequency is definitely linked to DNA replication.

Additionally, mutations induced on one strand of replicating DNA and fixed post-replicatively will result in chimeras which will probably express as wild types when subjected to selection for auxotrophy. The ideal mutagen causes lesions that are fixed in a pre-replicative process and will probably, therefore, have to be active on non-replicating DNA. The information simply is not available for plant cells as to which mutagens promote such mechanisms. In work with yeast (Nasim *et al.*, 1981), UV and methyl methane sulphonate produced mainly pure clones, while EMS and nitrous acid produced mainly mosaics.

III. Methods of Selection

A. Conditions and Media

Several important factors have to be taken into consideration when designing a selection system for auxotroph isolation. The "true" wild-type phenotype must first be established and tissue culture artifacts must be recognized when choosing the base line from which to look for auxotrophs. In a search for amino acid and vitamin auxotrophs, Gebhardt *et al.* (1981) used one white and one green wild-type line, which differed somewhat in their growth properties, as a means to reducing the potential number of "false" positives selected in the early stages of their experiments.

While a whole plant may be prototrophic for a given compound, it may not be required and/or synthesized by undifferentiated tissue growing in culture. If a search is to be made for fairly specific mutations, it may be wise to establish the presence of the compound in wild-type tissue before screening for auxotrophs.

It has long been recognized that the composition of the "rescue" medium can profoundly influence the phenotypes isolated. Gresshof (1978) pointed out several problematic factors such as antagonism of metabolites at high concentrations, or their supply in insufficient amounts. In early work with *Neurospora* Lein *et al.* (1948) found that three of seven auxotrophic mutants would not grow on complete medium. Histidineless mutants were always inhibited if certain other amino acids were present in the medium (Haas *et al.,* 1952). Isolating auxotrophs in a liverwort, Schieder (1976) also observed that the medium may influence what is isolated.

The choice of concentration of a given supplement in the medium may favour the isolation of certain kinds of mutants, but hinder the appearance of others. Thus dose response curves for histidine$^-$ and tryptophan$^-$ variants of *Hyoscyamus muticus* show optima very close to the concentrations used in the rescue medium for the original isolation experiments (Gebhardt *et al.,* 1983; Shimamoto and King, 1983).

Mutant isolation schemes frequently result in low density growth conditions either as a deliberate intention, e. g. to prevent coalescence of adjacent colonies before dilution, to limit cross feeding during enrichment (see below), and to limit selection pressures against auxotrophs growing more slowly than prototrophs in mixed culture (Roth and Lark, 1982), or as the inevitable consequence of high kill after mutagenesis. Many cultures do not survive at cell densities less than about 10^4 ml^{-1}, and various methods have been developed to either supply nutrients that may be leaking from viable cells into the medium, or to counteract putative toxic products released by dead or dying cells (although these could quite plausibly act as a feeder themselves). Advocated methods include feeder layers (Raveh *et al.,* 1973; Weber and Lark, 1979; Shneyour *et al.,* 1984; Horsch and King, 1984, 1985) and the use of enriched medium (Kao and Michayluk, 1975).

B. Positive Selection

Undoubtedly the easiest way to screen a population of cells for a mutant phenotype is to apply positive selection, i. e. design a scheme which kills wild-type cells and allows only mutants to survive. Theoretically, a vast number of genomes can be screened this way. Unfortunately, this is not an obvious technique for detecting auxotrophs — the auxotroph would need to express an "advantage" over the wild type under a given set of conditions, in addition to its growth requirement. So far in the field of *in vitro* auxotroph isolation there is only one phenotype where this technique has been fully exploited and yielded a large number of mutants at several loci: nitrate reductase-deficient (NR$^-$) mutants which cannot utilise nitrate as their sole nitrogen source but exhibit resistance to chlorate because they are unable to convert it to toxic chlorite as the wild type does (Aberg, 1947). Using this screen, NR$^-$ mutants and variants have been isolated *in vitro* in tobacco (Müller and Grafe, 1978; Buchanan and Wray, 1982), *Datura innoxia* (King and Khanna, 1980), *Rosa damascena* (Murphy and

Imbrie, 1981) and *Nicotiana plumbaginifolia* (Márton *et al.*, 1982; Negrutiu *et al.*, 1983). Chlorate resistance, however, is not always indicative of an inactive nitrate reductase enzyme, suggesting that other mechanisms, such as exclusion, can yield the resistant phenotype (Murphy and Imbrie, 1981; Buchanan and Wray, 1982; Márton *et al.*, 1982) (see also Chapter 5 in this volume).

Wildholm (1981) tried to develop a positive selection system with tobacco and carrot cells for tryptophan synthase-deficient mutants which would not convert non-toxic compounds to toxic tryptophan analogues. However, none of the EMS-treated cells which survived 5-fluoroindole or 6-fluoroindole selection had a tryptophan requirement for growth. The use here of diploid cell lines may have mitigated against finding recessive mutations.

When isolating sodium-dependent variants from haploid soybean cell cultures, Jia-Ping *et al.* (1981) commented upon the link between enhanced tolerance and raised requirement for a compound normally needed for growth but toxic in excess. This is a phenomenon which perhaps could be exploited more than it has been in auxotroph isolation.

D'Souza and Maher (1976) identified invertaseless mutants in *Arabidopsis* by spraying aseptically grown seedlings with a sucrose solution followed by a special reagent: putative mutants would not produce the red colour reaction typical of wild-type plants. To date, differential colour reactions have not been exploited for mutant identification in *in vitro* work — potential which perhaps has been underexploited.

C. Negative Selection

i) Total Isolation

It is extremely unfortunate that it has not been possible, so far, to adapt for the plant cell system the one technique — replica plating (Lederburg and Lederburg, 1952) — which has so simplified the requirement testing of putative auxotrophs in microbiology. Although there has been the occasional report of a replica plating method for plant cells (e. g. Schulte and Zenk, 1977) these have not found widespread applicability. Scientists searching for auxotrophs, therefore, have either spent considerable time and effort trying to develop mutant enrichment techniques or have resorted to the time and labour intensive task of individual clone testing which has, nevertheless, proved very effective (Savage *et al.*, 1979; Gebhardt *et al.*, 1981; Sidorov *et al.*, 1981; Roth and Lark, 1982).

ii) Enrichment

Mutant enrichment is a technique whereby conditions are imposed on a cell population which preferentially kill growing wild-type cells. The survivors of the treatment should contain an enhanced proportion of mutants over the original population thus reducing the subsequent time and labour spent testing clones.

For both microbiology and animal somatic genetics several enrichment procedures have been developed. For microorganisms these include substances probably specific for bacteria and fungi e. g. penicillin (Davis, 1948; Lederburg and Zinder, 1948). Others could have more general applicability, including tritium-labelled thymidine (Lubin, 1959) and nystatin (Snow, 1966). Ferenczy *et al.* (1975) developed a method of enriching for fungal mutants by the lysis of protoplasts selectively produced from growing colonies. The enrichment achieved was about 80-fold. An unsuccessful attempt was made in our laboratory to apply this procedure to *Rosa* cells, but it was not possible to reliably preferentially obtain protoplasts from dividing cells. Walton *et al.* (1979) developed an enrichment method for auxotrophic mutants of yeast based on the fact that stationary phase yeast are more resistant than exponentially growing cells to killing by heat.

The spectrum of successful "enrichment" compounds was extended by mammalian cell biologists and includes the nucleoside analogues BUdR (Puck and Kao, 1967) and FUdR (Basilico and Meiss, 1974). Thymine-less death of growing cells has also been used (De Mars and Hooper, 1960; Chu *et al.,* 1969) as a selection procedure.

A further compound, arsenate, was reported to have been successful as a counter-selective agent enriching for photosynthetic mutants of *Euglena gracilis* (Shneyour and Avron, 1975).

Of the agents listed above, nucleoside analogues and arsenate have proved most popular as counter-selective agents chosen by plant cell biologists. However, as two recent reviews point out (Blonstein and King, 1985; Horsch and King, 1985), there is no real evidence that either BUdR or arsenate have enriched for mutant cells. All too frequently data are not published for the mutant frequency before treatment by the agent, making a proper evaluation of the claims for enrichment impossible.

The possible reasons for the failure of several laboratories to establish reliable and repeatable enrichment techniques, even after several years of attempts, are numerous. That the processes by which the nucleoside analogues and arsenate preferentially kill growing plant cells have not been elucidated makes it difficult to identify those components of a system which might be altered to improve its efficacy. For example, with plant cells, light exposure is not needed as a post-treatment after BUdR to ensure killing as it is for animal cells, suggesting that its mode of action in plants may be different from that established in animal cells (Shillito *et al.,* 1981; Negrutiu *et al.,* 1985).

Another problem has been that in establishing enrichment conditions it has usually been necessary to use nitrogen-starved non-growing cells as models for auxotrophic mutants. Clearly the biochemistry of such cells may differ substantially from that of auxotrophs, and conditions which select for the former may be ineffective for the latter. However, now that a range of tissue-culture expressed auxotrophs are available these can be used profitably in reconstruction experiments to improve the protocols although, again, it must not be assumed that any one auxotroph will be a good model for all others.

Table 1. Amino-Acid, Vitamin and Purine Auxotrophs Isolated *In Vitro*.

Species	Mutagen	Requirement	Line code	Site of lesion	Fertile plants	Genetics[a]	Spontaneous reversion	Key reference(s)
Nicotiana tabacum	0.25 % EMS/ 1 h	Hypoxanthine			Yes	R?		Carlson *et al.* (1973) and
		Lysine			Yes	R?		Carlson (1970)
Hyoscyamus muticus	MNNG	Tryptophan	VIIIB9	*A* reaction of tryptophan synthetase	No	R (in fusion)	No ($<3.8 \times 10^{-8}$)	Gebhardt *et al.* (1983)
	MNNG	Histidine	VA5	imidazoleglycerolphosphate dehydratase	No	R (in fusion)	No ($<5.0 \times 10^{-8}$)	Gebhardt *et al.* (1983)
			XIIIB5	or histidinol-phosphate	No	R (in fusion)		
			IIIE3	aminotransferase	No	R (in fusion)		
	MNNG	Histidine	PO3	probably histidinol-phosphate aminotransferase	No	R (in fusion) allelic to VA5	Yes (3.2×10^{-7})	Shimamoto and King (1983)
	MNNG	Leucine	XIVE9 (also NR^-)		No			Gebhardt *et al.* (1982)
	MNNG	Nicotinamide	III_2F10	before quinilinic	No	R (in fusion) and not		Gebhardt *et al.* and
			IVH2	acid	No	allelic		P. J. King (pers. comm.)
			XE1	probably quinolinic acid transphosphoribosylase	No	R (in fusion) and not allelic to III_2F10 and IVH2		
Glycine max	UV	Asparagine or Glutamine	$Sln\text{-}asn^-$				($<4 \times 10^{-7}$)	Roth and Lark (1982)
Datura innoxia	0.5 % EMS/ 1.5 h	Isoleucine/ valine	IV-1	dihydroxyacid hydratase	No		Yes	Horsch and King (1983) and R. Wallsgrove (pers. comm.)

Species	Mutagen	Requirement	Line code	Site of lesion	Fertile plants	Genetics[a]	Spontaneous reversion	Key reference(s)
	0.05 % EMS/ 1 h	Adenine	Ad1	possibly in pathway leading to 5-aminoimidazole-4-carboxamide	No		$(<1.3 \times 10^{-7})$	King *et al.* (1980)
	0.05 % EMS/ 1 h	Pantothenate	Pn1	between α-ketoisovalerate and pantoic acid	No		No $(<2.5 \times 10^{-7})$	Savage *et al.* (1979) and King *et al.*(1980)
	0.25 % EMS/ 2.5 h	Threonine	JM3	homoserine kinase or threonine synthetase				Horsch and King (1984)
Nicotiana plumbaginifolia	UV 250 erg cm^{-2}	Isoleucine	6.5/10 6.8/7 4.1/9	threonine deaminase	Yes Yes Yes	R (in fusion) R (in fusion) R (in fusion)		Negrutiu *et al.* (1984, 1985)
	UV (as above)	Leucine	N12 7.2/7	2-OH-3-carboxy-isocaproate-isomerase or-dehydrogenase				as above
	UV (as above)	Histidine	Q16 2.1/12 3.1/14	histidinol-phosphate aminotransferase	Yes	R (in fusion)	2.0×10^{-7}	as above
	UV (as above)	Valine plus Isoleucine	K1 Q14 7.3/10	valine transaminase				as above
	UV (as above)	Methionine	10.1/9	β-cystathionase	Yes	R (in fusion)		as above
	UV (as above)	Tryptophan	10.1/10	tryptophan synthase	Yes	R (in fusion)		as above
	^{60}Co γ-rays	Uracil	URA401		No	R (in fusion)		Sidorov and Maliga (1982)
	^{60}Co γ-rays	Isoleucine	ILE401	L-threonine deaminase	No	R (in fusion)		Sidorov *et al.* (1981)
	^{60}Co γ-rays	Leucine	LEU403		Yes?	R (in fusion)		Sidorov und Maliga (1982)

[a] R = recessive

Other factors that have been cited as problematic in enrichment include density dependent effects (including cross feeding) (Horsch and King, 1984, 1985) and resistance to the enrichment agent. It has been suggested that BUdR may be poorly incorporated into plant DNA, but that in combination with FUdR and adenosine or uridine, uptake may be improved (Zryd, 1976; M. Suter, FMI, pers. comm.). Similarly, lowering phosphate in the medium may increase the uptake and effectiveness of arsenate (Horsch and King, 1985).

Mention should be made of the specialised method of enrichment which has been used successfully by two groups in the isolation of carrot cell lines showing temperature-sensitive impairment of embryogenesis (Terzi *et al.*, 1982; Breton and Sung, 1982). In this regime cells are grown under embryo-inducing conditions at the chosen restrictive temperature. Embryos are removed by filtration and the culture shifted to a permissive temperature. Regenerating embryos are cloned and retested for the temperature-sensitive embryogenic phenotype.

Despite the poor success so far with enrichment schemes for plant cell mutants, effort still needs to be expended in this field if the search at the cell level for rare auxotrophic phenotypes is ever to become a routine procedure.

IV. Phenotypes

A. Amino-Acid, Vitamin and Purine Auxotrophs

Table 1 summarises the amino-acid, vitamin and purine auxotrophs that have been isolated in the last fifteen years using *in vitro* techniques. Bearing in mind that until 1970 only thiamine and proline mutants had been identified in higher plants, it can be seen that plant tissue culture technology has been quite successful in increasing the range of mutant and variant material.

However, these achievements need to be viewed critically. Although the selection schemes were designed to screen for mutants requiring any of the 20 amino acids, so far lines with confirmed requirements for only 9 have been isolated — tryptophan, histidine, leucine, isoleucine, valine, threonine, methionine and asparagine and/or glutamine. The possible reasons for this shortfall have already been discussed. Tissue culture may not always provide the needs of the auxotroph seeker.

Even more disturbing has been the failure to convincingly demonstrate, in any one case, the Mendelian transmission of the variant phenotypes. Although the methionine$^-$ trait in *Nicotiana plumbaginifolia* was transmitted to all the progeny of selfed regenerants (Negrutiu *et al.*, 1985) it failed, as did isoleucine$^-$ and histidine$^-$, to segregate in fusion products. *Hyoscyamus muticus* nicotinamide$^-$ also failed to reappear in the progeny of a fusion hybrid with NR$^-$ *N. tabacum* (Potrykus *et al.*, 1984). As anther culture to rescue auxotrophic meiotic products is unlikely to be acceptable as standard procedure, it may never be possible to convincingly establish the genetic basis of these fundamental lesions, let alone map them physically, and conduct a linkage analysis.

In addition to these complications it can be seen from Table 1 that, with the exception of Carlson's possible tobacco variants, all the amino-acid mutations have been induced in a narrow range of so-called "model" species which, although amenable to tissue culture, have no prior established genetics or, of lesser import, economic value. The rather miserable spectre arises that when species which fulfil these conditions (e. g. tomato) have been sufficiently modified for tissue culturability, the whole mutant induction and screening procedure will take place all over again.

So far, the amino-acid auxotrophs have been isolated by tissue culture scientists intent on demonstrating that *in vitro* techniques can produce variation that had not yet been obtained at the whole plant level. They have also aimed to provide tissue-culture markers which would assist in such techniques as fusion and transformation. The mutant isolators have not been the biochemists intimately concerned with specific areas of plant metabolism. They would not only benefit most directly from the availablility of genetic variation within their field of study, but might also, from more profound chemical understanding, be better equipped to design effective, specific selection systems, especially positive schemes which could take advantage of the vast array of metabolite analogues that can be produced synthetically.

B. Nitrate Reductase Mutants

As discussed in Section III. B, great success has been achieved in the isolation of nitrate reductase-deficient auxotrophs by taking advantage of the positive chlorate resistance selection scheme. These mutants, and the role they have played in the elucidation of nitrogen metabolism in plants are fully discussed by John Wray in Chapter 5 of this volume and will not be treated further here.

C. Temperature-Sensitive Mutations

Temperature-sensitive (ts) mutations have contributed significantly to the understanding of many aspects of the regulation of development from micro-organisms (e. g. Edgar and Lielausis, 1964) to *Drosophila* (Suzuki *et al.,* 1976). Ts lines have also been obtained in mammalian somatic genetics (Naha, 1969; Thompson *et al.,* 1970) and a few in plant cell work, which are described below.

Two advantages accrue to ts mutations. First, they facilitate the resolution of the temporal activation of a series of related genes and second, the expression of a potentially very disruptive lesion can be suppressed while regeneration, crosses etc. are carried out. Bearing in mind the problems with regeneration and genetic analysis of the auxotrophic mutations that have been described in the previous section, the advantages of ts conditional lethals should be quite clear. However, even with an effective mutagenesis procedure one can expect ts auxotrophs to be quite rare emphasising further the urgent need for reliable enrichment procedures.

The most effective use to date of *in vitro* isolated temperature-sensitive mutants have been the studies by the groups of Z. R. Sung at Berkeley and M. Terzi in Pisa isolating ts mutants in carrot cell embryogenesis, which are helping them build up a time-sequence of the molecular and biochemical events which accompany and drive embryogenesis (Breton and Sung, 1982; Terzi *et al.*, 1982; Giuliana *et al.*, 1984). The enrichment by filtration procedure they used has already been described (see III. C. ii). Because these mutants are not strictly auxotrophs they will not be discussed in any detail here, although one *(ts 66)* isolated by Breton and Sung (1982), but subsequently lost, may have been an auxin auxotroph since it was unable to grow at all in the absence of 2,4-D in contrast to the wild type and other *ts-emb*⁻ lines.

A ts auxin auxotroph of *Hyoscyamus muticus,* XIIB2, which, unfortunately, has not yet been regenerated to fertile plants, is being studied in our laboratory with the aim of elucidating certain aspects of auxin production and action in plant cells. XIIB2 was originally isolated as an uncharacterised ts line in a large auxotroph isolation experiment (Gebhardt *et al.,* 1981). It is recessive in fusion, and plantlets that have been produced also exhibit the ts phenotype. It was subsequently found that the callus' phenotypic expression of browning followed by death 4 to 5 days after the temperature shift to 33° C could be corrected by the exogenous addition of a range of auxins, including indole-3-acetic acid. (The wild type is hormone autotrophic in culture.) It does not, however, appear to be a simple biosynthetic lesion because RIA has detected normal levels of IAA in XIIB2 at the high temperature (J. Oetiker, pers. comm.). We hope to produce more such ts auxotrophs to facilitate studies of hormone biosynthesis and action.

Malmberg (1979, 1980) also produced and partially characterized some *in vitro* derived ts mutants of *Nicotiana tabacum,* but this work has remained incomplete.

Temperature-sensitivity need not be the only approach to manipulating the expression of conditional lethals such as auxotrophs. Other factors that might be considered, for example, are pH and osmotic pressure.

V. Fusion and Transformation

From the start, attempts were not made to produce auxotrophs in culture solely to provide material for chemical analysis. It was also hoped that stable auxotrophic cell lines with low (preferably absent) reversion rates could be used as models in the refinement of fusion and transformation techniques.

So far, a great deal more has been done with fusion than with transformation, and for recent reviews on the technology, somatic hybridization, and segregation of organelles and cytoplasmic traits in fusion products see Lázár (1983), Harms (1983) and Fluhr (1983).

One use of fusion can be to attempt to regenerate a line which will not do so otherwise, because it has been found that fusion hybrids can often regenerate when only one of the two parents is so capable (Evans, 1983; Potrykus *et al.*, 1984).

Probably the most extensive fusion work has been done with the NR^- lines (see Chapter 5 in this volume). Where plant regeneration is not possible, fusion will provide the only means of determining dominance relationships and allele complementation (see Table 1).

Although for transformation non-revertant auxotrophs should have great advantages over leaky resistance phenotypes (Shillito *et al.*, 1983), there have been no confirmed reports yet of auxotroph complementation by transformation using genes of either microbial or plant origin.

VI. Conclusions and Future Prospects

In vitro culture techniques with cells have undoubtedly produced new plant variation which it might have been very difficult, if not impossible, to isolate at the whole plant level. In some, but not all, cases the new material has played an important role in the elucidation of certain biochemical processes in plants, and further isolation of auxotrophs seems justified if a number of criteria are met.

Auxotroph isolation should be "targeted". That is, cell populations should be screened for specific phenotypes intended as tools to study given processes.

In the absence of positive selection, enrichment procedures are essential to reduce effort and frustration, and to maximise the range of variation that can be isolated.

To ensure that a system can be properly analysed genetically it seems very advisable to work with variation that has additional conditionality (e. g. temperature-sensitivity) so that material can be handled with relative ease at the whole plant level.

Given these three objectives, targeting, enrichment and a conditionality additional to the auxotrophy, *in vitro* isolation of auxotrophs continues to hold promise as a technique contributing "Genetics" to the study of "Biochemistry".

Acknowledgements

I would like to thank Dr. Patrick J. King and other colleagues at the FMI for helpful discussion in the preparation of this chapter, and Dr. John King (Saskatoon) for providing me with unpublished results.

VII. References

Åberg, B., 1947: On the mechanism of the toxic action of chlorates and some related substances upon young wheat plants. Ann. R. Agric. Coll. Sweden **15,** 37—107.

Basilico, C., Meiss, H. K., 1974: Methods for selecting and studying temperature-sensitive mutants of BHK-21 cells. In: Prescott, D. M. (ed.), Methods of Cell Biology, Vol, 8, pp. 1—22. New York: Academic Press.

Behnke, M., 1976: Kulturen isolierte Zellen von einigen dihaploiden *Solanum-tuberosum*-Klonen und ihre Regeneration. Z. Pflanzenphysiol. **78,** 177—181.

Binding, H., 1974: Regeneration von haploiden und diploiden Pflanzen aus Protoplasten von *Petunia hybrida.* Z. Pflanzenphysiol. **74,** 327—356.

Binding, H., Nehls, R., Schieder, O., Sopory, S. K., Wenzel, G., 1978: Regeneration of mesophyll protoplasts isolated from dihaploid clones of *Solanum tuberosum.* Physiol. Plant. **43,** 52—54.

Blonstein, A. D., King, P. J., 1985: Plant mutant isolation via protoplasts. In: Fowke, L. C., Constabel, F. (eds.), Plant Protoplasts, pp. 151—168. Boca Raton, Flo.: CRC Press.

Bourgin, J.-P., Chupeau, Y., Missonier, C., 1979: Plant regeneration from mesophyll protoplasts of several *Nicotiana* species. Physiol. Plant. **45,** 288—292.

Breton, A. M., Sung, Z. R., 1982: Temperature-sensitive carrot variants impaired in somatic embryogenesis. Dev. Biol. **90,** 58—66.

van Breukelen, E. W. M., Ramanna, M. S., Hermsen, J. G. Th., 1975: Monohaploids (n = x = 12) from autotetraploid *Solanum tuberosum* (2n = 4x = 48) through two successive cycles of female parthenogenesis. Euphytica **24,** 567—574.

Buchanan, R. J., Wray, J. L., 1982: Isolation of molybdenum cofactor defective cell lines of *Nicotiana tabacum.* Mol. Gen. Genet. **188,** 228—234.

Carlson, P. S., 1970: Induction and isolation of auxotrophic mutants in somatic cell cultures of *Nicotiana tabacum.* Science **168,** 487—489.

Carlson, P. S., Dearing, R. D., Floyd, B. M., 1973: Defined mutants in higher plants. In: Srb, A. M. (ed.), Genes, Enzymes and Populations, pp. 99—107. New York: Plenum.

Chu, E. H. Y., Brimer, P., Jacobsen, K. B., Merriam, E. V., 1969: Mammalian cell genetics I. Selection and characterization of mutations auxotrophic for L-glutamine or resistant to 8-azaguanidine in Chinese hamster cells *in vitro.* Genetics **62,** 359—377.

Colijn, C. M., Kool, A. J., Nijkamp, H. J. J., 1979: An effective chemical mutagenesis procedure for *Petunia hybrida* cell suspension cultures. Theor. Appl. Genet. **55,** 101—106.

Davis, B. D., 1948: Isolation of biochemically deficient mutants of bacteria by penicillin. J. Am. Chem. Soc. **70,** 4267.

De Mars, R., Hooper, J. L., 1960: A method of selecting for auxotrophic mutants of HeLa cells. J. Exp. Med. **111,** 559—573.

D'Souza, E. S., Maher, E. P., 1976: An invertase mutant of *Arabidopsis.* Arabidopsis Inf. Serv. **13,** 79—84.

Dunwell, J. M., Sunderland, N., 1973: Anther culture of *Solanum tuberosum* L. Euphytica **22,** 317—323.

Durand, J., 1979: High and reproductive plating efficiencies of protoplasts isolated from *in vitro* grown haploid *Nicotiana sylvestris* Spegaz et Comes. Z. Pflanzenphysiol. **93,** 283—295.

Durand, J., Potrykus, I., Donn, G., 1973: Plantes issues de protoplastes de *Petunia*. Z. Pflanzenphysiol. **69,** 26—34.

Edgar, R. S., Lielausis, I., 1964: Temperature sensitive mutations of bacteriophage T4D: Their isolation and genetic characterization. Genetics **49,** 649—662.

Engel, P. P., 1968: The induction of biochemical and morphological mutants in the moss *Physcomitrella patens*. Am. J. Bot. **55,** 438—446.

Evans, D. A., 1983: Agricultural applications of plant protoplast fusion. Bio/Technology **1,** 253—261.

Feenstra, W. J., 1964: Isolation of nutritional mutants in *Arabidopsis thaliana*. Genetica **35,** 259—269.

Ferenczy, L., Sipiczki, M., Szegedi, M., 1975: Enrichment of fungal mutants by selective cell-wall lysis. Nature (London) **253,** 46—47.

Fluhr, R., 1983: The segregation of organelles and cytoplasmic traits in higher plant somatic fusion hybrids. In: Potrykus, I., Harms, C. T., Hinnen, A., Hütter, R., King, P. J., Shillito, R. D. (eds.), Protoplasts 1983. Lecture Proceedings (6th International Protoplast Symposium, Basel 1983), pp. 85—92. Basel: Birkhäuser.

Furner, I. J., King, J., Gamborg, O. L., 1978: Plant regeneration from protoplasts isolated from a predominantly haploid suspension culture of *Datura innoxia*. Plant Sci. Lett. **11,** 169—176.

Galbraith, D. W., Harkins, K. R., Maddox, J. M., Ayres, N. M., Sharma, D. P., Firoozabady, E., 1983: Rapid flow cytometric analysis of the cell cycle in intact plant tissues. Science **220,** 1049—1051.

Galun, E., Raveh, D., 1975: *In vitro* culture of tobacco protoplasts: Survival of haploid and diploid protoplasts exposed to X-ray radiation at different times after isolation. Radiat. Bot. **15,** 79—82.

Gavazzi, G., Nava-Racchi, M., Tonelli, C., 1975: A mutation causing proline requirement in *Zea mays*. Theor. Appl. Genet. **46,** 339—345.

Gebhardt, C., Schnebli, V., King, P. J., 1981: Isolation of biochemical mutants using haploid mesophyll protoplasts of *Hyoscyamus muticus*. II. Auxotrophic and temperature-sensitive clones. Planta **153,** 81—89.

Gebhardt, C., Fankhauser, H., King, P. J., 1982: Amino-acid auxotrophs of *Hyoscyamus muticus* and their characterisation. In: Fujiwara, A. (ed.), Plant Tissue Culture 1982. Proceedings of the 5th International Congress of Plant Tissue and Cell Culture, Tokyo, 1982, pp. 463—464. Tokyo: Maruzen Co.

Gebhardt, C., Shimamoto, K., Lázár, G., Schnebli, V., King, P. J., 1983: Isolation of biochemical mutants using haploid mesophyll protoplasts of *Hyoscyamus muticus* III. General characterisation of histidine and tryptophan auxotrophs. Planta **159,** 18—24.

Giuliano, G., Lo Schiavo, F., Terzi, M., 1984: Isolation and developmental characterization of temperature-sensitive carrot cell variants. Theor. Appl. Genet. **67,** 179—183.

Gowans, C. S., 1960: Some genetic investigations on *Chlamydomonas eugametos*. Z. Vererbungsl. **91,** 63—73.

Gresshof, P. M., 1978: Auxotrophic mutant isolation in higher plants — a thought experiment with *Arabidopsis thaliana*. Arabidopsis Inf. Serv. **15,** 88—91.

Haas, F., Mitchell, M. B., Ames, B. N., Mitchell, H. K., 1952: A series of histidineless mutants of *Neurospora crassa*. Genetics **37,** 217—226.

Harms, C. T., 1983: Somatic hybridization by plant protoplast fusion. In: Potrykus, I., Harms, C. T., Hinnen, A., Hütter, R., King, P. J., Shillito, R. D. (eds.), Protoplasts 1983. Lecture Proceedings (6th International Protoplast Symposium, Basel 1983), pp. 69—84. Basel: Birkhäuser.

Horsch, R. B., King, J., 1983: Isolation of an isoleucine-valine-requiring auxotroph from *Datura innoxia* cell cultures by arsenate counterselection. Planta **159,** 12—17.

Horsch, R. B., King, J., 1984: The isolation of auxotrophs from *Datura innoxia* Mill. cell cultures following recovery of arsenate-treated cells on feeder plates. Planta **160,** 168—173.

Horsch, R. B., King, J., 1985: Arsenate counterselective enrichment for auxotrophic plant cells works well in theory but not in practice. Can. J. Bot. **63,** 2115—2120.

Jia-Ping, Z., Roth, E. J., Terzaghi, W., Lark, K. G., 1981: Isolation of sodium dependent variants from haploid soybean cell culture. Plant Cell Rep. **1,** 48—51.

Kao, K. N., Michayluk, M. R., 1975: Nutritional requirements for growth of *Vicia hajastana* cells and protoplasts at very low population density in liquid media. Planta **126,** 105—110.

King, J., Khanna, V., 1980: A nitrate reductase-less variant isolated from suspension cultures of *Datura innoxia* (Mill.). Plant Physiol. **66,** 632—636.

King, J., Horsch, R. B., Savage, A. D., 1980: Partial characterization of two stable auxotrophic cell strains of *Datura innoxia* Mill. Planta **149,** 480—484.

Lamm, R., 1938: Notes on a haploid potato hybrid. Hereditas **24,** 391—396.

Langridge, J., Brock, R. D., 1961: A thiamine-requiring mutant of the tomato. Aust. J. Biol. Sci. **14,** 66—69.

Lázár, G. B., 1983: Recent developments in plant protoplast fusion and selection technology. In: Potrykus, I., Harms, C. T., Hinnen, A., Hütter, R., King, P. J., Shillito, R. D. (eds.), Protoplasts 1983. Lecture Proceedings (6th International Protoplast Symposium, Basel 1983), pp. 61—67. Basel: Birkhäuser.

Lederburg, J., Lederburg, E. M., 1952: Replica plating and indirect selection of bacterial mutants. J. Bacteriol. **63,** 399—406.

Lederburg, J., Zinder, N., 1948: Concentration of biochemical mutants of bacteria with penicillin. J. Am. Chem. Soc. **70,** 4267—4268.

Lein, J., Mitchell, H. K., Houlahan, M. B., 1948: A method for selection of biochemical mutants of *Neurospora*. Genetics **34,** 435—442.

Li, S. L., Rédei, G. P., 1969: Thiamine mutants of the crucifer, *Arabidopsis*. Biochem. Genet. **3,** 163—170.

Li, S. L., Rédei, G. P., Gowans, C. S., 1967: A phylogenetic comparison of mutation spectra. Mol. Gen. Genet. **100,** 77—83.

Lörz, H., Wernicke, W., Potrykus, I., 1979: Culture and plant regeneration of *Hyoscyamus* protoplasts. Planta Med. **36,** 21—29.

Lubin, M., 1959: Selection of auxotrophic bacterial mutants by tritium-labeled thymidine. Science **129,** 838—839.

Magnien, E., Devreux, M., 1980: A critical assessment of the protoplast system as a tool for radiosensitivity studies. In: Sala, F., Parisi, B., Cella, R., Ciferri, O. (eds.), Plant Cell Cultures: Results and Perspectives, pp. 121—126. Amsterdam: Elsevier.

Magnien, E., Dalschaert, X., Devreux, M., 1980: Different radiosensitivities of *Nicotiana plumbaginifolia* leaves and regenerating protoplasts. Plant Sci. Lett. **19,** 231—241.

Malmberg, R. L., 1979: Temperature-sensitive variants of *Nicotiana tabacum* isolated from somatic cell culture. Genetics **92,** 215—221.

Malmberg, R. L., 1980: Biochemical, cellular and developmental characterization of a temperature-sensitive mutant of *Nicotiana tabacum* and its second site revertant. Cell **22,** 603—609.

Márton, L., Dung, T. M., Mendel, R. R., Maliga, P., 1982: Nitrate reductase deficient cell lines from haploid protoplast cultures of *Nicotiana plumbaginifolia.* Mol. Gen. Genet. **182,** 301—304.

Müller, A. J., Grafe, R., 1978: Isolation and characterization of cell lines of *Nicotiana tabacum* lacking nitrate reductase. Mol. Gen. Genet. **161,** 67—76.

Murphy, T. M., Imbrie, C. W., 1981: Induction and characterization of chlorate-resistant strains of *Rosa damascena* cultured cells. Plant Physiol. **67,** 910—916.

Murr, S. M., Spurr, A. R., 1973: An albino mutant in *Plantago insularis* requiring thiamine pyrophosphate. J. Exp. Bot. **24,** 1271—1282.

Murr, S. M., Stebbins, G. L., 1971: An albino mutant in *Plantago insularis* requiring thiamine pyrophosphate. I: Genetics. Genetics **68,** 231—243.

Nagata, T., Takebe, I., 1971: Plating of isolated tobacco mesophyll protoplasts on agar medium. Planta **99,** 12—20.

Nagy, J. I., Maliga, P., 1976: Callus induction and plant regeneration from mesophyll protoplasts of *Nicotiana sylvestris.* Z. Pflanzenphysiol. **78,** 453—455.

Naha, P. M., 1969: Temperature sensitive conditional mutants of monkey kidney cells. Nature (London) **223,** 1380—1381.

Nasim, A., Hannan, M. A., Nestmann, E. R., 1981: Pure and mosaic clones — a reflection of differences in mechanisms of mutagenesis by different agents in *Saccharomyces cerevisiae.* Can. J. Genet. Cytol. **23,** 73—79.

Negrutiu, I., 1981: Improved conditions for large-scale culture, mutagenesis and selection of haploid protoplasts of *Nicotiana plumbaginifolia* Viviani. Z. Pflanzenphysiol. **104,** 431—442.

Negrutiu, I., Dirks, R., Jacobs, M., 1983: Regeneration of fully nitrate reductase-deficient mutants from protoplast culture of *Nicotiana plumbaginifolia* (Viviani). Theor. Appl. Genet. **66,** 341—347.

Negrutiu, I., Jacobs, M., Cattoir-Reynaerts, A., 1984: Progress in cellular engineering of plants: biochemical and genetic assessment of selectable markers from cultured cells. Plant Mol. Biol. **3,** 289—302.

Negrutiu, I., de Brouwer, D., Dirks, R., Jacobs, M., 1985: Amino acid auxotrophs from protoplast cultures of *Nicotiana plumbaginifolia* Viviani. I. BUdR enrichment selection, plant regeneration, and general characterisation. Mol. Gen. Genet. **199,** 330—337.

Nelson, O. E., Burr, B., 1973: Biochemical genetics of higher plants. Annu. Rev. Plant Physiol. **24,** 493—518.

Potrykus, I., Jia, J., Lázár, G. B., Saul, M., 1984: *Hyoscyamus muticus* + *Nicotiana tabacum* fusion hybrids selected via auxotroph complementation. Plant Cell Rep. **3,** 68—71.

Puck, T. T., Kao, F.-T., 1967: Genetics of somatic mammalian cells, V. Treatment with 5-bromodeoxyuridine and visible light for isolation of nutritionally deficient mutants. Proc. Nat. Acad. Sci., U.S.A. **58,** 1227—1234.

Racchi, M. L., Gavazzi, G., Monti, D., Manitto, P., 1978: An analysis of the nutritional requirements of the *pro* mutant in *Zea mays.* Plant Sci. Lett. **13,** 357—364.

Raveh, D., Hubermann, E., Galun, E., 1973: In vitro culture of tobacco protoplasts: use of feeder techniques to support division of cells plated at low densities. In Vitro **9,** 216—222.

Rédei, G. P., 1965: Genetic blocks in the thiamine synthesis of the angiosperm *Arabidopsis.* Am. J. Bot. **52,** 834—841.

Rédei, G. P., 1975: Induction of auxotrophic mutants in plants. In: Ledoux, L. (ed.), Genetic Manipulations with Plant Material, pp. 329—350. New York: Plenum.

Roth, E. J., Lark, K. G., 1982: Isolation of an auxotrophic cell line of soybean *(Glycine max)* which requires asparagine or glutamine for growth. Plant Cell Rep. **1,** 157—160.

Savage, A. D., King, J., Gamborg, O. L., 1979: Recovery of a pantothenate auxotroph from a cell suspension culture of *Datura innoxia* Mill. Plant Sci. Lett. **16,** 367—376.

Schieder, O., 1975: Regeneration von haploiden und diploiden *Datura innoxia* Mill. Mesophyll-Protoplasten zu Pflanzen. Z. Pflanzenphysiol. **76,** 462—466.

Schieder, O., 1976: The spectrum of auxotrophic mutants from the liverwort *Sphaerocarpos donnelli* Aust. Mol. Gen. Genet. **144,** 63—66.

Schulte, U., Zenk, M. H., 1977: A replica plating method for plant cells. Physiol. Plant. **39,** 139—142.

Shahin, E. A., 1985: Totipotency of tomato protoplasts. Theor. Appl. Genet. **69,** 235—240.

Shepard, J. F., 1980: Mutant selection and plant regeneration from potato mesophyll protoplasts. In: Rubenstein, I., Gengenbach, B., Phillips, R. L., Green, C. E. (eds.), Genetic Improvement of Crops. Emergent Techniques, pp. 185—219. Minneapolis: Univ. of Minnesota Press.

Shields, R., 1980: Getting out of the wood? Cell **19,** 303—304.

Shillito, R. D., Street, H. E., Schilperoort, R. A., 1981: Model system studies of the use of 5-bromo-2′-deoxyuridine for selection of deficient mutants in plant cell suspension and protoplast cultures. Mutat. Res. **81,** 165—175.

Shillito, R. D., Lázár, G., Paszkowski, J., Shimamoto, K., Nicola-Koukolikova, Z., Hohn, B., Hohn, T., Potrykus, I., 1983: Approaches to plant protoplast transformation using drug resistance and auxotroph complementation as selective markers. In: Lurquin, P. F., Kleinhofs, A. (eds.), Genetic Engineering in Eukaryotes, pp. 265—276. New York: Plenum.

Shimamoto, K., King, P. J., 1983: Isolation of a histidine auxotroph of *Hyoscyamus muticus* during attempts to apply BUdR enrichment. Mol. Gen. Genet. **189,** 69—72.

Shneyour, A., Avron, M., 1975: A method for producing, selecting, and isolating photosynthetic mutants of *Euglena gracilis.* Plant Physiol. **55,** 142—144.

Shneyour, Y., Zelcer, A., Izhar, S., Beckmann, J. S., 1984: A simple feeder-layer technique for the plating of plant cells and protoplasts at low density. Plant Sci. Lett. **33,** 293—302.

Sidorov, V. A., Maliga, P., 1982: Fusion-complementation analysis of auxotrophic and chlorophyll-deficient lines isolated in haploid *Nicotiana plumbaginifolia* protoplast cultures. Mol. Gen. Genet. **186,** 328—332.

Sidorov, V., Menczel, L., Maliga, P., 1981: Isoleucine-requiring *Nicotiana* plant deficient in threonine deaminase. Nature (London) **294,** 87—88.

Snow, R., 1966: An enrichment method for auxotrophic yeast mutants using the antibiotic "Nystatin". Nature (London) **211,** 206—207.

Sopory, S. K., Jacobsen, E., Wenzel, G., 1978: Production of monohaploid embryos and plantlets in cultured anthers of *Solanum tuberosum.* Plant Sci. Lett. **12,** 47—54.

Suzuki, D. T., Kaufman, T., Falk, D., and the U. B. C. *Drosophila* Research Group, 1976: 5. Conditionally expressed mutations in *Drosophila melanogaster.* In: Ashburner, M., Novitski, E. (eds.), The Genetics and Biology of *Drosophila,* Vol. 1A, pp. 207—263. London: Academic Press.

Terzi, M., Giuliano, G., Lo Schiavo, F., Nuti Ronchi, V., 1982: Studies on plant cell lines showing temperature sensitive embryogenesis. In: Burger, M. (ed.),

Embryonic Development, Part B: Cellular Aspects, pp. 521—534. New York: R. Alan Liss.

Thomas, E., 1981: Plant regeneration from shoot culture-derived protoplasts of tetraploid potato (*Solanum tuberosum* cv. Maris Bard). Plant Sci. Lett. **23,** 81—88.

Thompson, L. H., Mankovitz, R., Baker, R. M., Till, J. E., Siminovitch, L., Whitmore, G. F., 1970: Isolation of temperature-sensitive mutants of L-cells. Proc. Nat. Acad. Sci., U.S.A. **66,** 377—384.

Walton, E. F., Carter, B. L. A., Pringle, J. R., 1979: An enrichment method for temperature-sensitive and auxotrophic mutants of yeast. Mol. Gen. Genet. **171,** 111—114.

Weber, G., Lark, K. G., 1979: An efficient plating system for rapid isolation of mutants from plant cell suspensions. Theor. Appl. Genet. **55,** 81—86.

Widholm, J. M., 1981: Utilization of indole analogs by carrot and tobacco cell tryptophan synthase *in vivo* and *in vitro*. Plant Physiol. **67,** 1101—1104.

Zryd, J.-P., 1976: 5-bromodeoxyuridine as an agent in the selection of sycamore cell cultures. Plant Sci. Lett. **6,** 157—161.

Subject Index

Plant Gene Research

Edited by
E. S. Dennis, B. Hohn, Th. Hohn (Managing Editor), P. J. King, J. Schell, D. P. S. Verma

The first volume

Genes Involved in Microbe-Plant Interactions

Edited by **D. P. S. Verma,** Department of Biology, McGill University, Montreal, Canada, and **Th. Hohn,** Friedrich Miescher-Institut, Basel, Switzerland.

1984. 54 figures. XIV, 393 pages. ISBN 3-211-81789-1

Contents:

Springer-Verlag Wien New York

Plant Gene Research

Edited by
E. S. Dennis, B. Hohn, Th. Hohn (Managing Editor), P. J. King, J. Schell, D. P. S. Verma

The second volume

Genetic Flux in Plants

Edited by **Barbara Hohn,** Friedrich Miescher-Institut, Basel, Switzerland, and **Elisabeth S. Dennis,** CSIRO Division of Plant Industry, Canberra City, Australia.

1985. 40 figures. XII, 253 pages. ISBN 3-211-81809-X

This volume gathers together for the first time the most recent information on plant genome instability. The plant genome can no longer be looked upon as a stable entity. Many examples of change and disorder in the genetic material have been reported recently. Chloroplast DNA sequences have been found in nuclei and mitochondria. Mitochondrial DNA molecules can switch between various forms by recombination processes. Stress on plants or on cells in culture can cause changes in chromosome organization. DNA can be inserted into the plant genome by transformation with the Ti plasmid of *Agrobacterium tumefaciens,* and transposable elements produce insertions and deletions.

Contents:

Springer-Verlag Wien New York